T0332732

Mobile Multimedia Communications: Concepts, Applications, and Challenges

Gour Karmakar
Monash University, Australia

Laurence S. Dooley
Monash University, Australia

INFORMATION SCIENCE REFERENCE

Hershey · New York

Acquisitions Editor:	Kristin Klinger
Development Editor:	Kristin Roth
Senior Managing Editor:	Jennifer Neidig
Managing Editor:	Sara Reed
Copy Editor:	Amanda Appicello
Typesetter:	Amanda Appicello
Cover Design:	Lisa Tosheff
Printed at:	Yurchak Printing Inc.

Published in the United States of America by
Information Science Reference (an imprint of IGI Global)
701 E. Chocolate Avenue, Suite 200
Hershey PA 17033
Tel: 717-533-8845
Fax: 717-533-8661
E-mail: cust@igi-global.com
Web site: http://www.igi-global.com/reference

and in the United Kingdom by
Information Science Reference (an imprint of IGI Global)
3 Henrietta Street
Covent Garden
London WC2E 8LU
Tel: 44 20 7240 0856
Fax: 44 20 7379 0609
Web site: http://www.eurospanonline.com

Library of Congress Cataloging-in-Publication Data

Mobile multimedia communications : concepts, applications, and challenges / Gour Karmakar and Laurence S. Dooley, editors.

   p. cm.

   Summary: "This book captures defining research on all aspects and implications of the accelerated progress of mobile multimedia technologies. Topics include fundamental network infrastructures, modern communication features such as wireless and mobile multimedia protocols, personal communication systems, mobility and resource management, and security and privacy issues. This book will meet the needs of researchers in a variety of fields"--Provided by publisher.

   Includes bibliographical references and index.

   ISBN 978-1-59140-766-9 (hbk.) -- ISBN 978-1-59140-768-3 (ebook)

   1. Mobile communication systems. 2. Multimedia communications. I. Karmakar, Gour, 1970- II. Dooley, Laurence S., 1959-

   TK6570.M6M564 2007

   384.5'35--dc22

                              2007036431

British Cataloguing in Publication Data
A Cataloguing in Publication record for this book is available from the British Library.

All work contributed to this book set is original material. The views expressed in this book are those of the authors, but not necessarily of the publisher.

If a library purchased a print copy of this publication, please go to http://www.igi-global.com/reference/assets/IGR-eAccess-agreement. pdf for information on activating the library's complimentary electronic access to this publication.

Table of Contents

Detailed Table of Contents

Chapter I provides a brief description of the different generations of mobile communication networks including their standards, major services, and transmission rates. The characteristics of 4G from a users' point of view, as well as its major research challenges in terms of mobile terminals, network system, and mobile services are also presented in this chapter.

With the rapid development of mobile communication systems, demands for the transmission of multimedia information using portable devices are increasing day by day. Chapter II demonstrates the effectiveness of transport layer handoff schemes for multimedia transmission, and is compared with that achieved by Mobile Internet Protocol (IP).

Understating the parameters that determine the suitability of a communication standard for the transmission of multimedia information for a particular application is of paramount importance for mobile system designers, users, and researchers. Chapter III describes multimedia applications needing the services of mobile network systems, and presents the fundamentals of issues involved in the delivery of multimedia content with the desired Quality of Service (QoS). Current and future challenges in achieving successful mobile multimedia information transmission are also discussed.

The personalized communication system is gaining more importance with an increase of communication ways due to the social reform by the technological advancement. Chapter IV introduces a personalized system for device independent and position aware communication, which is secure, scalable, and an open architecture. This chapter also presents the further research in this field.

The application of multimedia information in treatment or medical diagnosis using portable devices is proliferating with the development of mobile multimedia communication. Chapter V provides an overview of the existing therapy methods based on paper materials and mobile electronic devices in the current literature, and proposes the software solutions for the patients, the careers, and the professional speech and language therapists (SLTs), which could be accessible by mobile devices such as personal digital assistants (PDA). Finally, the recommendations for the direction of future research and development are made in this chapter.

As mobile services play a role in the reforming of the social culture as well as creating a tremendous business opportunity, it is important to assess the acceptance of these services. The analysis and critical assessment of the existing theoretical acceptance models about the evolving mobile services and their underlying technologies is given in Chapter VI. Chapter VI also introduces an acceptance model for mobile services in the bedrock of the technology acceptance model and recommends the further research directions.

With the significant influence and increasing requirements of visual mobile communications in our day-to-day life, low bit-rate video coding to combat against the stringent bandwidth scarcity of mobile networks has become a prime research area in the recent time. Chapter VII presents a review of the image and video coding techniques suitable for mobile communications to provide the readers with a means of appreciating the wealth and radical advancement of the field, and also attempts to enlist and sketch the physical significance of the various important and popular quality metrics of the image and video coding techniques.

In the orthogonal frequency division multiplexing (OFDM) scheme, the total bandwidth can be split into many narrowband sub-channels. Therefore, it can avoid any multiple access interferences in the base station receiver, and use a subcarrier-specific adaptive modulation schemes. In Chapter VIII, a flexible frequency division multiple access scheme based on OFDM-FDMA and a radio resource management (RRM) employing dynamic channel allocation techniques are developed.

Ad-hoc network is another dimension of wireless networks, which has been a demanding field of research because of not having a predefined infrastructure. Wireless nodes are dynamically connected in an arbitrary manner to perform a particular emergency operation. This, therefore, makes the routing of a message promising and challenging. The state-of-the-art, the contrast and comparisons, and the related research challenges of ad-hoc routing protocols have been articulated in Chapter IX.

With the proliferation of network technologies, especially the wireless multimedia communication, Internet, and sensor networks, the demand for ubiquitous computing is progressively rising in every aspect of human life, ranging from remotely controlling the home appliances, to security and health monitoring. The overview of ubiquitous networking including its infrastructure, its influence on human lives, its applications and services, and its global evolution are presented in Chapter X.

As with social security and privacy, the security and privacy of all wireless communication services is of paramount importance. Chapter XI aims to present an overview of the security and privacy issues by highlighting the need to secure access to wireless networks and the loss that might accrue from the breach of a network. The vulnerabilities of the IEEE 802.11 and Bluetooth networks, as well as a paradigm for secure wireless networks are presented in this chapter.

Portable devices such as PDA and mobile sets are vulnerable to security threats because of their limited size and processing capabilities, openness, and access to air interface. The security architecture of the mobile devices including the security policy, access control, and physical and storage protection are described in Chapter XII. Chapter XII also articulates the security related to cellular communication networks, Wireless LAN, and Bluetooth.

Chapter XIII

The security of ad-hoc networks has been a difficult task to the recent research community due to a dynamic network topology, vulnerability of air interface to attacks, mobility of a node in any context and resource constraints. Chapter XIII provides the descriptions of various network-layer security attacks of ad-hoc networks, as well as the security protocols that protect the network layer from a number of attacks.

Chapter XIV

Pervasive computing provides the computing and communication technology to integrate the everyday works based on the philosophy of existence everywhere, but access at the same time. Therefore, this poses some inherent limitations that make it even more vulnerable to security attacks than mobile computing. In Chapter XIV, the state-of-the-art of pervasive security, privacy and trust, their challenges and requirements for pervasive applications, and some open issues related in this field have been described.

Preface

Mobile multimedia communication is increasingly in demand because of the basic need to communicate at any time, anywhere, using any technology. In addition, to voice communication, people have a desire to access a range of other services that comprise multimedia elements—text, image, animation, high fidelity audio and video using mobile communication networks. To meet these demands, mobile communication technologies has evolved from analog to digital, and the networks have passed through a number of generations from first generation (1G) to fourth generation (4G).

Chapter I provides a brief description of the different generations of mobile communication networks including their standards, major services, and transmission rates. The characteristics of 4G from a users' point of view, as well as its major research challenges in terms of mobile terminals, network system, and mobile services will also be presented in this chapter.

With the rapid development of mobile communication systems, demands for the transmission of multimedia information using portable devices are increasing day by day. Chapter II demonstrates the effectiveness of transport layer handoff schemes for multimedia transmission, and is compared with that achieved by mobile Internet protocol (IP).

Understating the parameters that determine the suitability of a communication standard for the transmission of multimedia information for a particular application is of paramount importance for mobile system designers, users, and researchers. Chapter III describes multimedia applications needing the services of mobile network systems, and presents the fundamentals of issues involved in the delivery of multimedia content with the desired quality of service (QoS). Current and future challenges in achieving successful mobile multimedia information transmission are also discussed.

The personalized communication system is gaining more importance with an increase of communication ways due to the social reform by the technological advancement. Chapter IV introduces a personalized system for device independent and position aware communication, which is secure, scalable, and an open architecture. This chapter also presents the further research in this field.

The application of multimedia information in treatment or medical diagnosis using portable devices is proliferating with the development of mobile multimedia communication. Chapter V provides an overview of the existing therapy methods based on paper materials and mobile electronic devices in the current literature, and proposes the software solutions for the patients, the careers, and the professional speech and language therapists (SLTs), which could be accessible by mobile devices such as personal digital assistants (PDA). Finally, the recommendations for the direction of future research and development are made in this chapter.

As the mobile services play a role in the reforming of the social culture as well as creating a tremendous business opportunity, it is important to assess the acceptance of these services. The analysis and critical assessment of the existing theoretical acceptance models about the evolving mobile services and their underlying technologies is given in Chapter VI. Chapter VI also introduces an acceptance model for mobile services in the bedrock of the technology acceptance model and recommends the further research directions.

With the significant influence and increasing requirements of visual mobile communications in our day-to-day life, low bit-rate video coding to combat against the stringent bandwidth scarcity of the mobile-networks has become a prime research area in the recent time. Chapter VII presents a review of the image and video coding techniques suitable for mobile communications to provide the readers with a means of appreciating the wealth and radical advancement of the field, and also attempts to enlist and sketch the physical significance of the various important and popular quality metrics of the image and video coding techniques.

In the orthogonal frequency division multiplexing (OFDM) scheme, the total bandwidth can be split into many narrowband sub-channels. Therefore, it can avoid any multiple access interferences in the base station receiver, and use a subcarrier-specific adaptive modulation schemes. In Chapter VIII, a flexible frequency division multiple access scheme based on OFDM-FDMA and a radio resource management (RRM) employing dynamic channel allocation techniques are developed.

Ad-hoc network is another dimension of wireless networks, which has been a demanding field of research because of not having a predefined infrastructure. Wireless nodes are dynamically connected in an arbitrary manner to perform a particular emergency operation. This, therefore, makes the routing of a message promising and challenging. The state-of-the-art, the contrast and comparisons, and the related research challenges of ad-hoc routing protocols have been articulated in Chapter IX.

With the proliferation of network technologies, especially the wireless multimedia communication, Internet and sensor networks, the demand for ubiquitous computing is progressively rising in every aspect of human life, ranging from remotely controlling the home appliances, to security and health monitoring. The overview of ubiquitous networking including its infrastructure, its influence on human lives, its applications and services, and its global evolution are presented in Chapter X.

As with social security and privacy, the security and privacy of all wireless communication services is of paramount importance. Chapter XI aims to present an overview of the security and privacy issues by highlighting the need to secure access to wireless networks and the loss that might accrue from the breach of a network. The vulnerabilities of the IEEE 802.11 and Bluetooth networks, as well as a paradigm for secure wireless networks are presented in this chapter.

Portable devices such as PDA and mobile sets are vulnerable to security threats because of their limited size and processing capabilities, openness, and access to air interface. The security architecture of the mobile devices including the security policy, access control, and physical and storage protection are described in Chapter XII. Chapter XII also articulates the security related to cellular communication networks, Wireless LAN, and Bluetooth.

The security of ad-hoc networks has been a difficult task to the recent research community due to a dynamic network topology, vulnerability of air interface to attacks, mobility of a node in any context and resource constraints. Chapter XIII provides the descriptions of various network layer security attacks of ad-hoc networks, as well as the security protocols that protect network layer from a number of attacks.

Pervasive computing provides the computing and communication technology to integrate the everyday works based on the philosophy of existence everywhere, but access at the same time. Therefore, this poses some inherent limitations that make it even more vulnerable to security attacks than mobile computing. In Chapter XIV, the state-of-the-art of pervasive security, privacy and trust, their challenges and requirements for pervasive applications, and some open issues related in this field have been described.

Finally, this book presents the state-of-the-art and research challenges of mobile computing and its next evolutionary step—ubiquitous and pervasive computing. In addition, the security issues related to these topics have also been elaborately depicted. Therefore, understanding these technologies help readers to have an in-depth idea of mobile multimedia communication infrastructure and its current and future research.

Gour C. Karmakar and Laurence S. Dooley

Acknowledgment

We would like to express our sincerest gratitude and greatest appreciation to the contributors of this book. Without their submissions this book would not have been published. All of the chapters are peer-reviewed by independent reviewers as well as editorial reviewed by one of the editors. We are also grateful to Mary Mathew and reviewers for their sincere help, support and excellent feedbacks that helped us so much in successfully completing this book.

We wish to give special thanks to all of the staff at Idea group Inc. for their whole hearted support starting from the proposal submission to the final publication. Without the help and support from Senior Academic Technology Editor Dr. Mehdi Khosrow-Pour, Managing Director Mrs. Jan Travers, Assistant Business Manager Corrina Chandler and Development Editor Ms. Kristin Roth, this book would not be possible to complete.

We would like to express our gratitude to the Gippsland School of and Information Technology, Monash University, Australia for providing logistic support and other necessary resources for completing this project.

Last but not the least, our special thanks to our families for their love, patience, and sacrifice for the whole project.

Gour C. Karmakar and Laurence S. Dooley

Chapter I
Introduction to Mobile Multimedia Communications

Gour C. Karmakar
Monash University, Australia

Laurence S. Dooley
Monash University, Australia

Michael Mathew
Monash University, Australia

ABSTRACT

In order to meet the ever increasing demand by people using mobile technology and its associated services based on multimedia elements in addition to voice, mobile communication technologies has since evolved from analog to digital and 1G to 4G. This chapter presents a contemporary review of all generations of mobile communication technologies, including their standards. 1G to 3G mobile communication technologies are mainly optimised for voice communication, using circuit switched networks. To provide high transmission mobile services at low cost in all levels of networks—personal, home, and cellular—it is imminent to exploit the merits of all existing technologies such as Bluetooth, WLAN, and HiperLAN, and use IP as a backbone network in 4G mobile communication standards. The key research challenges for mobile terminals, systems, and services for 4G networks are also presented in this chapter.

INTRODUCTION

Multimedia refers to the combination of different types of media elements such as text, audio, image, and video in a digital form which is represented and manipulated by a single electronic device or a single computing platform such as a PC (Chapman & Chapman, 2000). Interactive multimedia provides the interaction facility with users so they can access and exit the system as they wish.

Text comprises a string of alphabets from a particular character set. Image is a visual object consisting of a rectangular pattern of dots or primitive elements—lines, curves, circles, and so on (Halsall, 2001). Examples of images are computer generated graphics and digitized documents, pictures, and graphic arts. Images could be either 2D or 3D. Both text and images are inherently in a digital form. Bandwidth requirement for the transmission of text and image is less than that of high-fidelity audio and video. Audio is generally represented by the amplitude of sound waves, which includes low-fidelity speech during telephonic conversation, high-fidelity CD-quality audio and surround sound.

Video is a sequence of still images or frames displayed in a repaid succession so the human eye cannot pick up their transitions, and hence it creates an illusion of motion. Examples of video include movies, short films, and animation. Both audio and video are continuous time varying signals and analog in nature. For an integrated representation, all media elements must be represented in digital form.

With the rapid growth of the Internet, multimodal representation and interactive facilities of multimedia-based systems, the necessity of required information, accurate presentation, and quick perception, the applications of multimedia are burgeoning in every aspect of human life ranging from home, education, medicine, e-commerce and m-commerce, to airport security. The demands of these applications are met with the different types of multimedia-based services that are a combination of a number of media elements, but not limited to—interactive television (text, audio, and video), video phone and conference (speech and video), computer supported cooperative working (text and images), on-line education (text, images, audio and video), multimedia electronic mail (text, images, and audio,

for example), e-commerce (text, images, audio, and video) and Web and Mobile TV (text, audio, and video) (Halsall, 2001).

Due to the inherent high data rates, especially for audio and video, a number of compression techniques for both audio and video have been introduced, which has made it possible to transmit the video over broadband networks. Even with the advancement of compression technologies, the usual bit rate for a speech signal is 64kbps, while it is 384Kbps for low quality video, and up to 2Mbps for high quality video (Sawada, Tani, Miki, & Maruyama, 1998). The tentative bandwidth requirement for the future networks for enhanced-reality multimedia communications are projected from 1Mbps to 30Gbps for 3D audio and 3D video (Ohya & Miki, 2005).

Requirements for an on-demand real-time telephone network access for various purposes such as business, education, and social, cultural, and psychological factors, the scope of technology has been expanded from land-base fixed communications, to wireless and mobile multimedia communications. With the advent of wireless and mobile networks, people now have the opportunity to communicate with anybody, anywhere, and at any time. The number of mobile users is rapidly increasing all over the world (Rao & Mendoza, 2005; Salzman, Palen, & Harper, 2001). It is estimated the number of worldwide mobile users will be 1.87 billion by the end of 2007 (Garfield, 2004).

To meet the ever-increasing consumer demands and make the multimedia-based services as appealing as possible, the mobile networks have evolved from 1G to 4G. 1G mobile networks were commercially released in 1980. Examples of a 1G network include Advanced Mobile Phone Service (AMPS) and Total Access Communication System (TACS). An AMP is in American roaming, while TACS is European roaming. These are based on cellular analog technology. They were initially provided voice service at a rate of

2.4 kbps (Casal, Schoute, & Prasad, 1999), which had been extended to 19 kbps with the introduction of Cellular Digital Packet Data (CDPD) in the bedrock of 1G analog cellular networks for providing digital data service. Since all of these 1G networks are analog broadcasting, they are not secured and hence vulnerable to intercepting calls through a radio frequency scanner (McCullough, 2004).

For providing secured communication, higher transmission of up to 64 kbps, better quality of signal, as well as meeting ever increasing demand from users, 1G analog mobile standards evolved into 2G digital communication technologies around 1990. 2G mobile communication standards include Global System for Mobile Communication (GSM), Digital AMPS (D-AMPS), cdmaOne (IS-95), Personal Digital Cellular (PDC). Later, some of 2G standards have been enhanced to provide higher bit rate transmission up to 384 kbps, and data services including the Internet, which is known as 2.5G. An example of such an enhancement of GSM is a circuit switched network called GPRS, which is a packet switched network, and provides Internet facilities for mobile users (Andersson, 2001; Wikipedia, 2006).

To increase the capacity up to broadband communication, that is, 2 Mbps, and hence the number of subscribers, the specification of the 3G standard is outlined by IMT2000 considering the hierarchical cell structure, global roaming, and dynamic allocation of a radio frequency from a pool. This enables 3G to provide services such as fast Internet browsing and mobile multimedia communications—mobile TV, video phone, and video conferencing. Despite 3G being adopted as a current standard, its full-fledged implementation is being delayed because of the problems for the introduction of new hardware and dynamic frequency allocation—consequently making 3G obsolete and introducing 4G.

It is expected 4G mobile communication networks will be commercially available in the market within 2010. Its main purpose is to provide high transmission of multimedia data up to 1Gbps based on IP-core networks (Hui & Yeung, 2003; Ohya & Miki, 2005). It will consist of heterogeneous state-of-the-art technologies ranging from ad-hoc sensor networks, to intelligent home appliances, and to provide cutting-edge services including ubiquitous, reality, and personalised communications.

GENERATIONS OF MOBILE MULTIMEDIA NETWORKS

The first mobile telephone standard was introduced by AT&T in St. Louis, Missouri, USA, and was available for public users of this city in 1946. The mobile system (a radio telephone network) was installed on the top of a building in the metropolitan area. All users received their services through this system with a mobile operator, who routed both incoming and outgoing calls to the intended recipients. A high powered radio-based transmitter was used to cover the required area. However, the number of frequencies and hence the channels were very limited. It was difficult to have a dial tone and users had problems to complete calls. The system operated in a half-duplex communication mode, that is, one way communication at a time based on push-to-talk communication mode. The mobile set was very heavy and bulky.

An example of one of the initial mobile telephone standards is the Mobile Telephone Systems (MMT). Because of the demand beyond the capacity and previously mentioned disadvantages, the traditional analog mobile telephone technology was upgraded in 1960, and an Improved Mobile Telephone Service (IMTS) was introduced, which eliminates the push-to-talk mode by providing direct access to the dial telephone networks and full-duplex transmission. In this standard, the usage of high powered radio at 200 watts cre-

Figure 1. Organization of hexagonal cells for a coverage area

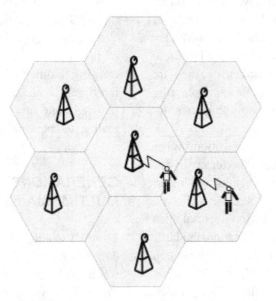

ated interference up to 100 miles, which allowed limited available channel capacity and frequency reuse. The limited channel capacity and excessive demands for communication at anytime from anywhere brought the birth of cellular mobile telephone networks in 1974 (Bates, 1995). In cellular technology, a coverage area is divided into small hexagonal zones known as cells shown in Figure 1.

Each cell has its base station consisting of a transmitter and two receivers per channel, a base station controller, an antenna system, and a data link to the base station controller or cellular office. The average size of a cell is three to five miles, which eventually resulted in a far less coverage area compared with non-cellular technology for a mobile phone, and hence reduced the required transmission power and interference between the neighbouring cells significantly. This made it possible to reuse the frequency and to develop a protection zone in the form of clustering of frequencies for a particular cell against the inter-

ference of its neighbouring cells, which paved the way of increasing the channel capacity of a cell, and hence the mobile communication networks. In order to meet the ever increasing demand for communication at anytime for anywhere using any technology, the cellular mobile communication technology has evolved so far into four generations shown in Table 1.

The description of each generation, including their related network standards are described in the following section.

First Generation (1G) Cellular Mobile Networks

Because of the increasing demands of users, it was necessary to increase the capacity and mobility of mobile telephone users. To meet this demand, the analog cellular technologies were designed and completed by AT&T in 1974 and commercialised in the 1980s. This analog service is still available around the world and played a vital role to increase

Table 1. *Generations of cellular mobile communication networks*

Year	Network Standards	Services	Capacity
1980's	1G Analog AMPS TACS NMT	Voice Small data	2.4 kbps – 9.6 kbps
1990's	2G Digital GSM D-AMPS PDC cdmaOne CSD iDEN	Voice SMS Some data	Up to 48.6 kbps
	2.5 G or 2G+ Digital GPRS HSCSD WiDEN EDGE CMDA2000 1xRTT	Voice Data Email Internet	Up to 384 kbps
2000's	3G Digital W-CDMA CDMA2000 1xEV TD-SCDMA	Fast Internet Mobile multimedia Video Phone Mobile TV Remote access	Up to 2Mbps
2010's	4G Digital Wi-Fi UMB UMTS Revision 8 WiMAX WiBro	Mobile Ubiquitous Reality communications (e.g., virtual reality and 3D applications) Multiple services at the same time Personalised communications	Up to 1Gbps

the popularity of mobile phone and the mobile users at 30 to 50% per year, which contributed to approximately 20 million users within 1990 (IFC, 2005). Examples of 1G generation mobile cellular networks are Advanced Mobile Phone Service (AMPS), Total Access Communication System (TACS), and Nordic Mobile Telephone (NMT).

Advanced Mobile Phone Service (AMPS)

AMPS is the first analog cellular system, which was designed and developed by AT&T in the 1970s in the Bell Labs as a research project, and commercially launched in America in 1984. It was one of the revolutionary achievements in the mobile communication technologies due to the smaller area coverage and low powered output through the use of the cellular technology. The low powered output produces lower interference, and hence it effectively allows frequency reuse in the form of cell clusters. The cells in a cluster may interfere with their neighbouring cells. To produce minimum interference, frequency reuse must be at least two cells apart. AMPS generally uses either 12 or 7-group cell patterns (Wesel, 1998). An example of a 7-group cell pattern with frequency repetition and separation among them is shown in Figure 2.

This directly yielded a higher capacity system compared with a traditional non-cellular analog mobile system. The division of area coverage into hexagonal cellular regions also provided the mobile users with a greater mobility. AMPS uses

Figure 2. 7-group cell pattern with frequency repetition and separation

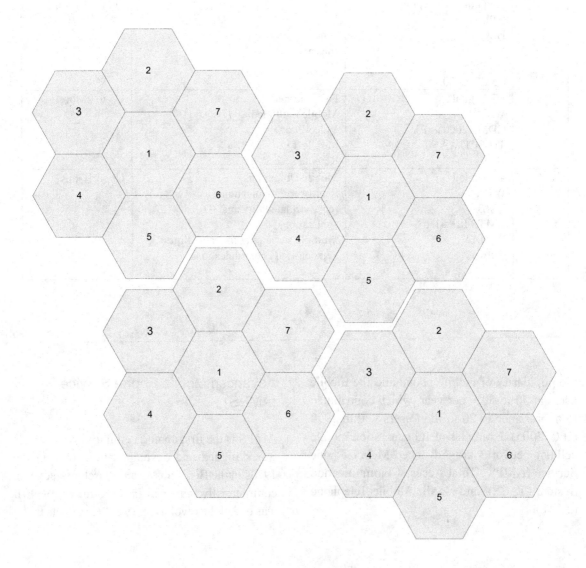

frequency division multiple accesses (FDMA) for voice channels and operates in 800-900 MHz. Initially in 1984, the Federal Communications Commission (FCC) allocated 666 channels (333 channels for each carrier) for two carriers—A and B. Lower frequencies were given to the license B, which was owned by wireline companies, while the upper frequencies were allotted to the license A, owned by non-wireline companies. Due to the rapid growth of the mobile cellular users, in 1989 the FCC had increased the channels from 333 to 416 for each carrier, that is, 832 in total (Wikipedia, 2006). Twenty-one of 416 channels are control channels that carry control data, and the remaining 395 channels are used for two way voice communications.

Despite its analog nature, AMPS has countrywide coverage, it is not possible to encrypt and decrypt messages, and therefore, anyone can intercept the call through a handheld radio frequency scanner. Hence, it is not secure at all. Other disadvantages are: the lifetime of the battery is not long enough, poor quality of the transmitted message, and call charge is costly (ePanorama, 2006). For these reasons, AMPS is being replaced with 2G standards such as Digital AMPS (D-APMS), GSM, and so on.

Total Access Communication System (TACS)

Total Access Communication System (TACS) was introduced in Europe in the late 1970s, and modified after AMPS using FDMA. Therefore, it is an analog mobile communication standard and is known as the European version of AMPS. It is used in the UK, China, and Japan (ePanorama, 2006). Extended TACS called ETACS was introduced in the UK by extending the channel capacity of TACS. Both TACS and ETACS are completely outdated in Europe and replaced by GSM (Wikipedia, 2006).

Nordic Mobile Telephone (NMT)

As with AMPS and TACS, Nordic Mobile Telephone (NMT) is another analog mobile telephone standard and was widely used in Nordic countries. It was introduced in the late 1970s in collaboration with Finland, Sweden, and Norway. There were two versions of NMT—NMT-450 and NMT-900. NMT-450 used 450 MHz frequency, while NMT-900 used 900 MHz. NMT is also completely outdated, and hence it has been replaced by GSM and W-CDMA (Wikipedia, 2006).

Second Generation (2G) Cellular Mobile Networks

The limitations of analog cellular networks have been alluded to in the previous section. To remove these limitations, meeting the ever increasing demand from users, and providing better quality, the countries and individual companies felt the importance for the introduction of digital cellular communication standards called 2G cellular mobile communication standards.

For digital encoding, the 2G standards are able to provide small data transmission in addition to voice communication. Some 2G standards can also provide a few extra advanced services: short messaging service (SMS) and enhanced messaging service (EMS). The digitisation of voice makes it possible to apply filtering or audio compression techniques to reduce the size of the transmitted voice, which leaves some room to multiplex more subscribers into the same radio frequency channel, and produce better quality compared with analog cellular standards by reducing noise. Examples of 2G cellular mobile communication standards include Global System for Mobile Communication (GSM), Digital AMPS (D-AMPS), cdmaOne (IS-95), and Personal Digital Cellular (PDC).

Global System for Mobile Communication (GSM)

Initially, to address many compatibility problems among many digital radio technologies of the European countries, the GSM-Group Special Mobile was formed by the Conference of European Posts and Telegraphs (CEPT) in 1982. After this, the full name of GSM was changed to Global System Mobile Communications, and its responsibility was handed over to the European Telecommunication Standards Institute (ETSI) in 1989 (ePanorama, 2006). The technical specification of GSM was completed in 1990, and it was commercially launched in 1991 from Finland.

From its initial journey, it evolved itself as the most popular standard, and took over 70% of the mobile market with 1.6 billions subscribers in 210 countries all over the world by 2005 (Wikipedia, 2006). The ever increasing popularity of GSM was due to the following main reasons:

- GSM is based on an open system architecture. Therefore, any operator from any country can provide its services using their available hardware.
- It was made compulsory by rule to use GSM in European countries to increase the interoperability.
- GSM first introduced SMS text messaging and pre-paid accounts which are very popular among students and teenagers. Pre-paid accounts also dominated the market of developing countries, since it does not require any accountability or personal verification.
- It can provide a good quality service with moderate level security and can work under a wide range of frequencies.
- The cost is low.
- It can provide international roaming, that is, if a user is overseas, which is not covered by his/her own operator, the user can still receive service by the operator of the visiting

location if there is an agreement between those two operators.

The majority of the GSM networks use 900 MHz or 1800 MHz frequency bands. There are some networks in USA and Canada that use 950 MHz or 1900 MHz frequency bands. In addition to these, GSM can also operate on 400 MHz and 450 MHz frequency bands. For multiplexing, it uses a combination of TDMA and FDMA, which allows the operator to allocate up to eight users, one for each time slot, into a single channel. The modulation technique used in GSM is Gaussian Minimum Shift Keying (GMSK), which effectively reduces the inter-channel interferences.

Like Integrated Services Digital Networks (ISDN), GSM can provide the data services—file or data transfer, and send a fax in up to a maximum speed of 9.6 kbps. GSM has been extended in order to increase its speed for providing faster data services. For example, High Speed Circuit Switched Data (HSCSD) is an expanded form of GSM, which can transmit up to 57.6 kbps by using multiple (e.g., eight) time slots for a single user or connection.

Digital AMPS (D-AMPS)

The digital form of the older analog AMPS standard is referred to as Digital AMPS (D-AMPS) or TDMA/IS-136. It is mainly used in the U.S. For transmitting information, it uses 824-849 MHz frequency band, while for receiving information, a frequency band of 869-894 MHz is utilised. The capacity of each radio frequency channel is 30 KHz. Each channel is divided into a number of time slots for squeezing several messages into a channel using the TDMA technique.

cdmaOne (IS-95)

cdmaOne is also known as IS-95. It is one of the 2G mobile communication standards and was originally introduced based on Code-Division

Multiple Access (CDMA) multiplexing technology in 1993 by Qualcomm. The two versions of IS-95 are IS-95A and IS-95B (Mobile, 2007). The former operates in a single channel at 14.4Kbps, while the latter can support up to eight channels and 115 kbps.

Personal Digital Cellular (PDC)

Personal Digital Cellular (PDC) is a 2G Japanese digital cellular standard. After its introduction, it was widely used in Japan. This standard enforced Japanese to look into another alternative standard, as their old standard was incompatible with all other existing standards. For this reason,

cdmaOne was gaining popularity in Japan in the late 1990s (Andersson, 2001).

Evolutions of GSM

As GSM is a circuit-switched mobile telephone network, it provides the limited amount of data services. In circuit-switched networks, dedicated connection is established for the entire life of a call and users usually need to pay per minute, whether they do or do not use the allocated connection. Another problem is users are charged as per usage time, and if it requires transmitting data through GSM, it needs to set up a dial-up connection. These issues acted as the driving force for the evolution

Figure 3. The architecture of the GPRS network

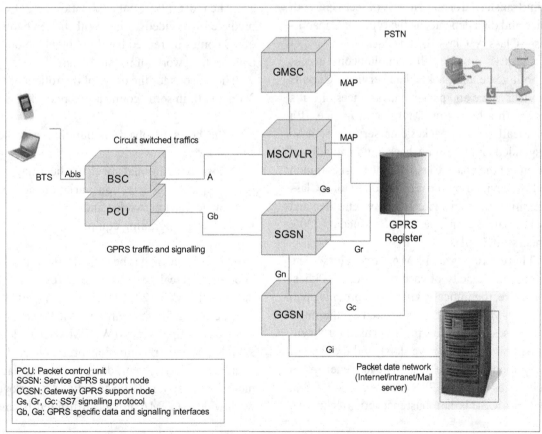

of GSM, and consequently, it evolved itself into a packet switched network, namely General Packet Radio Services (GPRS) for providing the facilities of a data network Internet. An extra network, namely GPRS core network is built on the bedrock of the GSM network, collectively called GPRS network. It allows GSM users to receive both vice and extra data services—mainly Internet browsing, e-mail, file and data transfer, and so forth. The architecture of the GPRS network is shown in Figure 3.

GPRS is regarded as 2.5 generation mobile network. GPRS is always online, that is, it provides the immediate data services whenever users require, just like a leased line or broadband Internet connection in which charges are based on data being transferred, or a flat rate provided by a particular operator. In contrast as mentioned previously, GSM usually charges based on the duration of a call, regardless of whether data is being transferred or not. The way of accessing both voice and data depends on the types of Capabilities—Class A, Class B, and Class C.

Class A provides the simultaneous access of both voice and packet data services. Class B provides voice and packet data services one at a time with a higher priority given to voice. If a voice call arrives, packet data services will be suspended, and resume back automatically at the end of the voice. Class B is the most common and is provided by most of the operators. Class C requires manual switching between voice and packet data by an operator (Wikipedia, 2006; Andersson, 2001).

There are eight TDMA time slots. The maximum capacity of each time slot is 20kbps. Therefore, theoretically GRPS can provide up to 160 kbps, however, this is limited by the multi slot classes (number of time slots used for downlinks, uplinks, and active slots), and the type of encryption techniques used. The four encryption techniques used in GPRS are CS-1, CS-2, CS-3, and CS-4. CS-1 is the robust encryption technique,

and hence it can accommodate 8kbps and cover 98% of the normal coverage, while CS-4 is the least robust encryption technique, which provides the data rate 20kbps and covers only 25% of the normal coverage area (Wikipedia, 2006).

Third Generation (3G) Cellular Mobile Networks

3G is a modern day mobile platform that has replaced the existing 2G standards. Some of the features of 3G articulated by IMT2000 include hierarchical cell structure, global roaming, and an expanding radio spectrum (Akyildiz, McNair, Ho, Uzunalioglu, & Wang, 1999). The claimed benefits of upgrading from the legacy networks mainly used for voice and SMS to 3G are endless. TV direct to your mobile, video calling, fast mobile Web browsing, and remote access are just the icing on the cake with the enormous potential of 3G. The most influential application for 3G was predicted to be video calling. With the deployment of 3G comes increased bandwidth, which opens a whole new world of possibilities.

The reasons for the delay of the rollout of the 3G network in some countries were due to:

- the costs involved in building new networks;
- costs of additional spectrum licensing;
- frequency distribution variation between 3G and 2G networks; and
- costs of upgrading equipment.

However, Japan has been the leader for 3G on a commercial scale, with an expected extinction of the 2G network by 2006 (Hui & Yeung, 2003).

One of the 3G standards is Wideband Code Division Multiple Access (WCDMA) technology, a higher speed transmission protocol used in an advanced 3G system, designed as a replacement for the aging 2G GSM networks deployed worldwide. WCDMA provides a different balance of cost in relation to capacity and performance. WCDMA promises to achieve reduced costs and provide more value for day-to-day applications (Wikipedia, 2006).

3G broadband mobile communications makes access to sophisticated workplace technology inside the 3G handset even faster, making working life more flexible and developing still further the "virtual office" complete with e-mails, video conferencing, and high speed access to services without the daily commute. WCDMA works on a higher frequency in comparison to GSM and therefore travel a shorter distance.

Obviously, the real applications for 3G are the real-time ones. They include video telephony (video conferencing), video streaming, remote wireless surveillance, multimedia real-time gaming, video on demand, and more. These applications will drive usage and increase service-provider revenue. Consequently, they also will raise equipment sales (Hackett, 2001; UMTS World, 2003).

Hierarchical Cell Structure (HCS)

This portrays the different environments and cell structures across a given area.

The capacity is conversely proportional to coverage. The hierarchical cell structure shown in Figure 4 will cover all areas of the mobile user, thereby giving the user wider coverage and mobility starting from the high capacity pico-cell to the large global cell. Since the satellite takes part into the congestion control and it is a global extension to micro and macro cells, the capacity of 3G networks will increase, and hence it is capable of providing the services with more users (Akyildiz et al., 1999).

Figure 4 highlights the different operating environments for the mobile user. The mobile user will access wireless networks using the mobile terminal. This terminal uses radio channels to communicate with the base stations. This gives the user access to the terrestrial networks. As the

Figure 4. The hierarchical cell structure

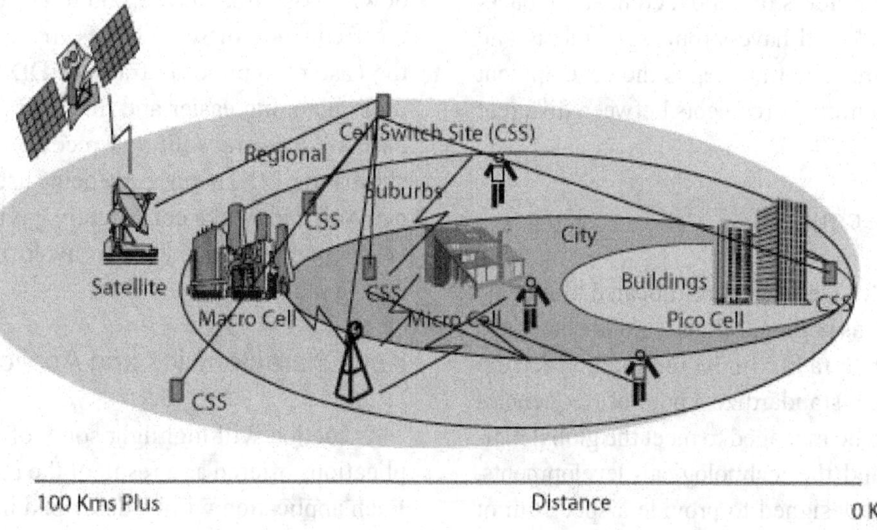

distance increases the bandwidth decreases, with the bandwidth at its maximum in the pico-cell. However, although the quality of service (QoS) drops with distance, the coverage is transformed from base stations to satellite using the fixed Earth stations in areas of remote access. Finally, there are a number of cell site switch (CSS) that govern a number of base stations. These CSSs manage the radio recourses and provide mobility management control, location updates, and manage global roaming.

Global Roaming

The third generation wireless networks begin to implement terminal and personal mobility as well as service provider portability. Terminal mobility is the ability of a network to forward a call to a mobile terminal irrespective of its attachment point to the network. This means if the user is in an area of poor coverage the mobile will automatically shift to a service provider with strong coverage in the area giving the user good QoS. Personal mobility provides the users with an access to their personal services regardless of their attachment points to the network or mobile terminals (Akyildiz et al., 1999).

This level of global mobility freedom will be dependent on the coordination of a wide range of service providers to build a compatible backbone network and have strong agreements and cost structure. The first step is the development of global roaming agreements between different countries.

Radio Spectrum

Previously frequencies were allocated to specific sectors such as paging, cellular, mobile data, and private mobile radio. In 3G the radio spectrum is designed to standardize a pool of frequencies which could be managed to meet the global market needs and the technological developments. IMT2000 is designed to provide a spectrum of

frequencies across different nations to meet with the demands of the mobile world, and hence it provides connections to heterogeneous backbone networks both wired and wireless.

Radio waves used to deliver 3G services are transmitted at a higher frequency than for 2G and travel a shorter distance. This may lead to coverage area or cell size for a 3G base station being smaller than 2G site. Also, with the increase in demand from users in a particular cell, the size of that cell shrinks; the only way to ensure QoS is to overlap between cells. Therefore, the QoS in a high usage environment will decrease due to factors such as conjunction, interference, and static.

The location of cell sites is critical with 3G networks to avoid interference between adjacent cells, which in turn is one of the major issues to be addressed. Researchers are always looking to develop ways to better share the frequency and increase the bandwidth (Ariyoshi, Shima, Han, Karlsson, & Urabe, 2002). One of the solutions implemented is time division duplex (TDD). This application is designed to separate outward and return signals on the frequencies being used. TDD has a strong advantage in the case where the asymmetry of uplink and downlink data speed is variable (Wikipedia, 2006).

Another technique that is used is frequency division duplex (FDD), in which the uplink and downlink sub-bands are said to be separated by the frequency offset. FDD is more efficient in the case of symmetric traffic. FDD also makes radio planning easier and more efficient, as the base stations transmit and receive on different sub-bands. FDD is currently being used by the 3G network and works consistently, giving the user two separate frequencies to avoid interference (Wikipedia, 2006).

Key Characteristics and Application

This section will highlight some of the key applications offered as a result of the rollout of 3G. Each application will be described in relation to practicality in the industry sectors, and critical feedback on the deployment and operations will be discussed. In areas where problems have been identified, a brief overview will cover the cause of the problem with recommended solutions (UMTS World, 2003).

Fixed-Mobile Convergence (FMC)

Fixed-mobile convergence (FMC) is based on the use of dual-mode handsets. However, it is a broad area. For the purposes of this section, FMC means the converging functionality between desk phones and mobiles—set the mobile as a dual identity where the mobile can be addressed as a desk phone and mobile at the same time.

FMC essentially is a system designed to ensure corporate users can, for example, expect the same functionality from their PBX as their mobile and vice versa—same address book, call redirection, call transfer, message bank, and so on.

Some key characteristics of FMC include:

- convergence of desk phone with mobile;
- mobile convergence integrated with VoIP over the PABX;
- functional transparency (address book, message bank, caller redirection, etc.);
- anytime, any place access to employees; and
- cost savings by organisation and clients.

The implementation of FMC has raised a few concerns in regards to employee security, PABX security, routing costs, and implementation of hardware that will enable a mobile to automatically convert to a part of the PABX and vice versa. The implementation of FMC is dependant on a few basic system characteristics such as:

1. Can a mobile have a dual identity?
2. What area of coverage will mobiles act as a part of the PABX and the security measure required?

3. What are the costs associated with implementation of hardware and software that will work over the two different systems?
4. Is it possible for the mobile to have the intelligence to recognize a call between the PABX and the mobile network?

The main advantage of FMC is the increased reach capacity of staff within the organisation. The other benefit is the cost saving structure for the firm routing calls via the PABX system.

Sales Force Automation

The sales force is the backbone of most organisations. 3G services are designed to introduce a new level of flexibility and convenience in the day to day life of the sales force. Figure 5 is an example of areas where 3G in conjunction with sales force automation (SFA) software can add value to the operations within organisation.

Some applications introduced to the sales force are:

1. direct link from the mobile device (lap top or PDA to the company server);
2. access to e-mail on the road;
3. video calls; and
4. wireless broadband service anywhere and anytime.

As 3G speeds increase in the future, and HSDPA rollouts take place, the true potential of such mobile applications is likely to grow. The issue of bandwidth will be history and the connection problem will be rectified with permanently connected devices (Gartner, 2005).

SFA software will provide an interface that will link the user to different departments within the firm accessing real-time updates on order, inventory management, virtual conferences, and client data.

Figure 5. Schematic diagram for mobile sales force automation

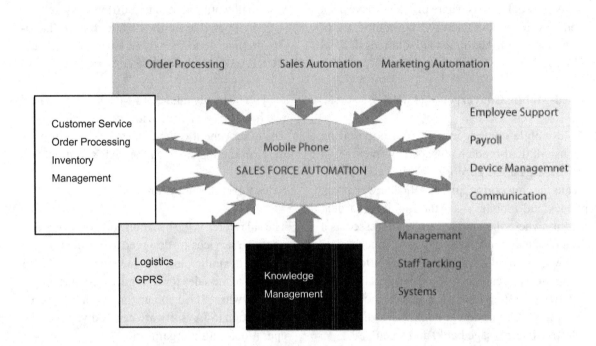

At the moment return on investment may be hard to forecast in terms of dollar value but as the systems develop the question will not be "how much does it cost to implement?" but rather "what are we missing by not have the application?" Some of the benefits driving the deployments are:

- speed of business: agents can have access to client data immediately;
- make changes in real time;
- improving accuracy for logistic and inventory management;
- real-time solutions and access to real-time information; and
- quick response time in coordinating within the organisation.

The current problem is the lack of custom SFA software in the marketplace. In cases where custom built SFA application are required, the cost seems to outweigh the benefits. There is also the case of overloading to many applications onto one small devices for it to be useful, which may turn it into a threat rather than a tool. The trick is getting companies to figure out what the key activities are and where the sales force will benefit from having mobile device access and focus on those.applications.

VoIP Mobile

VOIP integrated mobile services are still on the agenda. The use of the Internet as the sole platform to make calls is still not developed. The biggest benefit of mobile VoIP is having the ability to call over the Internet—in other words, creating huge cost savings. Unfortunately, this assumption is not true; the cost for organisations to set up mobile VoIP still seems to be expensive as it is in its infancy stages. That is not to say that the cost will gradually subside with developments in the field and time.

A number of companies are on the bandwagon trying to produce new developments to ensure quality of service. The two key factors in implementing successful VoIP is: (1) to increase the bandwidth; and (2) to ensure seamless flow of information from one point to another.

One in every five companies are looking into FMC—for most, it is cost savings rather than mobile VoIP, FMC is predicted to be the next big boom. It is believed that mobile VoIP will be an application running on 4G networks rather than the current 3G platforms.

Mobile TV

Mobile TV poses no direct benefit for the firms—companies are concerned it will be a distraction to the employees rather than a benefit. Mobile TV is targeted towards the mass audience rather than companies. For example, a certain target group would use this option to watch a football match or other similar programs.

The problem, however, is the end result—a fan or an enthusiast of a particular sport would prefer to watch the event on a big screen and with the ease of access to a number of venues. Mobile TV does not seem feasible. Businesses could incorporate mobile TV as a sales channel to run their advertising. For example, the GPRS on the phone might pick up a movie theater in the vicinity and automatically play the trailer of a movie.

Corporations will not spend money on unnecessary functionality such as mobile TV. To the general public, mobile TV is not a practical application due to a few reasons: (1) the end result

(looking at a show on such a small screen); (2) the cost of usage; and (3) ease of access to substitute options to view programs.

Video Calling

According to the Shosteck Group, the size of the market for video messaging and calling will reach an estimated US$10 to US$28 billion by 2010, while in 2005, the Yankee Group estimated that already in 2007, video services will generate revenue of more than US$3.5 billion. Video calling was set to be the key element that would differentiate 3G from 2G. The modern user can see the direct benefits of video images and real-time interaction on their phones. There is now no doubt services based on video are emerging as a major source of operator revenue.

Nokia's own estimate is that the market for mobile entertainment and media services, including music, ring tones, mobile TV, and browsing, will have a value of around 67 billion euros in 2010. This market will be flooded with a variety of video entertainment services (Le Maistre, 2002).

Until today, most users believe that 3G refers to video. This is because most service providers use the video calling as a flagship service in 3G launch, the person-to-person video call has already emerged as a key differentiator for 3G technology. Nokia studies show that as far as building 3G awareness among users go, the strategy has worked well in many markets. For example, according to research by Nokia in Hong Kong, Italy, Japan, and Korea in 2005, the vast majority of users in live networks named video calling as the key reason for subscribing to 3G (Orr, 2004).

However, the basic person-to-person video call is not available for EDGE networks, which may limit the service's appeal to operators pursuing joint 3G/EDGE deployment strategies. Lack of EDGE support will mean a lower user base for the service, which limits the value of the service to users as they will be able to contact fewer people.

Some users are also reluctant to subscribe to the service because of privacy concerns—they often do not like the idea of their face being seen on somebody else's terminal first thing in the morning (Lemay, 2006).

Video calling is real time and therefore dependant on the quality of service with demands on the bandwidth and consistent packet interchange without interference, congestion, or static. In reality, the technology of today still does not seem to have the quality of service for video calling because it is directly dependent on real-time transfer of information (UMTS World, 2003).

The Future of 3G

The ways of doing business is changing the usage of IT and communication departments of every business amalgamating into one unit unlike never before, with the IT sector being the driving force for the benchmarks set by communication demand.

The mobile phone is ever changing, providing the user with one key characteristic in today's timeless world—convenience. With the mobile phone integrated into our day-to-day life in more ways than one, the introduction of IP into the communication world will transform its existence into a compact and user-friendly "computer". This will bring in the introduction of 4G.

Fourth Generation (4G) Cellular Mobile Networks

Third generation (3G) mobile networks have given emphasis on the development of a new network standard and hardware such as IMT2000, and the strategy to dynamically allocate frequency spectrum. The major hindrances for the introduction of 3G are the development of new technologies, and how to take away all frequencies

from all countries and put them in a common pool. These problems are the main causes for the delay in the development of 3G networks. In spite of these, 3G can provide 2Mbps into a limited coverage area, that is, up to the macro (urban) cells. To solve these problems and to provide high-speed transmission and wider coverage, fourth generation mobile networks have been targeted to develop as IP-core heterogeneous networks. It is expected to be commercially launched by around 2010. There are a number of characteristics of 4G networks from users', operators', and service providers' point of views. From a user's point of view, 4G networks should have the following characteristics (Hui, 2003):

- The services of any networks ranging from body area to global communication networks including ad-hoc, sensors, and intelligent home appliances, should be available at anytime-anywhere irrespective of users' geographical locations and terminal devices. For example, users located anywhere in the world want an instantaneous access to their required service(s) provided by a number of networks with their existing terminals. This is possible since 4G networks are IP-core.
- High transmission of data including 3D enhanced real-time audio and video information with low transmission cost. The demands of users for multimedia data are rapidly increasing every day. However, concomitantly, users want affordable and lower usage cost. The multimedia information, especially audio and video, requires a high bandwidth, which inherently incurs a high transmission cost since the Internet provides high transmission services with lower cost. To address this catch-22, 4G networks have taken IP as a core or backbone.
- Personalised and integrated services: Personalised services mean all types of users

from different geographical locations and professions are able to receive the services they want in their customised ways using their available terminal devices. Integrated services mean users can simultaneously access multiple services from a number of service providers or networks using their same terminal devices. An example of integrated services is when a particular user with the same terminal device can concurrently browse the Internet using GPRS, make a phone call to their friend using

GSM and see cricket live on television, or watch a program using W-CDMA networks (Hui & Yeung, 2003).

A wide variety of 4G network architectures has been developed, which can be classified into the following two approaches:

1. Interworking: For the interworking among different networks such as ad-hoc, wireless LAN, cellular, digital video broadcast-terrestrial (DVB-T), digital audio broadcast

Figure 6. General architecture of the IP-based 4G network platform based on integration approach

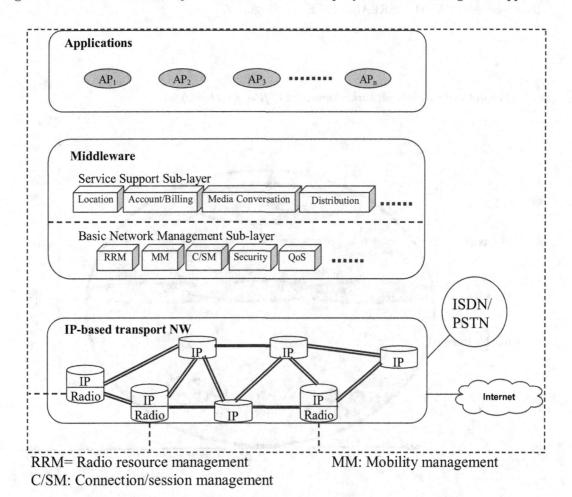

RRM= Radio resource management   MM: Mobility management
C/SM: Connection/session management

(DAB), satellite, and so on, new standard radio interfaces and technologies for 4G will be developed so they can provide seamless access and services to users. An example of 4G mobile networks architecture based on the interworking approach is presented in Muñoz and Rubio (2004), which comprises personal level (e.g., ad-hoc, wireless PAN), local level (e.g., Wireless LANs), national level (e.g., UMTS, GSM, and GPRS), regional level (e.g., DAB and DVB-T), and global coverage (e.g., satellite) networks.

2. Integration: During the development of 3G technologies, there are number of technologies which are being developed separately, such as Bluetooth, wireless local area networks (WLAN—IEEE 802.11), high performance radio LAN (HIPERLAN—IEEE 802.11a), and IMT2000. Each of these technologies has its own advantages and disadvantages. There is no such single technology which can provide the facilities of all these technologies. Therefore, researchers are giving emphasis on the development of 4G architecture by seamlessly integrating these existing technologies. A representative example of these architectures is presented in Figure 6 (Ruiz, 2002).

All existing wireless networks starting from wireless personal area networks (PANs) to global satellite networks are to be connected to an IP-based backbone in a hierarchical order. An example of a hierarchy of wireless networks connected to an IP-based backbone is shown in Figure 7.

Figure 7. Hierarchy of wireless networks connected to IP-based backbone

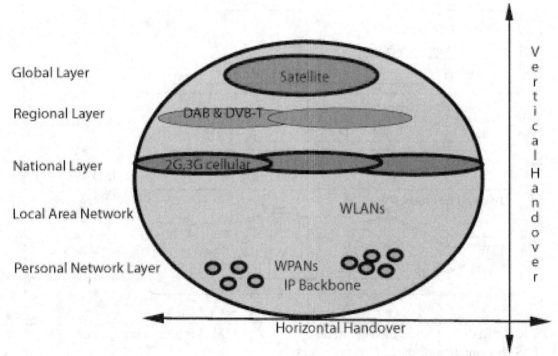

18

This hierarchical structure confirms that when a user is connected to a personal network layer (e.g., in an office, airport, shopping mall, hotel, visitor centre, or train station), the user will receive a high bit rate transmission. In comparison, when a user moves in the upper layers (e.g., highway, beach, and remote area), the user will automatically be connected to the upper layer networks (e.g., WLAN, 2G and 3G cellular, DAB and DVB-T, and satellite). For mobility management, this structure supports both horizontal and vertical handoffs. When a mobile terminal moves through the same network layer, a horizontal handoff takes place. If a mobile terminal moves across the network layers, a vertical handoff takes place. To meet the demands from users, operators, and service provides, and to seamlessly integrate the available network technologies, 4G networks possess a number of research challenges considering the aspects of mobile terminals, network systems, and mobile services that are described as in the following sections (Hui & Yeung, 2003; Ray, 2006).

Key Challenges of Mobile Terminals

Multimode Mobile Terminals

Single mobile terminals should be capable of accessing services from different mobile network technologies by automatically reconfiguring itself with them. The design of such a mobile terminal poses a number of problems related to high cost and power consumption, limitation in terminal size, and interpretational capability with downgraded technologies. An approach based on software radio has been proposed to adapt a mobile terminal with the interfaces of various network technologies (Buracchini, 2000).

Automatic Network System Discovery

A mobile terminal should be able to automatically detect its surrounding available wireless networks by receiving the signals broadcasted by them. For this, the multimodal terminal needs to scan the signal quickly as well as accurately from a number of diverse networks comprising different access protocols. This makes the job difficult and challenging. To address this issue, a software radio is suggested to use in scanning the signals sent from heterogeneous networks.

Automatic Selection of a Network System

Multimode mobile terminal is required to automatically select a suitable wireless network for each service. This may need an optimisation of each user requirement for each session based on available network resources, network service capability, and both application and network quality of service (QoS) parameters. While this can be a time-varying multivariable optimisation technique, it may involve complex intelligent processing of different network characteristics as alluded to earlier, including user profile and preference. Despite a number of solutions which have been introduced considering some of the previously-mentioned criteria, a number of issues are yet to be solved to select a suitable network (Hui & Yeung, 2003).

Key Challenges of a Network System

Terminal Mobility

Terminal mobility means the ability of a communication network to locate a mobile device anytime anywhere in the world, and hence route a call to that particular device irrespective of its network point of attachment. This system facility allows a mobile user to have a global roaming, that is, to roam the different geographical locations all over the world. For this, network systems need to have an efficient mobility management system, which includes both location and handoff managements. Location management mainly locates and updates the location of a mobile terminal, while handoff allows the continuation of a call when a call is in progress, regardless of the movement of a user or terminal across wireless networks. IP-based mobility management has been incorporated in mobile IPv6 (MIPv6) (Hui & Yeung, 2003).

In spite of its enhancement to reduce high handoff latency and packet loss, handoff management remains a fairly complex process due to an unacceptable degradation of QoS for real-time multimedia transmission, and not being able to measure the accurate handoff time. This is because, as shown in Figure 7, 4G requires both horizontal and vertical handoffs across wireless networks. To address this issue, researchers are investigating ways to develop an efficient handoff management scheme.

Adaptive and Light Weight Security System

Much of the existing mobile communication networks are mainly circuit switch based, that is, non-IP core. For this reason, the security standards which have been developed so far are mainly for non-IP based networks, and therefore are not directly applicable to IP-based mobile communication (4G) networks. In addition, to achieve the expected performance of QoS parameters, different standards have adopted different security schemes. For example, GSM has optimised its security for mainly voice delivery (Hui & Yeung, 2003). A particular security scheme is not suitable for 4G networks as it comprises a number of heterogeneous networks. Therefore, a light weight security system is needed to be developed, which can dynamically reconfigure itself considering the network and service types, operating environment, and user preference, and so forth.

Fault Tolerance and Network Survivability

Much effort has not been given to increase the fault tolerance and network survivability of wireless

networks compared with wired networks. While the wireless networks are more vulnerable than wired networks due to its tree type hierarchical structure, if damage occurs in a particular level, it may affect the whole level including all of its underlying levels. This becomes more serious for 4G networks because it consists of heterogeneous multiple tree types networks (Hui & Yeung, 2003).

Key Challenges of Mobile Services

Integrated and Intelligent Billing System

As mentioned in the beginning of this section, users' demand for 4G mobile communication networks to access any service from any network using any mobile device anywhere in the world. This enforces the system to keep record of who uses what service from which service provider, and its related rate information. Each service provided by a particular provider has its own charging scheme with negotiated QoS parameters. For example, in Australia, charges for 3G mobile cricket are different from that of browsing the Internet through a mobile device. Even for the same service, the billing scheme varies from provider to provider. For example for a particular service, namely the Internet, some service providers offer a flat rate, while others charge as per usage, such as time and data. In addition to these, some providers have introduced their charging scheme based on usage time zone, such as peak, non-peak, and free time. These factors make the billing system challenging and complicated, which is difficult enough for users to record and pay all bills they received from each service provider separately. This enforces users to demand a unified single billing system. However, to produce a single bill using all usage history requires an intelligent broker service (Hui & Yeung, 2003). In order to produce such a unified billing system, researchers are now delving into all possibilities.

Figure 8. Personal communication using a personalised video message

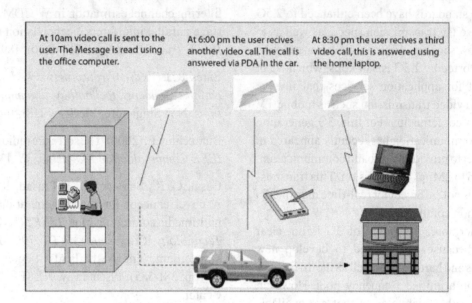

At 10am video call is sent to the user. The Message is read using the office computer.

At 6:00 pm the user recives another video call. The call is answered via PDA in the car.

At 8:30 pm the user recives a third video call, this is answered using the home laptop.

Personal Mobility

Personal mobility is one aspect of mobility management and global roaming. Personal mobility refers to the capability of a user to access any service from any network using any mobile terminal regardless of his network attachment point. Thus, this important characteristic of mobility management enables a user to receive a personalised service through his profile, that is, his preferred location, choice, and mobile devices. An example of a personalised service for a particular user, namely Mary for her video messages, is shown in Figure 8 (Hui & Yeung, 2003).

For this, a user is accessing her personalised video call at different times, and in different operating environments. Researchers have introduced a number of schemes for personal mobility management. Most of them are mainly based on mobile agent.

SUMMARY

The birth of mobile communication is mainly to fulfil people's expectation to communicate with anyone, anytime, anywhere. The transmission of multimedia through mobile communication networks and the competition to present equivalent services of wired communication networks is widely in demand. The elements of multimedia are text, image, audio, and video. The audio includes voice and music. The first generation mobile communication networks were analog, and mainly for voice services. There was no security scheme. The communication speed was also very low and covered a limited distance, and a limited number of users. To address these issues, there was a birth of 2G mobile communication standards which are in digital transmission. In addition, to afford higher voice transmission rates, 2G standards also provide SMS and data services. SMS services have been very popular with young people all over the world. Since 2G standards are unable to support e-mail and Internet services, some 2G standards have been enhanced to 2.5G to increase the transmission speed, as well as to afford those services. The maximum transmission rate supported by 2.5G is 384 kbps, which is not sufficient for application services that involve audio and video transmission such as mobile TV and video conferencing. For this, 3G generation mobile communication has recently appeared in the market to provide broadband communication, that is, up to 2Mbps transmission. This transmission rate is not sufficient for real-time multimedia information communication.

Furthermore, the future of 3G is not clear enough because of the need to develop new standards and hardware, as well as the difficulties to develop a unified frequency pool. However, there are other technologies that are available such as Bluetooth, Wireless LAN, HyperLAN, GPRS, and UMT2000. Each system has their own merits that are worthy to consider for 4G. These factors, including the demand for personalised communication services from any service provider all over the world, 4G mobile communications standards are projected to develop and include existing cutting-edge technology, and future technologies. Since the different standards have different types of services, security and billing systems, it poses a huge key research challenge for mobile terminals, network systems, and services that are also briefly described in this chapter.

REFERENCES

Akyildiz, I. F., McNair, J., Ho, J., Uzunalioglu, H., & Wang, W. (1999). Mobility management in next-generation wireless systems. *Proceedings of the IEEE, 87*(8), 1349-1351.

Andersson, C. (2001). *GPRS and 3G wireless applications.* NY: John Wiley & Sons, Inc.

Ariyoshi, M., Shima, T., Han, J., Karlsson, J., & Urabe, K. (2002). Effect of forward-backward filtering channel estimation in W-CDMA multi-stage parallel interference cancellation receiver. *IEICE Trans. Commun., E85-B*(10), 1898-1905.

Bates, R. J. (1995). *Wireless networked communications: Concepts, technology, and implementations.* NY; Singapore: McGraw-Hill, Inc.

Buracchini, E. (2000). The software radio concept. *IEEE Communication Magazine, 38,* 138-143.

Casal, C. R., Schoute, F., & Prasad, R. (1999). A novel concept for fourth generation mobile multimedia communication. *IEEE 50th Vehicular Technology Conference (VTC 1999-Fall), 1,* Amsterdam, The Netherlands, September 19-22 (pp, 381-385). Piscataway, NJ: IEEE Service Center.

Chapman, N., & Chapman, J. (2000). *Digital multimedia.* London: John Wiley & Sons.

ePanorama. (2006). Retrieved February 2007, from http://www.epanorma.net/links/tele_mobile.html

Garfield, L. (2004). *Infosync: Reporting from the digital.* Retrieved May 2007, from http://www.infosyncworld.com/news/n/5048.html

Gartner. (2005). *Smart phones are favoured as thin clients for mobile workers.* Retrieved May 2007, from http://www.nokia.com/NOKIA_COM_1/About_Nokia/Press/White_Papers/pdf_files/Whitepaper_TheMythsofMobility.pdf

Hackett, S. (2001). *Aglie communication: m.Net Australia consortium wins federal funding for 3G/WLL test bed and applications development environment.* Retrieved May 2007, from http://www.agile.com.au/press/press-29-05-2001.htm

Halsall, F. (2001). *Multimedia communications: Applications, networks, protocols and standards.* Harlow, England; NY: Addision-Wesley.

Hui, S. Y., & Yeung, K. H. (2003). Challenges in the migration to 4G mobile systems. *IEEE Communication Magazine, 41*(12), 54-59.

IFC. (2005). *IFC: Universal mobile telecommunications system (UMTS) proto.* Retrieved May 2007, from http://www.iec.org/online/tutorials/umts/topic01.html

Le Maistre, R. (2002). *US to top 3G chart in 2010 European editor, Unstrung.* Retrieved May 2007, from http://www.unstrung.com/document.asp?doc_id=24887

Lemay, R. (2006). *Perth, Adelaide get optus, Vodafone 3g.* Retrieved May 2007, from http://www.zdnet.com.au/news/communications/soa/perth-adelaide-get-optus-vodafone-3g/0,130061791,139261668,00.htm

McCullough, J. (2004). *185 wireless secrets: Unleash the power of PDAs cell phones, and wireless networks.* Indianapolis, IN: Wiley Publishing, Inc.

Mobile. (2007). *Mobile computing.* Retrieved May 2007, from http://searchmobilecomputing.techtarget.com/sDefinition/0,,sid40_gci506042,00.html

Muñoz, M., & Rubio, C. G. (2004). A new model for service and application convergence in B3G/4G networks. *IEEE Wireless Communications, 35*(5), 539-549.

Nokia. (2006). *State of workforce mobility.* Retrieved May 2007, from http://www.nokia.com/NOKIA_COM_1/About_Nokia/Press/White_Papers/pdf_files/Whitepaper_TheMythsofMobility.pdf

Ohya, T., & Miki, T. (2005). Enhanced-reality multimedia communications for 4G mobile networks. *1st International Conference on Multimedia Services Access Networks (MSAN '05),* Orlando, FL, June 13-15 (pp. 69-72). Piscataway, NJ: IEEE Service Center.

Orr, E. (2004). *3G-324M helps 3G live up to its potential.* Retrieved May 2007, from http://www.wsdmag.com/Articles/Print.cfm?ArticleID=7742

Rao, M., & Mendoza, L. (Eds.) (2005). *Asia unplugged: The wireless and mobile media boom in the Asia-Pacific.* New Delhi: Response Books (A Division of Sage Publications).

Ray, S. K. (2006). Fourth generation (4G) networks: Roadmap-migration to the future. *IETE Technical Review, 23,* 253-265.

Ruiz, P. M. (2002). *Beyond 3G: Fourth generation wireless networks.* Retrieved May 2007, from http://internetng.dit.upm.es/ponencias-jing/2002/ruiz/ruiz.PDF

Salzman, M., Palen, L., & Harper, R. (2001). *Mobile communications: Understanding users, adoption and design.* CHI 2001, Seattle, WA, March 31-April 5.

Sawada, M., Tani, N., Miki, M., & Maruyama, Y. (1998). Advanced mobile multimedia services and applied network techniques. *IEEE International Conference on Universal Personal Communications, 1,* 79-85.

UMTS World. (2003). *3G applications.* Retrieved May 2007, from http://www.umtsworld.com/applications/applications.htm

Wesel, E. K. (1998). *Wireless multimedia communications: Networking video, voice, and data.* Addison Wesley.

Wikipedia. (2006). *Wikipedia, the free encyclopedia.* Retrieved October 2006, from http://en.wikipedia.org/wiki/

Chapter II
Multimedia over Wireless Mobile Data Networks

Surendra Kumar Sivagurunathan
University of Oklahoma, USA

Mohammed Atiquzzaman
University of Oklahoma, USA

ABSTRACT

With the proliferation of wireless data networks, there is an increasing interest in carrying multimedia over wireless networks using portable devices such as laptops and personal digital assistants. Mobility gives rise to the need for handoff schemes between wireless access points. In this chapter, we demonstrate the effectiveness of transport layer handoff schemes for multimedia transmission, and compare with Mobile IP, the network layer-based industry standard handoff scheme.

I. INTRODUCTION

Mobile computers such as personal digital assistants (PDA) and laptop computers with multiple network interfaces are becoming very common. Many of the applications that run on a mobile computer involve multimedia, such as video conferencing, audio conferencing, watching live movies, sports, and so forth. This chapter deals with multimedia communication in mobile wireless devices, and, in particular, concentrates on the effect of mobility on streaming multimedia in wireless networks.

Streaming multimedia over wireless networks is a challenging task. Extensive research has been carried out to ensure a smooth and uninterrupted multimedia transmission to a mobile host (MH) over wireless media. The current research thrust is to ensure an uninterrupted multimedia transmission when the MH moves between networks or subnets. Ensuring uninterrupted multimedia transmission during handoff is challenging because the MH is already receiving multimedia from the network to which it is connected; when it moves into another network, it needs to break the connection with the old network and establish a connection with the new network. Figure 1 shows an MH connected to Wireless Network 1; when it moves, it has to make a connection with the new network, say Wireless Network 2. The re-establishment of a new connection takes a considerable amount of time, resulting in the possibility of interruption and resulting loss of multimedia.

The current TCP/IP network infrastructure was not designed for mobility. It does not support handoff between IP networks. For example, a device running a real-time application, such as video conference, cannot play smoothly when the user hands off from one wireless IP network to another, resulting in unsatisfactory performance to the user.

Mobile IP (MIP) (Perkins, 1996), from the Internet Engineering Task Force (IETF), addresses the mobility problem. MIP extends the existing IP protocol to support host mobility, including handoff, by introducing two network entities: home agent (HA) and foreign agent (FA). The HA and FA work together to achieve host mobility. The correspondent node (CN) always communicates with the mobile node (MN) via its home network address, even though MH may not dwell in the home network. For CN to have seamless access to MN, the MH has to be able to handoff in a timely manner between networks.

Handoff latency is one of the most important indicators of handoff performance. Large handoff latency degrades performance of real-time applications. For example, large handoff latency will introduce interruption in a video conference due to breaks in both audio and video data transmission. In addition to high handoff latency, MIP suffers from a number of other problems including triangle routing, high signaling traffic with the HA, and so forth. A number of approaches to reduce the MIP handoff latency are given next.

Mobile IP uses only one IP; a certain amount of latency in data transmission appears to be unavoidable when the MH performs a handoff. This is because of MN's inability to communicate with the CN through either the old path (because it has changed its wireless link to a new wireless network) or the new path (because HA has not yet granted its registration request). Thus, MH cannot send or receive data to or from the CN while the MH is performing registration, resulting in interruption of data communication during this time interval. This interruption is unacceptable in a real-world scenario, and may hinder the widespread deployment of real-time multimedia applications on wireless mobile networks. Seamless IP-diversity based generalized mobility architecture (SIGMA) overcomes the issue of discontinuity by exploiting multi-homing (Stewart, 2005) to keep the old data path alive until the new data path is ready to take over the data transfer, thus achieving lower latency and lower loss during handoff between adjacent subnets than Mobile IP.

The *objective* of this chapter is to demonstrate the effectiveness of SIGMA in reducing handoff latency, packet loss, and so forth, for multimedia transmission, and compare with that achieved by Mobile IP. The *contribution* of this chapter is to describe the implementation of a real-time streaming server and client in SIGMA to achieve seamless multimedia streaming during handoff. SIGMA *differs* from previous work in the sense that all previous attempts modified the hardware, infrastructure of the network, server, or client to achieve seamless multimedia transmission during handoff.

The rest of this chapter is organized as follows. Previous work on multimedia over wireless networks is described in the next section. The architecture of SIGMA is described in the third section, followed by the testbed on which video transmission has been tested for both MIP and SIGMA in the fourth section. Results of video over MIP and SIGMA and presented and compared

Figure 1. Illustration of handoff with mobile node connected to Wireless Network 1

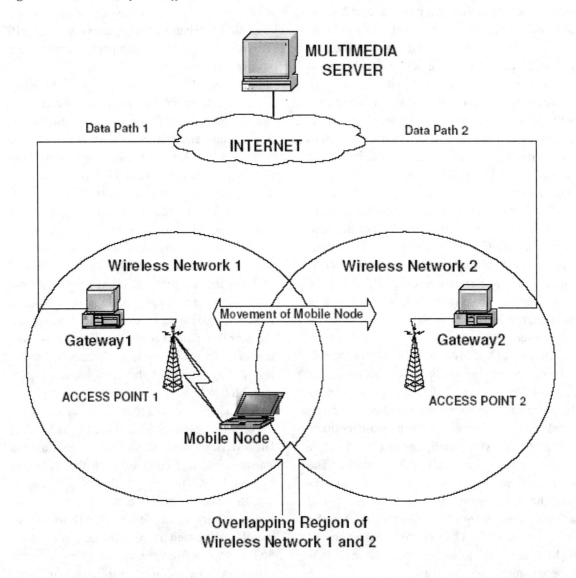

in the fifth section, followed by conclusions in the last section.

BACKGROUND

A large amount of work has been carried out to improve the quality of multimedia over wire-

less networks. They can be categorized into two types:

• Studies related to improving multimedia (e.g., video or audio) over wireless networks. They do not consider the mobility of the MN, but attempt to provide a high quality multimedia transmission within the same wireless network for stationary servers and clients.

- Studies related to achieving seamless multimedia transmission during handoffs. They consider mobility of the MH and try to provide a seamless and high quality multimedia transmission when the MH (client) moves from one network to another.

Although our interest in this chapter is seamless multimedia transmission during handoffs, we describe previous work on both categories in the following sections.

Multimedia over Wireless Networks

Ahmed, Mehaoua, and Buridant (2001) worked on improving the quality of MPEG-4 transmission on wireless using differentiated services (Diffserv). They investigated QoS provisioning between MPEG-4 video application and Diffserv networks. To achieve the best possible QoS, all the components involved in the transmission process must collaborate. For example, the server must use stream properties to describe the QoS requirement for each stream to the network. They propose a solution by distinguishing the video data into important video data and less important video data (such as complementary raw data). Packets which are marked as less important are dropped in the first case if there is any congestion, so that the receiver can regenerate the video with the received important information.

Budagavi and Gibson (2001) improved the performance of video over wireless channels by multiframe video coding. The multiframe coder uses the redundancy that exists across multiple frames in a typical video conferencing sequence so that additional compression can be achieved using their multiframe-block motion compensation (MF-BMC) approach. They modeled the error propagation using the Markov chain, and concluded that use of multiple frames in motion increases the robustness. Their proposed MF-

BMC scheme has been shown to be more robust on wireless networks when compared to the base-level H.263 codec which uses single frame-block motion compensation (SF-BMC).

There are a number of studies, such as Stedman, Gharavi, Hanzo, and Steele (1993), Illgner and Lappe (1995), Khansari, Jalai, Dubois, and Mermelstein (1996), and Hanzo and Streit (1995), which concentrate on improving quality of multimedia over wireless networks. Since we are only interested in studies that focus on achieving seamless multimedia transmission during handoff, we do not go into details of studies related to multimedia over wireless networks. Interested readers can use the references given earlier in this paragraph.

Seamless Multimedia over Mobile Networks

Lee, Lee, and Kim (2004) achieved seamless MPEG-4 streaming over a wireless LAN using Mobile IP. They achieved this by implementing packet forwarding with buffering mechanisms in the foreign agent (FA) and performed pre-buffering adjustment in a streaming client. Insufficient pre-buffered data, which is not enough to overcome the discontinuity of data transmission during the handoff period, will result in disruption in playback. Moreover, too much of pre-buffered data wastes memory and delays the starting time of playback. Find the optimal pre-buffering time is, therefore, an important issue in this approach.

Patanapongpibul and Mapp (2003) enable the MH to select the best point of attachment by having all the reachable router advertisements (RA) in a RA cache. RA cache will have the entire router's link whose advertisements are heard by the mobile node. These RAs are arranged in the cache according to a certain priority. The priority is based on two criteria: (1) the link signal strength, that is, signal quality and SNR level, and (2) the time since the RA entry was last updated. So the RAs with highest router priority are forwarded to the IP packet handler for processing. The dis-advantage of this method includes extra memory for the RA cache.

Pan, Lee, Kim, and Suda (2004) insert four components in the transport layer of the video server and the client. These four components are: (1) a path management module, (2) a multipath distributor module at the sender, (3) a pair of rate control modules, and (4) a multipath collector module at the receiver. They achieve a seamless video by transferring the video over multiple paths to the destination during handoffs. The overhead of the proposed scheme is two-fold: reduction in transmission efficiency due to transmission of duplicated video packets and transmission of control packets associated with the proposed scheme, and processing of the proposed scheme at the sender and receiver.

Boukerche, Hong, and Jacob (2003) propose a two-phase handoff scheme to support synchroni-zation of multimedia units (MMU) for wireless clients and distributed multimedia systems. This scheme is proposed for managing MMUs to deliver them to mobile hosts on time. The two-phase scheme consists of: setup handoff and end handoff. In the first phase, setup handoff procedure has two major tasks: updating new arrival BSs and maintaining the synchronization for newly arrived mobile hosts (MHs). If an MH can reach another BS, then MH reports "new BS arrived" to its primary BS. End handoff procedure deals with the ordering of MMUs and with the flow of MMUs for a new MH. Any base station can be a new primary base station. The algorithm notifies MHs, BSs, and servers, and then chooses the clos-est common node from the current primary base station and new base stations. This method suffers from the disadvantage of additional overhead of updating the base station (BS) with newly arrived BSs and ordering of MMUs.

SIGMA FOR SEAMLESS MULTIMEDIA IN MOBILE NETWORKS

Limitations of previously proposed schemes in achieving seamless multimedia transmission during handoff in a wireless environment have been discussed in the previous section. In this section, we will discuss our proposed handoff scheme, called SIGMA, which has been designed for seamless multimedia transmission during handoffs, followed by its advantages over previ-ous schemes.

Introduction to SIGMA

To aid the reader in getting a better understanding of SIGMA, in this section, we describe the various steps involved in a SIGMA handoff. A detailed description of SIGMA can be found in Fu, Ma, Atiquzzaman, and Lee (2005). We will use the stream control transmission protocol (Stewart, 2005), a new emerging transport layer protocol from IETF, to illustrate SIGMA.

Stream control transmission protocol's (SCTP) multi-homing (see Figure 2) allows an association between two endpoints to span across multiple IP addresses or network interface cards. One of the addresses is designated as the primary while the other can be used as a backup, in the case of failure of the primary address, or when the upper layer application explicitly requests the use of the backup. Retransmission of lost packets can also be done over the secondary address. The built-in support for multi-homed endpoints by SCTP is especially useful in environments that require high-availability of the applications, such as Signaling System 7 (SS7) transport. A multi-homed SCTP association can speedup recovery from link failure situations without interrupting any ongoing data transfer. Figure 2 presents an example of SCTP multi-homing where two nodes,

CN and MH, are connected through two wireless networks, with MH being multi-homed. One of MN's IP addresses is assigned as the primary address for use by CN for transmitting data packets; the other IP address can be used as a backup in case of primary address failure.

STEP 1: Obtain New IP Address

Referring to Figure 2, the handoff preparation procedure begins when the MH moves into the overlapping radio coverage area of two adjacent subnets. Once the MH receives the router advertisement from the new access router (AR2), it should initiate the procedure of obtaining a new IP address (IP2 in Figure 2). This can be accomplished through several methods: DHCP, DHCPv6, or IPv6 Stateless Address Autoconfiguration (SAA) (Thomson & Narten, 1998). The main difference between these methods lies in whether the IP address is generated by a server (DHCP/DHCPv6)

or by the MH itself (IPv6 SAA). For cases where the MH is not concerned about its IP address but only requires the address to be unique and routable, IPv6 SAA is a preferred method for SIGMA to obtain a new address since it significantly reduces the required signaling time.

STEP 2: Add IP Addresses to Association

When the SCTP association is initially setup, only the CN's IP address and the MH's first IP address (IP1) are exchanged between CN and MH. After the MH obtains another IP address (IP2 in STEP 1), MH should bind IP2 into the association (in addition to IP1) and notify CN about the availability of the new IP address (Fu, Ma, Atiquzzaman, & Lee, 2005).

SCTP provides a graceful method to modify an existing association when the MH wishes to notify the CN that a new IP address will be added

Figure 2. An SCTP association featuring multi-homing

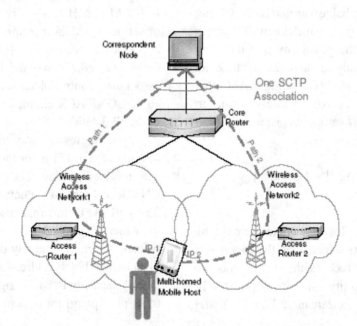

to the association and the old IP addresses will probably be taken out of the association. The IETF Transport Area Working Group (TSVWG) is working on the "SCTP Address Dynamic Reconfiguration" Internet draft (Stewart, 2005), which defines two new chunk types (ASCONF and ASCONF-ACK) and several parameter types (Add IP Address, Delete IP address, Set Primary Address, etc.). This option will be very useful in mobile environments for supporting service reconfiguration without interrupting on-going data transfers.

In SIGMA, MH notifies CN that IP2 is available for data transmission by sending an ASCONF chunk to CN. On receipt of this chunk, CN will add IP2 to its local control block for the association and reply to MH with an ASCONF-ACK chunk indicating the success of the IP addition. At this time, IP1 and IP2 are both ready for receiving data transmitted from CN to MH.

STEP 3: Redirect Data Packets to New IP Address

When MH moves further into the coverage area of wireless access network2, data path2 becomes increasingly more reliable than data path1. CN can then redirect data traffic to the new IP address (IP2) to increase the possibility of data being delivered successfully to the MH. This task can be accomplished by the MH sending an ASCONF chunk with the Set-Primary-Address parameter, which results in CN setting its primary destination address to MH as IP2.

STEP 4: Updating the Location Manager

SIGMA supports location management by employing a location manager that maintains a database which records the correspondence between MH's identity and current primary IP address (Reaz, Atiquzzaman, & Fu, 2005). MH can use any unique information as its identity, such as the home address (as in MIP), domain name, or a public key defined in the public key infrastructure (PKI).

Following our example, once the Set-Primary-Address action is completed successfully, MH should update the location manager's relevant entry with the new IP address (IP2). The purpose of this procedure is to ensure that after MH moves from the wireless access network1 into network2, further association setup requests can be routed to MH's new IP address IP2. This update has no impact on existing active associations.

We can observe an important difference between SIGMA and MIP: the location management and data traffic forwarding functions are coupled together in MIP, whereas they are *decoupled in SIGMA to speedup handoff and make the deployment more flexible*.

STEP 5: Delete or Deactivate Obsolete IP Address

When MH moves out of the coverage of wireless access network1, no *new* or *retransmitted* data packets should be directed to address IP1. In SIGMA, MH can notify CN that IP1 is out of service for data transmission by sending an ASCONF chunk to CN (Delete IP Address). Once received, CN will delete IP1 from its local association control block and reply to MH with an ASCONF-ACK chunk indicating the success of the IP deletion.

A less aggressive way to prevent CN from sending data to IP1 is for the MH to advertise a zero receiver window (corresponding to IP1) to CN (Goff, Moronski, Phatak, & Gupta, 2000). This will give CN an impression that the interface (on which IP1 is bound) buffer is full and cannot receive any more data. By deactivating instead of deleting the IP address, SIGMA can adapt more gracefully to MH's zigzag (often referred to as ping pong) movement patterns and reuse the previously obtained IP address (IP1), as long as the lifetime of IP1 has not expired. This will reduce the latency and signaling traffic that would have otherwise been caused by obtaining a new IP address.

Timing Diagram of SIGMA

Figure 3 summarizes the signaling sequences involved in SIGMA. Here we assume IPv6 SAA and MH initiated Set-Primary-Address. Timing diagrams for other scenarios can be drawn simi-

larly, but are not shown here because of space limitations. In this figure, the numbers before the events correspond to the step numbers in the previous sub-sections, respectively.

Advantages of SIGMA over the Previous Works

A number of previous work have considered seamless multimedia transmission during handoff, as mentioned in the second section, which have their own disadvantages. Here, we discuss the

Figure 3. Timeline of signaling in SIGMA

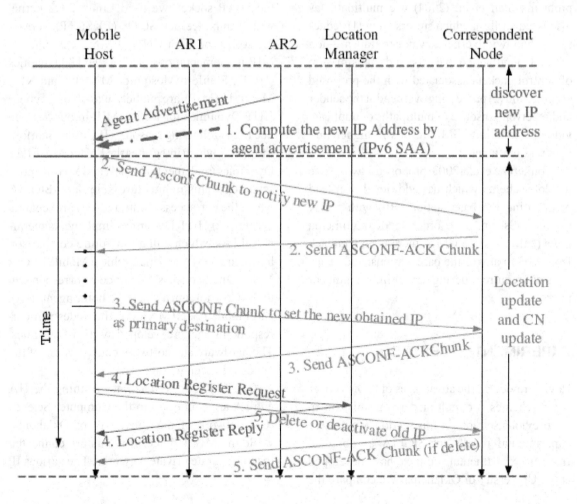

advantages of SIGMA over previous work. Lee et al. (2004) performed pre-buffering adjustment in client. Playback disruption may occur if the pre-buffered data is not enough to overcome the discontinuity of data transmission that occurs during handoff. Moreover, excessive pre-buffered data wastes memory usage and delays the starting time of playback. Find the optimal pre-buffering time is an important issue in this approach. Since SIGMA does not pre-buffer any data in the client, such optimization issues are not present in SIGMA.

Patanapongpibul et al. (2003) use the router advertisement (RA) cache. The disadvantage of this method is that it needs extra memory for RA cache; SIGMA does not involve any caching and hence does not suffer from such memory problems. Pan et al. (2004) use multipath (as discussed earlier), which suffers from (1) reduction in bandwidth efficiency due to transmission of duplicated video packets and transmission of control packets associated with the proposed scheme, and (2) processing overhead at the sender and receiver. Absence of multipaths or duplicate video packets in SIGMA results in higher link bandwidth efficiency.

Boukerche et al. (2003) proposed a two-phase handoff scheme which has additional overhead of updating the base station (BS) with newly arrived BSs, and also ordering of multimedia units (MMUs). In SIGMA, there is no feedback from MH to any of the base stations, and hence does not require ordering of multimedia units or packets.

EXPERIMENTAL TESTBED

Having reviewed the advantages of SIGMA over other schemes for multimedia transmission in the previous section, in this section, we present experimental results for SIGMA as obtained from an experimental setup we have developed at the University of Oklahoma. We compare the results of handoff performance during multimedia transmission over both SIGMA and Mobile IP. To make a fair comparison, we have used the same test bed for both MIP and SIGMA. Figure 4 (to be described later) shows the topology of our test bed, which has been used by a number of researchers—Seol, Kim, Yu, and Lee (2002), Wu, Banerjee, Basu, and Das (2003), Onoe, Atsumi, Sato, and Mizuno (2001)—for measurement of handoff performance. The difference in data communication between the CN and the MH for MIP and SIGMA lies in the lower layer sockets: the file sender for MIP is based on the regular TCP socket, while that for SIGMA is based on SCTP socket. We did not use the traditional *ftp* program for file transfer because it was not available for the SCTP protocol. To obtain access to the SCTP socket, we used Linux 2.6.2 kernel with Linux Kernel SCTP (LKSCTP) version 2.6.2-0.9.0 on both CN and MN. A number of MIP implementations, such as HUT Dynamics (HUT), Stanford Mosquito (MNET), and NUS Mobile IP (MIP), are publicly available. We chose HUT Dynamics for testing MIP in our test bed due to the following reasons: (1) Unlike Stanford Mosquito, which integrates the FA and MN, HUT Dynamics implements HA, FA, and MH daemons separately. This architecture is similar to SIGMA where the two access points and MH are separate entities. (2) HUT Dynamics implements hierarchical FAs, which will allow future comparison between SIGMA and hierarchical Mobile IP. Our MIP testbed consists four nodes: correspondent node (CN), foreign agent (FA), home agent (HA), and mobile node (MN). All the nodes run corresponding agents developed by HUT Dynamics. The hardware and software configuration of the nodes are given in Table 1.

The CN and the machines running the HA and FA are connected to the Computer Science (CS) network of the University of Oklahoma, while the MH and access points are connected to two separate private networks. The various IP

Table I. Mobile IP and SIGMA testbed configurations

Node	Hardware	Software	Operating System
Home Agent(MIP) Gateway1 (SIGMA)	Desktop, two NICs	HUT Dynamics 0.8.1 Home Agent Daemon (MIP)	Redhat Linux 9 kernel 2.4.20
Foreign Agent (MIP) Gateway2 (SIGMA)	Desktop, two NICs	HUT Dynamics 0.8.1 Foreign Agent Daemon (MIP)	Redhat Linux 9 kernel 2.4.20
Mobile Node	Dell Inspiron- 1100 Laptop, one Avaya 802.11b wireless card	HUT Dynamics 0.8.1 Mobile Node Daemon (MIP), File receiver	Redhat Linux 9 kernel 2.4.20
Correspondent Node	Desktop, one NIC	File sender	Redhat Linux 9 2.6.20

Table 2. Mobile IP and SIGMA network configurations

Node	Network Configuration
Home Agent (MIP) Gateway1 (SIGMA)	eth0: 129.15.78.171, gateway 129.15.78.172; eth1:10.1.8.1
Foreign Agent (MIP) Gateway2 (SIGMA)	eth0: 129.15.78.172 gateway 129.15.78.171; eth1: 10.1.6.1
Mobile Node	Mobile IP's Home Address: 10.1.8.5 SIGMA's IP1: 10.1.8.100 SIGMA's IP2 : 10.1.6.100
Correspondent Node	129.15.78.150

addresses are shown in Table 2. IEEE 802.11b is used to connect the MH to the access points.

The network topology of SIGMA is similar to the one of Mobile IP except that there is no HA or FA in SIGMA. As shown in Figure 4, the machines which run the HA and FA in the case of MIP act as gateways in the case of SIGMA. Table 1 shows the hardware and software configuration for the SIGMA experiment. The various IP addresses are shown in Table 2. The experimental procedure of Mobile IP and SIGMA is given next:

1. Start with the MH in Domain 1.
2. **For Mobile IP:** Run HUT Dynamics daemons for HA, FA, and MN. **For SIGMA:** Run the SIGMA handoff program, which has two functions: (1) monitoring the link layer signal strength to determine the time to handoff, and (2) carrying out the signaling shown in Figure 4.
3. Run file sender/video server and file receiver/ video client (using TCP sockets for Mobile

Figure 4. SIGMA and Mobile IP testbed

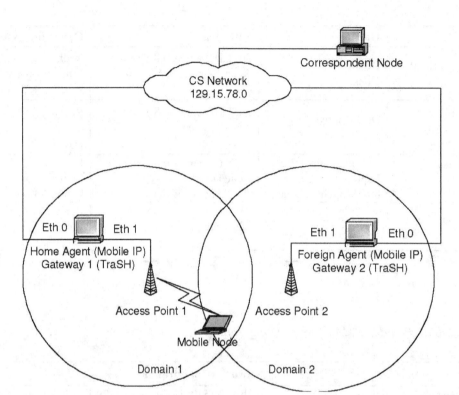

IP, using SCTP sockets for SIGMA) on CN and MN, respectively.

4. Run Ethereal (ETHEREAL) on the CN and MH to capture packets.

5. Move MH from Domain 1 to Domain 2 to perform handoff by Mobile IP and SIGMA. Capture all packets sent from CN and received at MN.

RESULTS

Various results were collected on the experimental setup and procedure described earlier. In this section, we present two kinds of results: file transfer and multimedia transmission. The reason for showing the results of file transfer is to prove that SIGMA achieves seamless handoff not only for multimedia but also for file transfers.

Results for File Transfer

In this section, we present and compare the results of handoffs using MIP and SIGMA for file transfer. For comparison, we use throughput, RTT, and handoff latency as the performance measures. *Throughput* is measured by the rate at which packets are received at the MN. *RTT* is the time required for a data packet to travel from the source to the destination and back. We define *handoff latency* as the time interval between the MH receiving the last packet from Domain

Figure 5. Throughput during MIP handoff

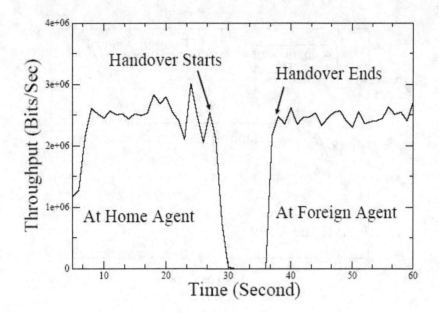

Figure 6. Packet trace during MIP handoff

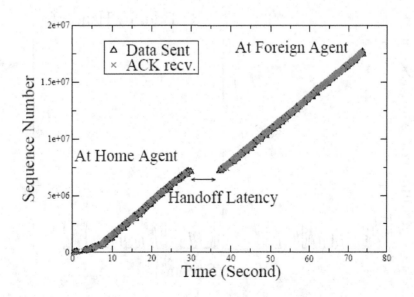

Figure 7. Zoomed in view during MIP handoff instant

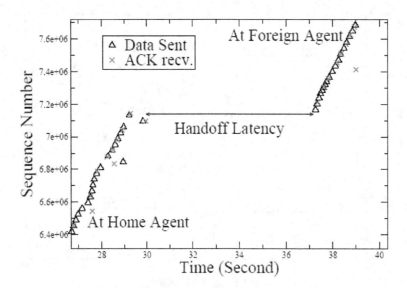

Figure 8. RTT during MIP handoff

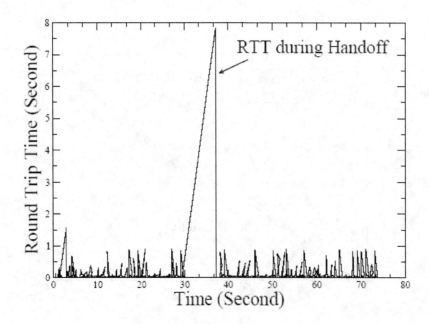

1 (previous network) and the first packet from Domain 2 (the new network). The experimental results are described next.

Results from Mobile IP Handoff

Figure 5 shows the throughput during Mobile IP handoff between Domain 1 and Domain 2. The variations in throughput within HA (from 20 second to 30 second) and within FA (from 37 second to 60 second) are due to network congestion arising from cross traffic in the production CS network.

The average throughput before, during and after handoff are 2.436 Mbps, 0 Mbps and 2.390 Mbps, respectively. Figure 6 shows the packet trace during MIP handoff. The actual handoff latency for MIP can be clearly calculated by having a zoomed-in view of the packet trace graph. Figure 7 shows a zoomed-in view of the packet trace, where the calculated handoff latency is eight seconds for Mobile IP. Figure 8 shows the RTT for the MIP handoff. As we can see, the RTT is high for eight seconds (the handoff latency time), during the handoff.

The registration time (or registration latency) is also a part of the handoff latency. Registration latency, the time taken by the MH to register with the agent (HA or FA), is calculated as follows. Ethereal capture showed that the MH sent a registration request to the HA at time t = 14.5123 second and received a reply from the HA at t = 14.5180 second. Hence, the calculated registration time for registering with HA is 5.7 milliseconds. Similarly, during MIP handoff, Ethereal capture showed that the MH sent a registration request to FA at time t =7.1190 second and received a reply from the FA at t =7.2374, resulting in a registration time of 38.3 milliseconds. This is due to the fact that after the MH registers with the HA, it can directly register with the HA. On the other hand, if it registers with the FA, the MH registers each new care-of-address with its HA possibly

through FA. The registration latency is, therefore, higher when the MH is in the FA.

Results from SIGMA Handoff

Figure 9 shows the throughput during SIGMA handoff where it can observed that the throughput does not go to zero. The variation in throughput is due to network congestion arising from cross traffic in the production CS network. Although we cannot see the handoff due to it being very small, it should be emphasized that the ethereal capture showed the handoff starting and ending at t = 60.755 and t = 60.761 seconds, respectively, that is, a handoff latency of six milliseconds.

Figure 10 shows the packet trace during SIGMA handoff. It can be seen that packets arrive at the MH without any gap or disruption; this is also a powerful proof of SIGMA's smoother handoff as compared to handoff in Mobile IP. This experimentally demonstrates that *a seamless handoff can be realized with SIGMA*. Figure 11 shows a zoomed-in view of the packet trace during the SIGMA handoff period; a handoff latency of six milliseconds can be seen between the packets arriving at the old and new paths.

Figure 12 shows the RTT during SIGMA handoff. A seamless handoff is evident from the absence of any sudden RTT increase during handoff.

Result of Multimedia Data Transfer

To test the handoff performance for multimedia over SIGMA, we used a streaming video client and a streaming server at the MH and CN, respectively (details in the fourth section). Apple's Darwin Streaming Server (DARWIN) and CISCO's MPEG4IP player (MPEG) were modified to stream data over SCTP. A seamless handoff, with no interruption in the video stream, was achieved with SIGMA.

Figure 9. Throughput during SIGMA handoff

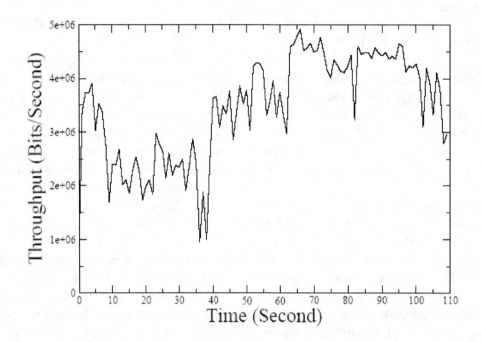

Figure 10. Packet trace during SIGMA handoff

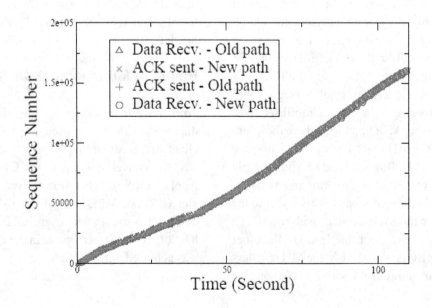

Figure 11. Zoomed in view during SIGMA handoff

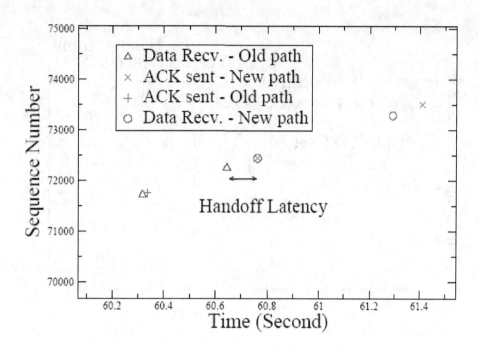

Figure 12. RTT during SIGMA handoff

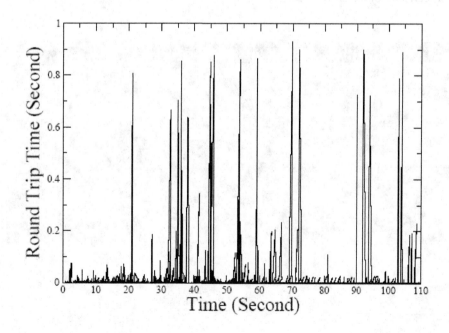

Figure 13. Throughput of video during SIGMA handoff

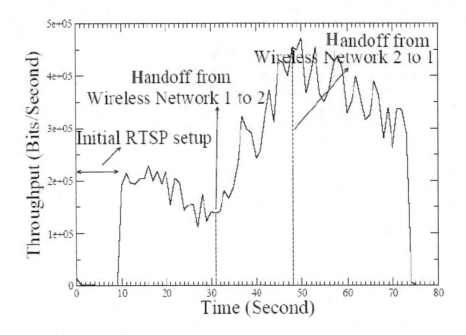

Figure 14. Screen shot of MPEG4-IP player

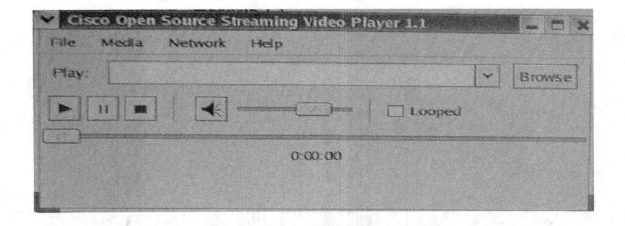

Figure 15. Screen-shot of MPEG4-IP player playing streaming video

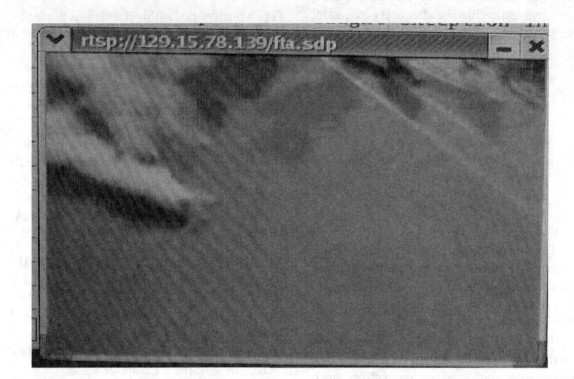

Figure 13 shows the throughput of multimedia (video) data, when the MH moves between subnets. The connection request and setup between the client and server is carried out during the first 10 seconds. It can be seen that the throughput does not drop during handoff at time = 31 second when MH moves from wireless network 1 to 2. A second handoff takes place when the MH moves from network 2 to network 1 at time = 48. It is seen that seamless handoff is achieved by SIGMA for both the handoffs.

Figure 14 shows a screen capture of the MPEG4IP player used in our experiment. Figure 15 shows the video playing in the player during handoff, where "rtsp://129.15.78.139/fta.sdp" rep-

resents the server's IP address and the streaming format (SDP).

Comparison of SIGMA and MIP Handoffs

We observed previously that the registration time of MIP was only 0.1 second, and the handoff latencies of MIP and SIGMA were eight seconds and six milliseconds, respectively. We describe the reasons for the MIP handoff latency being much longer than its registration time in the following:

1. In HUT Dynamics, the MIP implementation used in this study, the MH obtains a registration lifetime after every successful registration. It originates another registration on expiry of this lifetime. So it is possible for the MH to postpone registration even after it has completed a link layer handoff and received FA advertisements. This may introduce some delay which can be up to the duration of a life time.

2. As mentioned in the previous section, the registration of MH also costs some time, measured as 38.3 milliseconds in our testbed.

The handoff latency in MIP comes from three factors: (1) remaining home registration lifetime after link layer handoff which can be from zero to a lifetime, (2) FA advertisement interval plus the time span of last time advertisement which is not listened by MN, and (3) registration latency. During these three times, the CN cannot communicate through either the previous path because it has completed link layer handoff, or the new path because MH has not yet completed the registration. As a result, the throughput was zero during this time. Obviously, such shortcoming has been eliminated in SIGMA through multi-homing and decoupling of registration and data transfer. Consequently, data continue to flow between the CN and MH during the handoff process.

CONCLUSION AND FUTURE TRENDS

We have shown that SIGMA achieves seamless multimedia transmission during handoff between wireless networks. As future work, video streaming can be tested over SIGMA during vertical handoffs, that is, between wireless LANs, cellular, and satellite networks.

ACKNOWLEDGMENT

The work reported in this chapter was funded by National Aeronautics and Space Administration (NASA) grant no. NAG3-2922.

REFERENCES

Ahmed, T., Mehaoua, A., & Buridant, G. (2001). Implementing MPEG-4 video on demand over IP differentiated services. *Global Telecommunications Conference, GLOBECOM*, San Antonio, TX, November 25-29 (pp. 2489-2493). Piscataway, NJ: IEEE.

Boukerche, A., Hong, S., & Jacob, T., (2003). A two-phase handoff management scheme for synchronizing multimedia units over wireless networks. *Proc. Eighth IEEE International Symposium on Computers and Communication*, Antalya, Turkey, June-July (pp. 1078-1084). Los Alamitos, CA: IEEE Computer Society.

Budagavi, M., & Gibson, J. D. (2001, February). Multiframe video coding for improved performance over wireless channels. *IEEE Transactions on Image Processing, 10*(2), 252-265.

DARWIN. Retrieved June 23, 2005, from http://developer.apple.com/darwin/projects/streaming/

ETHEREAL. Retrieved June 30, 2005, from www.ethereal.com

Fu, S., Atiquzzaman, M., Ma, L., & Lee, Y. (2005, November). Signaling cost and performance of SIGMA: A seamless handover scheme for data networks. *Journal of Wireless Communications and Mobile Computing, 5*(7), 825-845.

Fu, S., Ma, L., Atiquzzaman, M., & Lee, Y. (2005). Architecture and performance of SIGMA: A seamless mobility architecture for data networks. *40th IEEE International Conference on Com-*

munications (ICC), Seoul, Korea, May 16-20 (pp. 3249-3253). Institute of Electrical and Electronics Engineers Inc.

Goff, T., Moronski, J., Phatak, D. S., & Gupta, V. (2000). Freeze-TCP: A true end-to-end TCP enhancement mechanism for mobile environments. *IEEE INFOCOM*, Tel Aviv, Israel, March 26-30 (pp. 1537-1545). NY: IEEE.

Hanzo, L., & Streit, J. (1995, August). Adaptive low-rate wireless videophone schemes. *IEEE Trans. Circuits Syst. Video Technol.*, *5*(4), 305-318.

HUT. Retrieved June 1, 2005, from http://www.cs.hut.fi/research/dynamics/

Illgner, R., & Lappe, D. (1995). Mobile multimedia communications in a universal telecommunications network. *Proc. SPIE Conf. Visual Communication Image Processing*, Taipei, Taiwan, May 23-26 (pp. 1034-1043). USA: SPIE.

Khansari, M., Jalai, A., Dubois, E., & Mermelstein, P. (1996, February). Low bit-rate video transmission over fading channels for wireless microcellular system. *IEEE Trans. Circuits Syst. Video Technol.*, *6*(1), 1-11.

Lee, C. H., Lee, D., & Kim, J. W. (2004). Seamless MPEG-4 video streaming over Mobile-IP enabled wireless LAN. *Proceedings of SPIE, Multimedia Systems and Applications*, Philadelphia, Pennsylvania, October (pp. 111-119). USA: SPIE.

LKSCTP. Retrieved June 1, 2005, from http://lksctp.sourceforge.net

MIP. Retrieved June 1, 2005, from opensource.nus.edu.sg/projects/mobileip/mip.html

MNET. Retrieved June 1, 2005, from http://mosquitonet.stanford.edu/

MPEG. Retrieved June 1, 2005, from http://mpeg4ip.sourceforge.net/faq/index.php

Onoe, Y., Atsumi, Y., Sato, F., & Mizuno, T. (2001). A dynamic delayed ack control scheme on Mobile IP networks. *International Conference on Computer Networks and Mobile Computing*, Los Alamitos, CA, October 16-19 (pp. 35-40). Los Alamitos, CA: IEEE Computer Society.

Pan, Y., Lee, M., Kim, J. B., & Suda, T. (2004, May). An end-to-end multipath smooth handoff scheme for streaming media. *IEEE Journal on Selected Areas in Communications, 22*(4), 653-663.

Patanapongpibul, L., & Mapp, G. (2003). A client-based handoff mechanism for Mobile IPv6 wireless networks. *Proc. Eighth IEEE International Symposium on Computers and Communications*, Antalya, Turkey, June-July (pp. 563-568). Los Alamitos, CA: IEEE Computer Society.

Perkins, C. (1996). IP mobility support. *IETF RFC 2002*, October.

Reaz, A. S., Atiquzzaman, M., & Fu, S. (2005). Performance of DNS as location manager. *IEEE Globecom*, St. Louis, MO, November 28-December 2 (pp. 359-363). USA: IEEE Computer Society.

Seol, S., Kim, M., Yu, C., & Lee., J. H. (2002). Experiments and analysis of voice over MobileIP. *13th IEEE International Symposium on Personal, Indoor and Mobile Radio Communications (PIMRC)*, Lisboa, Portugal, September 15-18 (pp. 977-981). Piscataway, NJ: IEEE.

Stedman, R., Gharavi, H., Hanzo, L., & Steele, R. (1993, February). Transmission of subband-coded images via mobile channels. *IEEE Trans. Circuit Syst. Video Technol.*, *3*, 15-27.

Stewart, R. (2005, June). *Stream control transmission protocol (SCTP) dynamic address configuration*. IETF DRAFT, draft-ietf-tsvwgad-dip-sctp-12.txt.

Thomson, S., & Narten, T. (1998, December). *IPv6 stateless address autoconfiguration.* IETF RFC 2462.

Wu, W., Banerjee, N., Basu, K., & Das, S. K. (2003). Network assisted IP mobility support in wireless LANs. *Second IEEE International Symposium on Network Computing and Applications, NCA'03*, Cambridge, MA, April 16-18 (pp. 257-264). Los Alamitos, CA: IEEE Computer Society.

Chapter III
Quality of Service Issues in Mobile Multimedia Transmission

Nalin Sharda
Victoria University, Australia

ABSTRACT

The focus of this chapter is on the quality of service (QoS) aspects involved in transmitting multimedia information via mobile systems. Multimedia content and applications require sophisticated QoS protocols. These protocols need to manage throughput, delay, delay variance, error rate, and cost. How errors are handled in a multimedia session can have significant impact on the delay and delay variance. Coding and compression techniques also influence how the final presentation is transformed by the impediments encountered on a mobile network. Providing the user with the ability to negotiate between cost, quality, and temporal aspects is important, as this allows the user to strike a balance between these factors. In moving from 2G to 3G, and, over the next decade to 4G mobile networks, the ability to transmit multimedia information is going to improve constantly. Nonetheless, providers must develop viable economic models and user interfaces for providing differentiated QoS to the users.

INTRODUCTION

Transmission of multimedia information over mobile networks to portable devices, such as laptops, mobile phones, and PDAs (personal digital assistants), is leading to the development of new applications. However, successful transmission of multimedia information over mobile networks cannot be taken for granted. Understating the impediments to successful transmission of multimedia information is of paramount importance. This chapter focuses on multimedia applications that use mobile networks, and issues involved in the delivery of multimedia content with the desired quality of service (QoS). Current and future challenges in achieving successful mobile multimedia information transmission are also discussed.

Multimedia applications require more sophisticated QoS protocols than those for simple data transmission. The main parameters that underpin QoS are throughput, delay, delay variance, error rate, human perception of quality, and cost (Sharda, 1999). The interplay between these factors is rather complex, therefore, some simplifying assumptions must be made in developing methodologies for delivering multimedia content with the desired QoS.

For the delivery of desired QoS, one of the most promising concepts developed over the last few years is that of resource reservation. This entails reserving resources such as bandwidth on interconnects, and buffer space and processing power on switching nodes.

Packet switching networks embody the idea of statistical time division multiplexing (STDM); that is, resources are allocated to a communication session based on the demands of the traffic. This leads to more efficient, and therefore, more economical usage of the resources. However, the need to allocate resources dynamically adds complexity to the communication system's operation and management. Mobile multimedia communications are further complicated due to their variable transmission quality, the need to keep track of end system location, restrictions placed due to limited battery life, reduced screen size, and the cost of the connection.

Over the last decade, some progress has been made in establishing mobile multimedia transmission systems. However, much research and development is still required before we can take it for granted that a multimedia application, such as videoconferencing, would run with the desired QoS over a mobile communication infrastructure on a hand-held device as we zoom down a freeway at high speed, and, all this at a reasonable cost.

The next section of this chapter presents the challenges introduced by the mobile multimedia content, applications, and communication systems. It begins with an overview of mobile multimedia systems, and then presents the implications of coding and compression techniques for transmitting multimedia. Requirements of various multimedia applications and their relationship to mobile communication systems are also presented.

The third section presents QoS issues in transmitting multimedia content over mobile systems. Fundamentals of QoS concepts and different QoS models are introduced, and a novel model for managing QoS in real time is presented.

The fourth section presents directions for future research, and the final section gives the conclusions.

MOBILE MULTIMEDIA SYSTEMS

Overview

This section presents an overview of coding methods used for various media types, multimedia applications, and current mobile communication systems. QoS issues related to each of these are also discussed.

Multimedia communication systems combine different types of media contents, such as text, audio, still images, and moving images, to achieve the overall objective of a communication session. Therefore, the network needs to provide a service which works well for all media types.

The requirements for successfully transmitting a particular media type depend upon its coding and compression techniques, and the application in which it is being used. Media content that must be transmitted live, or processed in real time, poses more stringent requirements. Consequently, live video conferencing is one of the most challenging multimedia applications.

The network infrastructure and the communications protocols used for transmission play a vital role in satisfying the demands of a given application. In general, multimedia transmission requires high bandwidth, low error rate, low delay, and very low delay variance. To date, we have not solved all of these challenges for even wired media. Fulfilling these requirements for achieving high-quality multimedia transmission over wireless connections is even more challenging.

The transition from the 2nd generation (2G) mobile systems to the 3rd generation (3G) mobile communication infrastructure presents new opportunities; however, still there are many challenging problems that need to be overcome. One of the key features missing in the current systems is the facility for the user to negotiate with the system and strike a compromise between the three key service aspects—quality, cost, and time—just as any market-oriented goods or services have to strike a balance between the quality, cost, and its delivery time.

Errors encountered in any transmission system can be either ignored, or detected and corrected. Errors can be ignored only if the received message is usable even with some errors. If errors in the received message are not acceptable, then these errors must be detected and corrected. Reverse error correction protocol requests retransmission of packets received with errors. This not only adds delay to the final reception of packets, it also adds delay jitter, as different packets encounter different delays. Forward error correction protocols include additional error correction bits, so that some of the errors can be corrected from the received data; this adds to the total data traffic. The choice of error handling method depends upon the type of data, its coding methodology, and the application.

Multimedia Content

By definition, a multimedia system combines different media types: text, audio, still and mov-

ing images. Each of these content types can be further categorised into sub-types. For example, still images can be bi-tonal, greyscale, or full-colour; furthermore, these can have continuous variation in tone—as in a photograph, or have sudden variation in the intensity—as in a printed page. A variety of techniques are used for digitally coding still images, depending upon the image type and application. Similarly, many text representation techniques and associated digital coding techniques are used. Audio and video are even more complex, as these are time varying quantities and involve continuous sampling over time. Errors and delays introduced at any stage of sampling, encoding, transmitting, and decoding of audio and video can lead to reduction in the quality of the final presentation.

Most multimedia content needs to be compressed to reduce the storage space and transmission bandwidth. Uncompressed multimedia content has in-built redundancy, and a few corrupted bits do not change the contents dramatically. Conversely, compressed media is compact, and has much less redundancy. Consequently, any errors during transmission affect compressed content more severely.

Mobile transmission systems are inherently more error-prone than wired transmission systems. The requirements for successfully transmitting a particular media type over a network depend not only on its coding and compression techniques, but on its application as well. However, all multimedia content is for human consumption, therefore, the criteria for acceptable quality of presentation ultimately depends upon human perception. For example, streamed video can accept a few seconds of delay, but live video conferencing becomes rather ineffective if the round-trip delay exceeds even a tenth of a second.

Text Coding

Despite the move towards graphical information, text remains a vital part of any multimedia presentation. One of the most enduring text codes is the American Standard Code for Information Interchange (ASCII). ASCII began its life as a 7-bit code designed for use with teletypes. Today, if someone talks of an ASCII document, they essentially refer to a text document with no formatting. Applications such as Notepad create ASCII text, and word processors can save a file as "text only". Extended ASCII codes were designed for computers to be able to handle additional characters from other languages. It took some time to get a single standard for these additional characters, and there are a few Extended ASCII sets.

Unicode provides a text code that is independent of platform, program, or language. In Unicode, a unique 16-bit number is reserved for every character. The Unicode standard aims to provide a universal repertoire with logical ordering that is efficient. The latest version of Unicode Standard is Unicode 4.0.1, and supports around a hundred international scripts.

ASCII and Unicode have been used extensively over wire-line communication systems, and can be used over wireless media as well. In general, transmission of text codes does not require high bandwidth or stringent limits on delay and delay variance. Hence maintaining QoS in transmitting text is often not much of a problem. Nonetheless, a new code set was designed for sending short text messages over mobile systems.

Short message service (SMS) uses a 7-bit code set that enables one to send and receive text messages of up to 160 characters on mobile phones. Some 8-bit messages are used for sending smart messages (such as images and ring tones) and for changing protocol settings. For Unicode-based text messages, 16-bit codes of maximum 70 characters can be used. These are viewable by most phones, and some appear as a flash SMS, that is, appear on the screen immediately upon arrival, without pressing any button. The SMS code was originally developed for the 2G technology, and therefore works well with 2G as well as 3G systems. The only possible issue with respect to QoS can be errors; bandwidth, delay, and delay jitter do not impede the transmission of SMS messages.

Non-Textual Information

The standard developed to transmit multimedia information over the Internet is the multipurpose Internet mail extensions (MIME). This standard was developed by the Internet Engineering Task Force (IETF) to support the transmission of mixed-media messages across TCP/IP networks. This also became the standard for transmitting foreign language text which the ASCII code could not represent.

Multimedia messaging service (MMS) provides the ability to send messages that combine text, sounds, images, and video over wireless networks. This requires handsets that are MMS capable. MMS is an open wireless standard specified by the WAP (wireless application protocol) forum—which has now been consolidated into the Open Mobile Alliance (OMA). In the WAP protocol, a notification message triggers the receiving terminal to start retrieving the message automatically using the WAP GET command. This retrieval may be modified by applying filters defined by the user. The content that can be transmitted with the WAP protocol can use a variety of media types and encoding standards.

Audio Coding

The basic technique for digitising analog audio signals is called pulse code modulation (PCM). In this technique, an analog audio signal is sampled at a rate double that of the maximum frequency that needs to be captured, and each sample is stored using 8-bit or 16-bit words.

Phone quality audio signals are sampled at 8,000 samples per second, and stored with 8-bit resolution; this generates 64 Kbps data rate. CD quality audio has two channels; it is sampled at 44,000 samples per second and saved with 16-bit resolution, giving a 1.4 Mbps data rate. A variety of compression techniques are used to reduce the bandwidth required to transmit audio signals. Compression becomes particularly important for CD quality stereo music, as the required 1.4 Mbps bandwidth is not economically available even in wire-line networks, much less so in wireless networks.

The MP3 (MPEG audio Layer 3) compression format has become one of the most widely used standards for transmitting high quality stereo audio. MP3 is one of three audio coding schemes associated with the MPEG video compression standard. The MP3 standard provides the highest level of compression and uses perceptual audio coding and psychoacoustic compression to remove all redundant and irrelevant parts of a sound signal that the human ear does not hear. MP3 uses modified discrete cosine transform (MDCT) and improves the frequency resolution 18 times with respect to that of the MPEG audio Layer 2 coding scheme. It manages to reduce the CD bit rate of 1.4 Mbps down to 112-128 Kbps (a factor of 12) without sacrificing sound quality. Since MP3 files are small, they are easily transferred across the Internet, and are also suitable for transmission over wireless networks.

The next generation of MP3 standard is called mp3PRO. It is fully compatible with MP3, while halving the storage and bandwidth requirements. With this standard CD quality stereo can be transmitted at 64 Kbps. Furthermore, it can be used with digital rights management software, and can be ported transparently to any MP3-friendly application.

Advanced audio coding (AAC) is a wideband audio coding algorithm that exploits two main coding strategies to reduce the amount of data needed to encode high-quality digital audio. First, it removes signal components that are not important from a human perception point of view, and second, it eliminates redundancies in the coded audio signal. The MPEG-4 AAC standard incorporates MPEG-2 AAC, for data rates above 32 Kbps per channel. Additional techniques increase the effectiveness of the AAC technique at lower bit rates, and are able to add scalability and/or error resilience. (These techniques extend AAC into its MPEG-4 version: ISO/IEC 14496-3, Subpart 4.) The MPEG-4 aacPlus standard combines advanced audio coding techniques such as spectral band replication (SBR), and parametric stereo (PS). The SBR techniques deliver the same audio quality at half the bit rate, while the PS techniques (optimised for the 16-40 Kbps range) provide high audio quality at bit rates as low as 24 Kbps (Dietz & Meltzer, 2002).

The aacPlus codec family includes two versions. Version 2 of aacPlus is the high quality audio codec targeted for use in the 3GPP (3rd Generation Partnership Project). The aacPlus version 1 standard is adopted by 3GPP2 and ISMA (Internet Streaming Media Alliance) for digital video broadcasting (DVB).

The relationship between the aacPlus codec family members is shown in Figure 1 (Dietz & Meltzer, 2002). To compress the incoming stereo audio, the encoder extracts parametric representation of the stereo aspect of the audio. The stereo parametric information takes 2-3 Kbps and is transmitted along with the mono signal. Based on the parametric representation of the stereo information, the decoder regenerates the stereo signal from the received mono audio signal.

To be able to transmit high quality stereo audio, it is necessary to compress it to reduce the bandwidth, otherwise it may not be possible to obtain the desired QoS, especially over wireless networks. However, high level of compression makes the transmitted signal highly susceptible to errors, especially if the audio is being transmitted in real time. Any loss in the parametric information will severely degrade the quality of the reproduced stereo signal. Stereo is often used for music, and the slightest imperfection in music gets noticed by even non-experts.

Figure 1. Relationship between aacPlus audio codecs v1 and v2 (Dietz & Meltzer, 2002)

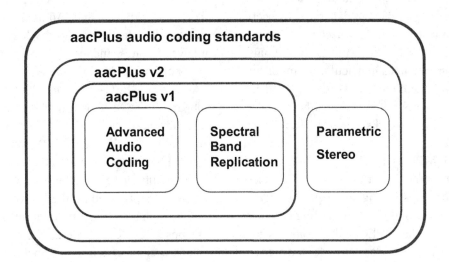

Human ears are more sensitive to errors than human eyes. Human hearing faculties behave like differentiators, accentuating any variations, while human eyes behave like integrators, smoothening out variations (Sharda, 1999). Therefore, an audio stream should be given higher priority as compared to text or an image data stream.

Still Image Coding

Still image coding depends upon the type of image and its compression algorithm. Standards such as JPEG (Joint Photographic Experts Group), GIF (Graphics Interchange Format), and PNG (Portable Network Graphics) have dominated the field so far. JPEG is generally used for lossy compression of continuous tone images, such as photographs. GIF is a bitmap image format for pictures with 256 colours. PNG is a lossless bitmap image format. PNG improves upon the GIF format and is freely available.

The newer JPEG 2000 image compression standard uses a wavelet transform instead of the discrete cosine transform used in JPEG (Taubman & Marcellin, 2002). Therefore, JPEG 2000 can give higher compression ratio without generating the blocky and blurry artefacts introduced by the original JPEG standard. It also allows progressive downloads to extract various image resolutions, qualities, components, or spatial regions, without having to decompress the entire image. Distortion performance is also improved over the original JPEG standard, especially at low bit rates and at extremely high quality settings. JPEG 2000 is more error resilient as compared to the original JPEG standard (Secker & Taubman, 2004). This makes JPEG 2000 much better suited for applications requiring image transmission over wireless networks, as errors and delays introduce fewer observable artefacts in the displayed image.

Wireless networks experience higher error rates, and have lower bandwidth. Therefore, they are more severely challenged when transmitting digital images. Since the JPEG 2000 standard provides higher compression ratios, it is more suitable for the low bandwidth wireless networks; however, some additional issues need to be addressed (Santa-Cruz, Grosbois, & Ebrahimi, 2002). Issues such as error resilience over wireless networks are being addressed by the JPEG 2000 Wireless (JPWL) team. Their aim is to standardise tools and methods for efficient transmission of JPEG 2000 images over error-prone wireless networks. One of the techniques being developed by JPWL make the code stream more error resilient by adding redundancy, or by interleaving data (Dufaux & Nicholson, 2004). The decoder not only detects errors, but also corrects some, where possible. Another technique changes the sensitivity of different parts of the code stream to errors. More sensitive sections of the code stream are more heavily protected than the less sensitive sections. The third technique describes the locations of the remaining errors in the code stream; the decoder then uses this information to exclude the corrupted parts of the code stream from the decoding process.

By standardising these techniques in JPWL, JPEG 2000 is being made more resilient to transmission errors, making it an ideal choice for the transmission of digital images and video over wireless media.

Moving Image Coding

Moving image coding can entail storing up to 20-30 image frames in every second. This demands very high bandwidth for high quality uncompressed video. Development of video coding standards that provide low resolution and low frame rate video suitable for transmission over networks began with the H.261 standard published by the ITU (International Telecom Union) in 1990, with data rates in multiples of 64 Kbps. The H.263 version provided a replacement for H.261 (in 1995)

to work at all bit rates. It was further enhanced as H.263v2 (in 1998) and H.263v3 (in 2000). H.263 is similar to H.261, with improved performance and error recovery, and supports CIF,[1] QCIF, SQCIF, 4CIF, and 16CIF images. As these standards are designed for multiples of 64Kbps rates these are sometimes called px64 (where p can be 1-30). Originally these data rates were expected to suit ISDN (integrated services digital network) lines, nonetheless, these standards are useful in transmitting video over other wire-line and wireless networks also. H.263 is the baseline standard for the new 3G-324M standard, which targets the 3G wireless networks (Smith & Jabri, 2004).

Another option within the 3G-324M specification is the next generation video coding standard MPEG-4 AVC. It was approved in 2003 and called MPEG-4 AVC or ITU-T H.264, or simply advanced video coding (AVC).

MPEG-4 AVC doubles the compression efficiency of earlier standards for the same picture quality, which leads to 50% lower bandwidth (Navakitkanok & Aramvith, 2004). Therefore, it is far better than the earlier standards for wireless transmission. It offers improved resilience to transport errors, improved bit rate scalability, and stream switching for transmission over less reliable network infrastructure, such as wireless networks.

Motion JPEG 2000 (like Motion JPEG) can perform video compression applying only intra-frame compression. This makes Motion JPEG 2000 well suited for video transmission over wireless networks. It has been demonstrated that Motion JPEG 2000 outperforms MPEG-4 in terms of coding efficiency, error resilience, complexity, scalability, and coding delay (Tabesh, Bilgin, Krishnan, & Marcellin, 2005).

The JPWL work has taken into consideration the general principle underpinning networking protocols, with particular attention given to 3G networks (3GPP/3GPP2), wireless LANs (WLAN based on the IEEE 802.11 standards family), and

Digital Radio Mondiale (DRM), making motion JPEG 2000 particularly suitable for wireless networks.

Multimedia Applications

Applications which have so far been bound to wire-line networks and desktop computers, now want to be let loose. The only option is to use wireless networks and portable devices. Areas for such mobile multimedia applications include both personal and business communications. E-learning, marketing, travel, and tourism are just a few of the burgeoning application areas that can make good use of mobile multimedia systems. Some of the potential killer applications based on the JPEG 2000 Wireless (JPWL) methods include video streaming and video conferencing (Liu & Choudary, 2004).

Mobile systems offer new opportunities and challenges as they become capable of transmitting multimedia information. Such applications need to transmit not only the core information, but also some associated meta-information. Most electronic systems use multi-tier information transmission processes, which include: intimation of arrival (bell, ring, beep, and vibrate); abbreviated information (subject, caller ID); textual information (text message, SMS); multimedia information, and meta-information for layered retrieval of the information.

How a particular information type and associated meta-information is used depends upon the application, the user preference, user device, and the required QoS (Cheng & Shang, 2005).

Text Applications

Text is very useful for communication. It is often said that a picture is worth a thousand words; nonetheless, we should not forget that a few well chosen words can be worth scores of pictures. Additionally, text requires much lower bandwidth,

and has greater certainty of meaning. It is more reliable in the face of transmission errors, especially if we use either reverse or forward error correction protocols.

Text is easy to transmit asynchronously or synchronously. One can send an SMS to a friend during work, without the fear of disturbing her in an import meeting. The receiver can reply in her own time, or the two can engage in a brief chat session to fix their evening rendezvous. The runaway success of SMS follows the predicate that "brevity is the soul of wit," as SMS allows succinct messages that convey the meaning quickly. Coded messages based on SMS have also become prevalent, further reducing the time taken to enter and read the message.

Some commonly used SMS codes include: *ATB—All the best; BRB—Be right back; GR8—Great; LUV—Love; PCM—Please call me; TTYL—Talk to you later; 2DAY—Today; and WER R U—Where are you?* SMS codes are also being used to download information to mobile phones, such as snow photos to check the condition on ski slopes.

In Japan, codes called Emoji have been developed. These are colourful, and often animated inline graphics used for mobile messaging. However, these are not standardised or interoperable between carriers. Emonji's are treated as characters, and each carrier has its own set.

In conclusion, text or text-like messages are, and will remain, an important aspect of mobile communications, especially because these are inexpensive, highly expressive, and are least problematic with respect to delivery with the desired QoS.

Audio Applications

Transmission of voice was the original motivation for developing the mobile communications technology. However, digital radio is also coming online, and integrating digital radio in mobile phones is in the offing. An Austrian company Livetunes has developed UMTS-enabled handset with digital radio. SIRIUS Satellite Radio can transmit commercial-free music and other audio entertainment to cars and homes. Mobile audio commercials over such digital radio channels allow advertisers to send audio commercials to their customers' mobile phone. The customer receives a phone call; upon answering the call, the audio commercial is played. It can include new offers, promotions, and announcements. To avoid spamming, companies have to provide their own subscriber database and the audio clip.

The QoS requirements for audio are different for bi-directional conversation than those for uni-directional digital radio transmission. For digital radio, buffering can be used to remove any delay jitter; however, excessive buffering can add unacceptable delay to conversational applications (Sharda, 1999).

Human hearing is very sensitive to any distortion in audio. For conversational audio, we can tolerate some errors, as long as the meaning of the spoken words is clear. If there is a problem in understanding the meaning, then the listener can always ask the speaker to repeat what was said. This is like reverse error correction working at the highest communication layer, that is, the user layer. However, this cannot work for stereo music; as human hearing works like a differentiator, and any distortion gets accentuated. Furthermore, our hearing is capable of picking slightest variation between the two channels of stereo music. Therefore, QoS issues are very important when high quality stereo music is transmitted, but not so important for conversational audio.

Still Image Applications

Applications needing still image transmission can use multimedia messaging service (MMS). Examples of MMS based applications include: weather reports giving images, stock prices dis-

played as graphs, football goals displayed as a slide show, and many more. An extension of still image transmission is animated text messages. The main QoS factor that affects still images is delay. As delay jitter does not effect still image transmission, any errors can be overcome by using reverse error correction protocols. If such additional delays are not acceptable, then images can be displayed with errors. Uncompressed images can tolerate a high level of errors; however, the ability to tolerate errors reduces for compressed images. The original JPEG type compression techniques lead to blocky images when errors occur, as they use discrete cosine transform. However, JPEG 2000 compression standard overcomes this problem by using wavelet transform. Images compressed with JPEG 2000 degrade "gracefully" in face of errors. As wireless communication systems are inherently more error prone, image-based applications will benefit from the use of JPEG 2000 standard for their compression (Dufaux & Nicholson, 2004).

Content repurposing is also becoming important, so that the content creator can compile content only once, and the system can vary image size and resolution depending upon the display screen size and the communications channel bandwidth (Rokou & Rokos, 2004)

The aim of content repurposing is to push the content with the most appropriate resolution, so that it can be transmitted over the available network to meet the QoS goals. In general, this would imply pushing lower resolution images over wireless networks. However, with JPEG 2000 and JPWL, the system can push a rough image to begin with, which keeps improving as more data bits are transmitted.

Video Applications

Video phones are a natural extension of the current audio telephony. Wire-line based video phones were demonstrated decades ago, however, these never became popular. Mobile video telephony is likely to become popular once the cost of transmitting acceptable quality video becomes affordable. In the meanwhile, look-at-this (LAT) applications will generate demand for mobile Internet and 3G wireless networks, as these will create large amount of real-time mobile video. Some possible LAT application areas include:

a.   **Retail:** Before purchasing an item, the consumer sends an image of the item to their partner for comment or approval.

b.   **Real Estate:** An agent sends images of the building and its surrounding areas to the prospective customer.

c.   **General Business:** A worker sends live video to colleague(s) at other location(s) while holding a voice conversation. This could be applied to developing new ideas; designing new products; repairing faulty equipment; maintaining, installing, or inspecting a system.

QoS is of great importance in video transmission. Video conferencing is the most challenging multimedia application for transmission over mobile systems. Much effort has gone in to migrating from 2G networks to 3G networks to provide the desired QoS for video transmission. However, the cost of transmission is still high enough for it to be an impediment in its large-scale adoption.

Mobile Communication Systems

The desire to communicate over long distances has been an innate need for human beings since time immemorial. We can reflect that the earliest telecommunications systems devised by human beings were wireless systems, namely, smoke signals, semaphore flags, drums, and yodelling and so forth. Therefore, it is not surprising that electronic communications are also moving towards wireless systems.

Evolution of Telecommunications

Electric telecommunications began with the telegraph demonstrated by Morse in 1837 and the telephone developed by Bell in 1876. Marconi began his experiment with radio transmission in 1895. Automation of circuit switching systems began in 1919 with the Strowger exchange. The era of satellite communications dawned with the Telstar satellite in 1950. Saber became the first major data network in 1962.

Evolution of Mobile Systems

An early landmark in the development of wireless communications was the patent for the spread spectrum concept, proposed in 1941 by Hedy Lamarr. The first mobile telephone service was setup in St. Louis by AT&T as far back as 1946. Some theoretical breakthroughs also occurred around this time. In 1948, Claude Shannon published the Shannon-Hartley equation, and in 1949 Claude Shannon and Robert Pierce develop the underlying concepts for CDMA (code-division multiple access). In 1950, Sture Lauhrén made the world's first cellphone call, and by 1956, Swedish PTT Televerket operated a mobile telephone service. In 1969, the Nordic Mobile Telephone Group started a mobile service. CDMA was deployed for military systems in the 1970s. In 1973, Motorola vice presidents Marty Cooper and John Mitchell demonstrated the first public call from a handheld wireless phone.

Evolution of Digital Mobile Systems

First Global System for Mobile Communications (GSM) technology based networks were implemented by Radiolinja in Finland in 1991. In 1992, the Japanese Digital Cellular (JDC) system was introduced. By 1993, the IS-95 CDMA standard got finalised. First meetings of the 3GPP (3rd Generation Partnership Project) Technical

Specification Group was held in December 1998. In 2000, Siemens demonstrated the world's first 3G/UMTS (3rd Generation Universal Mobile Telecommunications System) call over a TD-CDMA (time division-CDMA) network.

In 2000, commercial GPRS (general packet radio service) networks were launched. These networks supported data rates up to 115 Kbps, as compared to GSM systems with 9.6 Kbps data rates. In 2001 NTT (Nippon Telegraph & Telephone Corp.) produced commercial WCDMA (wide-band CDMA) 3G mobile network. In 2003, Ericsson demonstrated the transmission of IPv6 traffic over 3G UMTS using WCDMA technology.

Fixed Wireless vs. Mobile Communications

We need to distinguish between fixed wireless communication systems and mobile communication systems. A mobile communication system frees the end systems from the tyranny of being connected to a wall socket, and provides the ability to communicate anytime and anywhere. It allows the freedom to roam outside the home or the office.

Fixed wireless communication systems are local alternatives to wired communication systems. These do not provide mobility outside the home or the office, nonetheless, they provide a cost effective telecommunications connection for a given location with the ability to move around within a specified boundary.

For remote locations, satellite-based communication systems may be the only means of establishing a connection; however, these can be expensive. Satellite connections add about half a second round trip delay, making full-duplex audio or video connections rather difficult. While there is appreciable delay in a satellite connection, the delay variance is not very high, as the number of

hops is fixed at two—transmitter to the satellite, and satellite to the transmitter. Error rates can be high on a satellite connection, especially burst errors—in case of atmospheric disturbances.

Universal Mobile Telecommunications System (UMTS)

Today, there are more than 60 3G/UMTS networks using WCDMA technology. Over 25 countries have adopted this technology, and there is a choice of over 100 terminal designs in Asia, Europe, and the U.S. The 3G mobile technologies identified by ITU for 3G/UMTS offer broadband capabilities to support a large number of voice and data customers, and offer much higher data rates at a lower incremental cost than the 2G technologies (Myers, 2004).

One of the issues driving the development and proliferation of 3G technologies is the recognition that there is a need for guaranteeing the QoS for multimedia traffic. Without guaranteed QoS, many applications fail to perform as per the users' expectations. Until the users are confident of getting the quality they need for running mobile multimedia applications effectively, they will not shift from their current mode of operation and adopt the new wireless networking technologies for multimedia information transmission.

QUALITY OF SERVICE IN MOBILE SYSTEMS

This section gives an overview of the various approaches being trailed for the provision of QoS in mobile networks. While much work has been done in providing QoS guarantees at the network infrastructure level, a holistic approach to providing end-to-end QoS has been missing to some extent. We begin by presenting a QoS model that focuses on the user, and develops a methodology for allowing the user to negotiate with the system to find a compromise between cost, quality, and temporal aspects.

QoS Concepts and Models

The three layer quality of service (TRAQS) model shown in Figure 2 comprises three layers for QoS management in multimedia communications (Sharda & Georgievski, 2002). These three layers are: user perspective layer, application perspective layer, and transmission perspective layer. Each layer performs QoS processing for a set of QoS parameters that are related to the specific perspective. The main functions of the three perspective layers are:

- **User Perspective Layer (UPL)** interacts and performs QoS negotiations with the user and then transfer the QoS request to the APL.
- **Application Perspective Layer (APL)** first assesses the QoS request received from the UPL, and aims to satisfy the needs of the multimedia application by requesting the required services from the TPL.
- **Transmission Perspective Layer (TPL)** is responsible for negotiating with the network infrastructure to obtain appropriate communication services that can guarantee QoS.

Similarly, the various QoS protocols developed at the network infrastructure level need to be

Figure 2. Three layer QoS (TRAQS) model (Sharda, 1999)

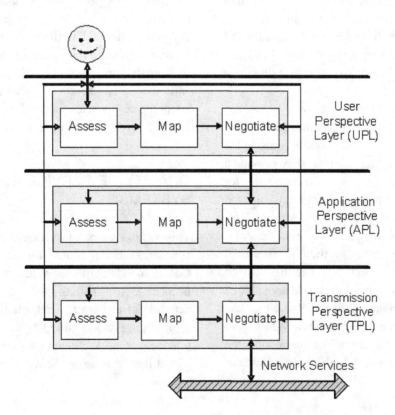

able to communicate with the TPL to allow the user to specify the desired compromise between cost, quality, and temporal issues such as delay and jitter.

Quality, Cost, Temporal Triangle (QCTT)

In purchasing any goods or services one needs to find a compromise between three important factors: cost, quality, and time. While one would like to get the best quality at the least cost and in the shortest time, in practice, this is not possible. One must strike a compromise between these three factors. So far mobile communications systems have not come to grips with this reality. Future communication systems must provide users the ability to specify what quality and temporal aspects (such as delay and jitter) they want, and then systems should respond with the cost it would charge to provide that quality.

Figure 3. Quality, cost, temporal triangle (QCTT) model (Georgievski & Sharda, 2005a)

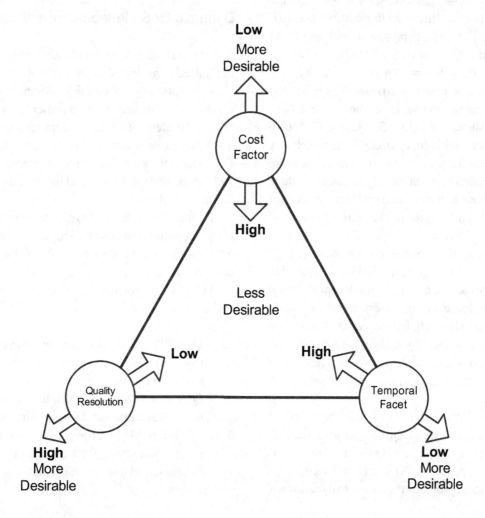

Only with a differentiated cost can the telecommunications service providers afford to deliver the required QoS. If the cost is too low, the network may be overwhelmed with traffic, and none of the users can then obtain the desired QoS. Over and above this, the network services provider may not be able to make profit. On the other hand, if the cost is too high, there will not be enough consumers using the service, once again making it difficult for the service provider to get return on investment. Whereas, by providing the user the ability to negotiate, the consumer and the service provider can both have a win-win situation; some consumers pay high cost as they need higher quality, while other consumers can pay lower cost, as they have lower QoS requirements.

In the following sections, we first explain the concepts involved in the quality, cost, temporal triangle (QCTT), and then present an implementation of the same (Georgievski & Sharda, 2005a).

The three performance aspects—quality, cost, and time—are bound by a tri-partite dependency and thus can be modelled as a triangular relationship, as shown in Figure 3. The QCTT model embodies an inherent restriction on the delivery of QoS, that is, it is possible to achieve the more desirable parameter values only for two of the three performance aspects, while the third aspect must be forced to the less desirable value (Georgievski & Sharda, 2005b).

For example, if a user chooses to have high quality resolution (e.g., large image size, high frame rate), and, the more desirable, low temporal facet (e.g., low delay and jitter), then the cost factor has got to be high. By embedding the quality, cost temporal (QCTT) model in a user interface, we can provide the ability to dynamically manage QoS even while a multimedia session is in progress.

A multimedia communication session first needs to enter static QoS specifications, and then carry out dynamic QoS management as the session proceeds. An interface based on the QCTT model provides the ability to dynamically manage QoS. Such interfaces are described in the following sections.

Static QoS Specification

Figure 4 shows the user interface developed for negotiating static QoS prior to initiating a multimedia communication session. Using this interface, the user is able to specify the desired QoS, and then interactively negotiate with the system. It uses intuitive GUI elements such as a four colour system, a user status response, and a system status signalling system. These GUI elements allow the user to request the desired QoS, and get feedback if the network can deliver the same (Georgievski & Sharda, 2005a).

Dynamic QoS Management with QCTT

A dynamic QoS management interface is shown in Figure 5. This interface uses the QCTT model for re-negotiating QoS while a communication session is taking place. This is achieved by using three GUI elements: three sliders, buttons, and pivot point displacement. The system feedback GUI elements include: system QoS provision ring and values, and QCT threshold line (Georgievski & Sharda, 2005a).

To specify the desired QoS, the user moves the pivot point in the QCT triangle to a location which indicates the desired values for quality, cost, and temporal parameters.

The system provides visual feedback as follows:

1. **QoS Provision Ring** displays the current QoS parameter values that the system is able to provide.
2. **QoS Provision** values display the current numerical values set for QoS parameters.
3. **QCT Threshold Line** uses a three-colour scheme to provide feedback for displaying desirable and non-desirable values for each aspect.

Figure 4. Static QoS negotiation user interface (Georgievski & Sharda, 2005a)

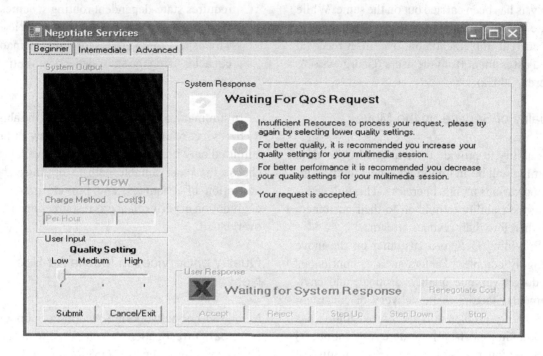

Figure 5. Dynamic QoS management user interface (Georgievski & Sharda, 2005a)

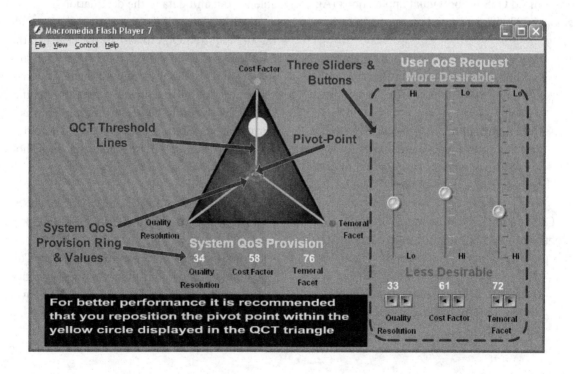

This system has been tested and a usability analysis has been carried out on the same. While some improvements have been stipulated in the current implementation, overall, it received good assessment from the users (Georgievski & Sharda, 2005a).

Quality of Service on the Move

The ability to provide the requested QoS while roaming will be an important aspect in differentiating various mobile operators. This will determine their ability to hold on to their customers and therefore their revenue stream.

Providing QoS to a customer on the move is highly complex. Factors such as continuous handover, variable quality, dropout, and environmental factors make delivery of consistent QoS highly problematic. QoS provisioning has three main aspects: (1) resource reservation, (2) QoS routing protocol, and (3) Call admission control policy.

The integrated services (IntServ) framework developed under the RFC 1633 aims to provide customised QoS to individual applications (Aggélou, 2003). This is based on two aspects:

1. **Resource Reservation:** Each router needs to know the amount of buffer space and link bandwidth it needs to reserve for a session.
2. **Call Admission:** Each router determines the resources already committed to current sessions it is serving, before accepting the request from a new session.

QoS routing is the most important protocol for mobile networks, the main objective specified for this protocol in the RFC 2386 are:

1. **Dynamic Determination of Feasible Paths:** This is based on policy and cost constraints.

2. **Optimisation of Resource Usage:** This requires state-dependent routing schemes.
3. **Graceful Performance Degradation:** This aspect compensates for transient inadequacies using the state-dependent routing scheme.

In summary, a mobile network needs the ability to reserve resources, ensure that a new call is admitted only if enough resources are available, choose the most suitable path to optimise the utilisation of resources, and provide graceful degradation in performance as resources become overloaded.

Quality of Services in Mobile Ad-Hoc Networks

Mobile ad-hoc networks are becoming an important area of investigation. As routing paths are not fixed in an ad-hoc network, QoS routing becomes an even more dynamic problem (Aggélou, 2004). In any ad-hoc network, a variety of routes with differing node capacity and power may be available to transmit data to the destination.

In general, not all routes are capable of providing the required QoS to satisfy the needs of the mobile users. Even when a route is selected that initially meets the user requirements, its error characteristics will not remain constant with time, due to the dynamic nature of routing and node placement in mobile ad-hoc networks. Therefore, ongoing re-routing will be required in an ad-hoc mobile network.

MOSQUITO: Mobile Quality of Service Provision in the Multi-Service Network

The MOSQUITO project, at the University College London, explored a microeconomic approach to resource allocation for providing QoS over multi-service network.

In this protocol, a base station sells bandwidth and QoS guarantees in small auctions to mobile terminals. A simple price setting/bidding function is used to determine the outcome of the auction. This research project aims to explore if:

- Microeconomics can be used for resource allocation.
- The performance of such a system can be measured.
- The algorithm creates a stable system or a chaotic one.
- Chaos can be characterised and controlled.

To use microeconomics for QoS provisioning, such questions need to be answered. Additionally, pricing functions need to be established using some simplifying assumptions; because, without simplifying heuristics, the juxtaposition of a myriad of factors such as pricing, routing, and quality selection will make real-time negotiations impossible.

FUTURE DIRECTIONS

There is no doubt that the future is heading towards mobile communications. And multimedia information will increasingly become the main traffic being transmitted, or blocked, on these networks. One solution to this problem is the so called "brute force" method: that is, "throwing" more bandwidth at the multimedia applications. However, experience shows that as more resources are made available without a viable economic model, the system ultimately gets overloaded. Therefore, developing QoS systems that provide the user with the ability to negotiate with the network infrastructure are going to be of paramount importance.

Some of the developments in this area point to the following:

1. In the near future, mobile computing and communication systems will suffer from low bandwidth and low performance due to battery limitations.
2. Increasingly, mobile systems will provide higher bandwidth and combine different wireless technologies, such as high performance local wireless networks and wide area networks.
3. Third generation mobile systems will combine IP-traffic with traditional voice traffic.

The next generation of mobile networking technology is called 4G, or "3G and beyond" by IEEE (Aggélou & Tafazolli, 2001). In Japan, NTT DoCoMo is conducting tests under the 4G banner for 100 Mbps speeds with moving terminals, and 1 Gbps for stationary terminals. The first commercial release by NTT DoCoMo is expected in 2010. This technology aims to provide on demand high quality video and audio. 4G will use OFDM (orthogonal frequency division multiplexing), and also OFDMA (orthogonal frequency division multiple access) to better allocate network resources, and service multiple users simultaneously. Unlike the 3G networks, which use both circuit switching and packet switching, 4G will use packet switching only. Additionally, many QoS issues will be handled by developing new protocols. Nonetheless, the author contends that providing the ability to negotiate a compromise between cost, quality, and temporal aspects will remain an important issue.

CONCLUSION

Transmission of multimedia information over mobile networks is becoming increasingly important. New applications are in the offing if such multimedia information can be transmitted with the desired QoS. Text and still images do not pose much problem when transmitting these over mobile networks, as delay and delay variance do not adversely effect the operation of applications using text or still images. JPEG 2000 standard provides a marked improvement over the current standards such as JPEG, GIF, and PNG for still image transmission. Audio and video transmission, especially for full-duplex applications requiring real-time operation, poses the most demanding requirements for providing the desired QoS. While 3G networks, and 4G networks of the future, are capable of providing the required infrastructure for delivering multimedia content with the desired QoS, their user interfaces need to provide the ability to strike the desired balance between quality, cost, and temporal aspects.

ACKNOWLEDGMENTS

The author would like to thank Dr. Mladen Georgievski for his useful suggestions and other contributions towards the preparation of this chapter.

REFERENCES

Aggélou, G. (2004). *Mobile ad hoc networks.* New York: McGraw-Hill Professional.

Aggélou, G., & Tafazolli, R. (2001). QoS support in 4th generation mobile multimedia ad hoc networks. *Proceedings of the Second International Conference on 3G Mobile Communication Technologies,* London, March 26-28 (pp. 412-416). London: Institute of Electrical Engineers.

Aggélou, G. N. (2003). An integrated platform for quality-of-service support in mobile multimedia clustered ad hoc networks. In M. Ilyas (Ed.), *The handbook of ad hoc wireless networks* (pp. 443-465). Boca Raton, FL: CRC Press, Inc.

Cheng, A., & Shang, F. (2005). Priority-driven coding of progressive JPEG images for transmission in real-time applications. *11th IEEE International Conference on Embedded and Real-Time Computing Systems and Applications (RTCSA'05),* Hong Kong, August 17-19 (pp. 129-134). Washington, DC: IEEE Computer Society.

Dietz, M., & Meltzer, S. (2002, July). CT-aacPlus: A state-of-the-art audio coding scheme. *EBU Technical Review,* (291), 1-7. Retrieved from http://www.ebu.ch/en/technical/trev/trev_291-dietz.pdf and http://www.ebu.ch/en/technical/trev/trev_index-digital.html

Dufaux, F., & Nicholson, D. (2004). JPWL: JPEG 2000 for wireless applications. Photonic devices and algorithms for Computing VI. In K. M. Iftekharuddin, & A. A. S. Awwal (Eds.), *Proceedings of the SPIE, 5558,* 309-318.

Georgievski, M., & Sharda, N. (2005a). Enhancing user experience for networked multimedia systems. *Proceedings of the 4th International Conference on Information Systems Technology and its Applications (ISTA2005),* Massey University, Palmerston North, New Zealand, May 23-25 (pp. 73-84). Bonn: Lecture Notes in Informatics (LNI), Gesellschaft für Informatik (GI).

Georgievski, M., & Sharda, N. (2005b). Implementation and usability of user interfaces for quality of service management. *Tencon'05: Proceedings of the Annual technical Conference of IEEE Region 10,* Australia, November 21-24. New Jersey: IEEE.

Liu, T., & Choudary, C. (2004). Content-aware streaming of lecture videos over wireless networks. *IEEE Sixth International Symposium on Multimedia Software Engineering (ISMSE'04),* Miami, FL, December 13-15 (pp. 458-465). Washington, DC: IEEE Computer Society.

Myers, D. (2004). *Mobile video telephony.* New York: McGraw-Hill Professional.

Navakitkanok, P., & Aramvith, S. (2004). Improved rate control for advanced video coding (AVC) standard under low delay constraint. *International Conference on Information Technology: Coding and Computing (ITCC'04), 2*, Las Vegas, NV, April 5-7 (p. 664). Washington, DC: IEEE Computer Society.

Rokou, F. P., & Rokos, Y. (2004). Integral laboratory for creating and delivery lessons on the Web based on a pedagogical content repurposing approach. *Fourth IEEE International Conference on Advanced Learning Technologies (ICALT'04)*, Joensuu, Finland, August 30-September 1 (pp. 732-734). Washington, DC: IEEE Computer Society.

Santa-Cruz, D., Grosbois, R., & Ebrahimi, T. (2002). JPEG 2000 performance evaluation and assessment. *Signal Processing: Image Communication, 17*(1), 113-130.

Secker, A., & Taubman, D. S. (2004). Highly scalable video compression with scalable motion coding. *IEEE Transactions on Image Processing, 13*(8), 1029-1041.

Sharda, N. (1999). *Multimedia information networking*. New Jersey: Prentice Hall.

Sharda, N., & Georgievski, M. (2002). A holistic quality of service model for multimedia communications. *International Conference on Internet and Multimedia Systems and Applications (IMSA2002)*, Kaua'i, Hawaii, August 12-14

(pp. 282-287). Calgary, Alberta, Canada: ACTA Press.

Smith, J. R., & Jabri, M. A. (2004). The 3G-324M protocol for conversational video telephony. *IEEE MultiMedia, 11*(3), 102-105.

Tabesh, A., Bilgin, A., Krishnan, K., & Marcellin, M. W. (2005). JPEG2000 and motion JPEG2000 content analysis using codestream length information. *Proceedings of The Data Compression Conference (DCC'05)*, Snowbird, UT, March 29-31 (pp. 329-337). Washington, DC: IEEE Computer Society.

Taubman, D., & Marcellin, M. (2002). *JPEG2000: Image compression fundamentals, standards and practice*. Netherlands: Kluwer Academic Publishers.

ENDNOTE

[1] CIF: Common Intermediate Format. A video format used in videoconferencing systems. It is part of the ITU H.261 videoconferencing standard, and specifies a data rate of 30 frames per second (fps), with each frame containing 288 lines and 352 pixels per line. Other CIF based standards include: QCIF - Quarter CIF (176x144), SQCIF - Sub quarter CIF (128x96), 4CIF - 4 x CIF (704x576), and 16CIF - 16 x CIF (1408x1152).

Chapter IV
On a Personalized System for Device Independent and Position Aware Communication

Odej Kao
Technische Universität Berlin, Germany

Stefan Lietsch
University of Paderborn, Germany

ABSTRACT

This chapter introduces a personalized system for device independent and position aware communication. It combines the advantages of mobile universal messaging and location aware architectures, and supports its users by providing additional and helpful information. One focus is the integration into the well-known session initiation protocol to ease the implementation into existing applications. We think the combination of both approaches can support the user and help him to get more information. The chapter also focuses on how to design a modular, secure, and scalable communication system that could easily be ported to other fields of application and gives an outlook on future trends in this field.

INTRODUCTION

Since the early attempts to support human communications by technological media such as the telegraph or telephone, a long time has passed. Many new technologies of communication have been developed and people became accustomed to them. Nowadays, we have so many ways to pass messages to each other that it becomes a complex task to maintain all these different systems.

Additionally new methods to communicate not only with human beings but also with machines arise. This includes a range of applications from simple Web-based software up to the completely voice-controlled household. The speed of development brings benefits together with problems. Two main issues can be identified: on the one hand, a user needs several devices to use the different methods of communication, and, on the other hand, the exact address of the recipient needs to be known to establish a connection via a particular channel of communication. In addition, the user needs to know or at least needs to try out the media-related address on which his partner is reachable. These problems get more severe the more different channels of communication a user utilizes. To tackle these issues a system that eases the management of communication is needed. This system should provide an extended phonebook, where the addresses of all users are stored as well as the functionality to translate between potentially incompatible media. Furthermore, the system should be easy to use and should offer various services to its users. There already are systems that are capable of these two functionalities as shown in this chapter. But in respect of supporting the user with the given data much more can be done. This chapter describes a system that provides the usual functionalities and also supports the user with all information it has. Additionally input data from positioning devices, either native ones (e.g., GPS) or passive ones (e.g., cell phone positioning) is used to connect the existing information with locations to extend the support of the users. This system combines the two aspects: uniform communication and user support in a modular, scalable, secure, and user-centered communication platform.

After introducing three example communication systems and giving a short comment on location systems, the concept of the communication and location system (CoLoS) is presented. To show the potential of the system a prototypical implementation is described.

BACKGROUND

As mentioned in the *Introduction*, there already are systems that enable a communication independent of the underlying media. The first of its kind was the mobile people architecture (MPA) (Appenzeller et al., 1999) developed by the Mosquito Group of the Stanford University. It is typical for device independent communication systems. The main goal of all such systems is to provide access to all its users independent of the actual communication devices used. In the mobile people architecture, this is mainly achieved by extending the traditional ISO-OSI reference model by a personal layer. This layer represents an individual person by managing all the owner's devices and their reachability. Thus communication requests no longer go directly to the devices but to the user itself, that is, to his representative, the so-called personal proxy. This proxy decides how to proceed with the incoming request, routes it to the appropriate device, and manages the communication. This so-called personal level routing is the main achievement of the MPA, but it brings one bottle-neck namely the personal proxy. All the communication has to go through the proxy although there may be an alternative, faster way. This problem is solved in the iceberg architecture (Wang et al., 2000). The system is based on the MPA but has one major difference: it no longer has decentralized proxies for every user but concentrates many proxies in centralized units called iceberg points of presence (IPoPs). These IPoPs have interfaces to many access networks (e.g., telephony, cellular, and Internet) and are interconnected by fast network connections. This ensures that all communication can be routed in a fast and direct way. There is also a billing unit in the system to charge the users for certain services. The iceberg architecture is therefore a highly developed system that enables device independent communication but it does not support the user beyond this functionality and is limited to communication services.

The integrated personal mobility architecture (IPMoA) (Thai, Wan, Seneviratne, & Rakotoarivelo, 2003) takes a slightly different approach to ensure the device independency and the mobility of the user. It does not primarily focus on the reachability but on the mobility of its users. The users are able to access all their data and applications from every remote location and with every available device. By including communication applications, a device independent communication is possible. The whole system is based on agents that commute between the home and the foreign network and exchange the data between those networks. A high level of personalization can thereby be reached, but since nearly all data must be fetched from the home network, there may be problems for instance with time-sensitive and synchronous applications.

All systems presented here have different kinds of information about their users, for example, reachability, different addresses, and so forth. This information is used to provide the functionalities of the communication systems. Our approach is to take all the data gathered by the different communication systems, add some additional data, for example, from location systems and generate information that supports the user far more than what was possible with previous systems. A platform for device independent communication and its personalization is therefore provided which combines the advantages of both approaches.

A variety of localization techniques form the basis of current research. For instance localization by determining the positions of all kind of mobile communication devices (e.g., GSM phones or WIFI devices) has been proposed by Youssef, Agrawala, and Shankar (2003) and Zimmermann, (2001). Another research direction is to use proprietary short range radio techniques based on Bluetooth or infrared to locate the users of the system as proposed by Gonzalez-Castano and Garcia-Reinoso (2002). Since support for as many different localization systems as possible was desired, committing to one technique or direction was not an option. The integration of the positioning applications was done independently from underlying mechanisms. This is why only a few example systems will be integrated in the prototype, but the main focus will be on supplying an extendable communication and localization platform.

CONCEPT AND ARCHITECTURE

This section describes the concept of a device independent and location aware communication system and gives a glimpse on possible fields of application. For a better understanding of what each component of the system does, four exemplary services are presented and some requirements are defined.

Services from the User's Perspective:

- **Find communication partners in your proximity.** One self-evident service is to announce possible communication partners or friends which are detected nearby the location of the user. He can then decide whether he wants to contact the person or perform other actions. Possible scenarios for this service could be exhibitions where interesting exhibitors nearby are indicated to the user or the sign-posting to one specific person.
- **Discover the cheapest or fastest connection.** Another possible service is to suggest the best and/or cheapest network connection for the user's current position. Afterwards, he can decide on his own if he wants to change to the proposed connection. Furthermore, automatic connection handover mechanisms are possible.
- **Hints on services close to the user.** The system can point out services nearby to the user. These could be communication services like

a locally-bounded NetMeeting conference or non-communication related services like a public printer or fax machine.

- **Support the user with additional information.** This is achieved by a configurable information generator. It can access all the data gathered by the different systems and search for useful information to support the user by applying rules on a database. In this way, all information available in the common database can be used to find new information that supports the user. This approach removes the barrier of incompatible databases used for the different communication services and allows it to use all available resources to provide valuable information to the user.

To make sure that these services and other new ideas work properly, are accepted by the users, and do not cause security problems, some requirements must be fulfilled.

Technical Requirements

- **Extendibility and Integratability:** The system needs to be flexibly extendable (regarding the supported communication services) and easily integrated into the users' working environment. This can be achieved best by a modular design and flexible interfaces. These features allow the seamless addition of future communication services and limit the components requiring installation on the user's device to a minimum.
- **Hardware Independency and Portability:** The client side of the system must be independent of the user's hardware or at least be available in different versions for several devices. Although differences in computational power, graphical abilities, and other hardware dependent features need to be considered.

- **Transparency:** The system should be transparent to the user as far as possible. The user should therefore be able to continue working with their well-known communication and location services without recognizing whether or not the CoLoS system is active.
- **Ease of Use:** The user should profit from the system without having additional work to do. It should provide useful information and choices and support the user during communication. In addition the system should recognize existing applications independently and integrate them into the platform.
- **Optionality and Privacy:** The user must be able to decide whether or not he wants to submit information about him at any time. This is extremely important, since unwanted surveillance is easily possible through localization systems. Although, if the user decides to turn off the system, it must be possible to use existing applications for communication and localization as usual.
- **Data Security:** Since communication and location systems gather many security relevant data about their users, it is very important to protect this data against unauthorized access. They must be safely stored and transported. That demands the utilization of secure mechanisms for authorization and encryption in the system.

In addition to the services described, we also want to integrate the standard functionalities of other communication systems. One function already mentioned is the communication with users whose addresses are not immediately determined. The system finds the desired user by any given information (e.g., name, e-mail, preferences) and gives choices if more than one person matches. Another feature is the communication with incompatible devices. That is that two people with totally different devices (e.g., a mobile phone and an instant messenger) can communicate without recognizing the incompatibility.

Concept

The following sector briefly introduces the communication and location system (CoLoS) and its main components.

Figure 1 shows the two sides of the developed system and their main components. The CoLoS client allows the service utilization and notifies the user about new, incoming information. This is provided by a GUI fitted to the particular device. Existing applications are integrated through the so-called client interface which can be seen as a universal interface. The data exchange between server and client is handled by the CoLoS connection on the client and the controller on the server side whereas the data is transmitted in packets of the CoLoS/SIP protocol. One of the main server components is the user register where all the data is stored in a fast database. This data can be accessed by the decision engine to combine

it, following pre-specified rules in order to create useful information. The last main component is the communication dispatcher and its communication and translator modules. This enables the system to translate between incompatible types of media transparently. In the following, we describe the main components of the CoLoS server and client.

Server Modules

- **Controller:** The controller is the main server component. It decides by the type of an incoming message how to proceed with it. It passes the information of a message including a location update to the user register in order to update the database. The controller also decides by means of the data from the user register by which device a user is reachable and if an incoming communication request can be fulfilled (with or without translation).
- **User Register:** The user register can be seen as an extended phone book. It stores all user

Figure 1. Schematic representation of the CoLoS components

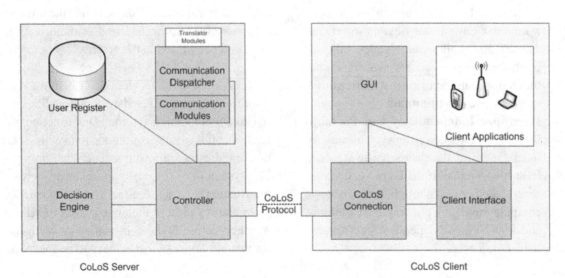

data irrespective of its origin. Some example data are: addresses for each communication device, availability and preference for communication ways, current location, profile of the user, a buddy list, public key, and so forth. The list of stored data can be extended arbitrarily to cover all useful information about the system's users.

- **Communication Dispatcher:** The communication dispatcher is invoked by the controller if a user requests a communication with another user who is not reachable under a compatible device. That is, for example, if user A wants to use an IP-Phone and user B only has his Instant Messenger enabled. The controller recognizes this incompatibility and passes all messages concerning this communication to the communication dispatcher. It determines the corresponding communication and translation modules (in this case, an IP-Phone, a messenger interface, and a text-to-speech/speech-to-text translator), translates the messages, and passes them back to the controller. Existing systems, as for example the SoftBridge System (Lewis, Tucker, & Blake, 2003) can be utilized. Afterwards the messages are sent to the receiver under the address of the initial sender so that the whole translation action is transparent for the users.

- **Decision Engine:** The decision engine tries to find out which services can be offered to the users. This is a major functionality of the CoLoS since all available data is taken into account. It therefore applies rules on the user register after every change in the database to look for new and useful information. One possible rule could be: "Check if a user in the user's buddy list is in his proximity after he changed his position". If one or more matches are found the result is passed to the controller. The controller then generates messages containing this information and sends them to the correspondent

users. While the database is growing and is changed more often with an increasing number of users, new mechanisms must be found to restrict the search to relevant fields of data. For example, a strategy based on the current location of the user is possible. Additionally rules can be defined and added while the system is running.

Client Modules

- **CoLoS Connection:** The CoLoS connection is the correspondent to the controller on the server side. It gets commands and communication data from the GUI or the client interface, packs them into CoLoS protocol packets and sends them to the server. When a new packet is received the CoLoS connection decides by its type what to do with it. For example, an incoming message with the type "user information" is passed to the GUI where it is displayed correspondingly.

- **Client Interface:** The client interface is the bridge between existing communication and location applications and the CoLoS. The interface notices which of the registered applications are started and announces this through the CoLoS connection to the server. Furthermore, it intercepts connection requests to pass them to the server for further processing. Incoming messages are handed over to the corresponding application, and the whole communication process is monitored for faults and interruptions to enable a quick solution. Since the client interface acts transparently, the user can continue to work with his applications as usual but also has the advantages of the CoLoS. The interface can also be utilized to get location data from positioning applications. This functionality must be controllable by the user at all time to avoid an unwanted surveillance.

- **Graphical User Interface (GUI):** By the GUI, a user can access all functionalities offered by the CoLoS easily and quickly. It for example alerts him on incoming requests or offers a "buddy list" where he can save contacts he often uses. Also searching and security functions are integrated. All settings made by the GUI are sent to the server and stored in the user registry.

CoLoS Protocol

The data exchanged between the CoLoS server and the clients is, as mentioned previously, encapsulated in packets of the CoLoS protocol. This ensures an efficient and secure data transmission. The CoLoS protocol is implemented as an overlay protocol based on TCP/IP/SIP. It has several control fields (e.g., sender and receiver address, type of the message, additional options) and one optional data field.

Example

For a better understanding of the system and the interaction of its components, we describe one

exemplary process in the CoLoS. The scenario is that user A has the CoLoS client and a positioning device enabled and changes his location. Another user B, who is on A's friends list, is nearby A's new location. User A is notified and can choose between different options.

This process is depicted in Figure 2 and takes place as follows: The localization device realizes the change of positions and passes this information to the client interface (1). The client interface gathers it and generates location update messages in fixed intervals (2). This avoids an overload of the system due to too fast or too many location updates. The messages are passed to the CoLoS connection where they are packed in CoLoS protocol packets and transmitted to the controller on the server side (3). The controller receives the packet and extracts the type and the information included. Due to the type of the message, the controller decides to initiate a query through the decision engine (4a). Simultaneously, the position update is send to the user register (4b). The decision engine selects the rules corresponding to the received information and applies them to the user

Figure 2. Exemplary processing of a request

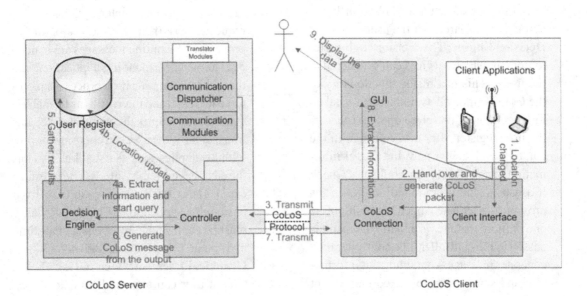

register to search for helpful information (5). In this case, the rule "Find friends of User A that are closer than 50 m" could be selected and used to start a query. If the query has one or more results, this information is passed back to the controller (6). There a CoLoS protocol packet is generated and transmitted to the client (7). The ColoS connection extracts the data and analyzes the type (8). Afterwards the information is passed to the GUI where it is visualized and different choices of actions are presented to the user (9).

This process is already implemented for demonstration purpose in a CoLoS prototype, and it works well in a closed environment. It is exemplary for the interaction in the CoLoS system.

FUTURE TRENDS

It is likely that a variety of new ways of improving the quality of message exchange systems will be developed in the near future. The most challenging task in foreseeable time will be to integrate all the different systems. Therefore work must be done on common protocols and universal interfaces. There already are first efforts such as the session initiation protocol (SIP) (Camarillo et al., 2002) that is already used by the CoLoS prototype or the mobile IP system (Perkins, 2002). The possibilities, problems, and a comparison of both approaches are discussed in Kwon, Gerla, Das, and Das (2002) and Wang and Abu-Rgheff (2003). However most of the systems currently available only focus on the reachability. In future developments, users will become the main issue because they need to be supported and do not want to put additional effort in controlling another difficult application. A desirable scenario is that no matter what device a user currently utilizes he is reachable by a uniform address and can reach any other user without knowing which device he got. Additionally, a user should be able to set

preferences and get information that helps him. To get to this scenario, standardization needs to be done and a uniform, scalable, secure, and fast framework must be developed. The CoLoS is a concept for such a framework and may be used as a starting point for building up a productive system in the near future.

The other field of rapid development is the area of location detection. Currently, there are considerable efforts to locate devices that are not natively designed for localization such as cell phones or WLAN devices. Since accuracy and speed of localization of these devices steadily increase these techniques become more and more popular.

CONCLUSION

Regarding the fast development in communication technologies today, a system was designed that is not limited to existing applications. In fact, an open platform was developed which is able to integrate existing and future technologies and to link them seamlessly. This is achieved by a modular structure and the use of standardized and application independent protocols. Thus the system can be extended by arbitrary applications to serve its users as a transparent, easy-to-use, and secure communication base.

In the future, the simplification and personalization of communication will gain even more importance since the number of ways to communicate increases steadily and many users do not want or simply cannot take care of the maintenance of all the media. Additionally, negative aspects of the expanding communications world like unwanted spam messages, and so forth, could be effectively countered by intelligent communication platforms as proposed in this chapter.

REFERENCES

Appenzeller, G., Baker, M., Lai, K., Maniatis, P., Roussopoulos, M., Swierk, E., & Zhao, X. (1999, July). The mobile people architecture. *Mobile Computing and Communications Review, 3*(3), 36-42.

Camarillo, G., Handley, M., Johnston, A., Peterson, J., Rosenberg, J., Schooler, E., Schulzrinne, H., & Sparks, R. (2002). RFC 3261 – SIP: Session initiation protocol. *IETF Internet Standard.* Retrieved September 18, 2007, from www.ietf. org/rfc/rfc3261.txt

Gonzalez-Castano, F. J., & Garcia-Reinoso, J. (2002). Bluetooth location networks. *IEEE Global Telecommunications Conference (GlobeComm) 2002,* Taipei, Taiwan, November 17-21 (pp. 233-237). New York: IEEE.

Kwon, T. T., Gerla, M., Das, S., & Das, S. (2002). Mobility management for VoIP service: Mobile IP vs. SIP. *IEEE Wireless Communications, 9,* 66-75.

Lewis, J., Tucker, W., & Blake, E. (2003). Soft-Bridge: A multimodal instant messaging bridging system. *Proceedings of the Southern African Telecommunication Networks and Applications Conference (SATNAC) 2003,* Southern Cape, South Africa, September 7-10 (pp. 255-256). South Africa: SATNAC.

Perkins, E. C. (2002). RFC 3220 – IP mobility support for IPv4. *IETF Internet Standard.* Retrieved September 18, 2007, from www.ietf. org/rfc/rfc3220.txt

Thai, B., Wan, R., Seneviratne, A., & Rakotoarivelo, T. (2003, February). Integrated personal mobility architecture: A complete personal mobility solution. *ACM Mobile Networks and Applications (MONET) Special Issue: Personal Environment Mobility in Multi-Provider and Multi-Segment Network, 8,* 27-36.

Wang, H., Raman, B., Biswas, R., Chuah, C., Gummadi, R., Hohlt, B., Hong, X., Kiciman, E., Mao, Z., Shih, J., Subramanian, L., Zhao, B., Joseph, A., & Katz, R. (2000, August). Iceberg: An Internet-core network architecture for integrated communications. *IEEE Personal Communications (2000): Special Issue on IP-based Mobile Telecommunication Networks, 7*(4), 10-19.

Wang, Q., & Abu-Rgheff, M. A. (2002). Integrated mobile IP and SIP approach for advanced location management. *4th International Conference on 3G Mobile Communication Technologies, 2003,* London, UK, June 25-27 (pp. 205-209). New York: IEEE.

Youssef, M., Agrawala, A., & Shankar, U. (2003). WLAN location determination via clustering and probability distributions. *IEEE International Conference on Pervasive Computing and Communications (PerCom) 2003,* Dallas, Texas, March 23-26 (pp. 143-150). New York: IEEE.

Zimmermann, R. (2001). Lokalisierung Mobiler Geraete (german). *Seminar Mobile Computing ETH Zuerich 2001.* Technical report. Retrieved September 18, 2007, from www.vs.inf.ethz.ch/ edu/SS2001/MC/beitraege/07-location-rep.pdf

TERMS AND DEFINITIONS

Bluetooth: Bluetooth provides a way to connect and exchange information between devices like personal digital assistants (PDAs), mobile phones, laptops, PCs, printers, and digital cameras via a secure, low-cost, globally available short range radio frequency.

Global Positioning System (GPS): The GPS is a satellite navigation system used for determining one's precise location and providing a highly accurate time reference almost anywhere on Earth or in Earth's orbit. It uses an intermediate circular orbit (ICO) satellite constellation of at least 24 satellites.

Instant Messenger: A program which hooks up to a service for conversation in real time. Involved parties see each line of text right after it is typed (line-by-line), thus making it more like a telephone conversation than exchanging letters. Popular instant messaging services on the public Internet include AOL Instant Messenger, Yahoo! Messenger, .NET Messenger Service, and ICQ.

ISO/OSI Reference Model: The OSI model divides the functions of a protocol into a series of layers. Each layer has the property that it only uses the functions of the layer below, and only exports functionality to the layer above. A system that implements protocol behavior consisting of a series of these layers is known as a "protocol stack" or "stack". Protocol stacks can be implemented either in hardware or software, or a mixture of both. Typically, only the lower layers are implemented in hardware, with the higher layers being implemented in software.

Overlay Protocol: Protocol that uses the services of underlying protocols to either transport its data or combine the services and offer more complex ones.

Session Initiation Protocol (SIP): SIP is a text-based protocol, similar to HTTP and SMTP, for initiating interactive communication sessions between users. Such sessions include voice, video, chat, interactive games, and virtual reality.

Software Agent: Software that acts as an agent for another, for example, a user of a system. It autonomously fulfills the tasks that are given to it. For example, an agent can find data by predetermined rules in a given environment (e.g., the Internet).

Chapter V
Mobile Multimedia for Speech and Language Therapy

Nina Reeves
University of Gloucestershire, UK

Sally Jo Cunningham
University of Waikato, New Zealand

Laura Jefferies
University of Gloucestershire, UK

Catherine Harris
Gloucestershire Hospitals NHS Foundation Trust, UK

ABSTRACT

Aphasia is a speech disorder usually caused by stroke or head injury (Armstrong, 1993). Related communication difficulties can include word finding, speaking, listening, writing, and using numbers (FAST, 2004). It is most commonly acquired by people at middle age or older, as a result of stroke or other brain injury. Speech and language therapy is "the process of enabling people to communicate to the best of their ability" (RCSLT, 2004). Treatment, advice, and support are provided based on assessment and monitoring activities that conventionally are carried out in face-to-face sessions. This chapter considers issues in providing technology to continue to support aphasic patients between therapy sessions, through multimedia applications for drill-and-practice in vocalizing speech sounds. Existing paper therapy aids are generally designed to be used under the guidance of a therapist. Multimedia applications enable people with aphasia to practise spoken language skills independently between sessions, and mobile multimedia speech and language therapy devices offer still greater promise for blending treatment and support into an aphasic person's daily life.

INTRODUCTION

Current trends in the demography of the developed world suggest that increased longevity will lead to a larger population of patients needing rehabilitation services after a stroke (Andrews & Turner-Stokes, 2005). An essential part of these services is speech and language therapy (SLT) (NHS, 2004) to enable the patient with aphasia to return to the community and live as independently as possible. At present, even in countries where SLT is a well-developed profession, resources in terms of staff and mobile communication devices for loan are limited (Harris, 2004). Therapy generally cannot offer a "cure" for aphasia; instead, the goals of therapy are to support the person in capitalizing on remaining language ability, regaining as much of their prior language skills as possible, and learning to use compensatory methods of communication.

This chapter describes the existing therapy methods based on paper materials and mobile electronic devices commonly called augmentative and alternative communication (AAC) devices and proposes the development of software solutions which could be delivered flexibly via readily available mobile devices such as personal digital assistants (PDA) used in a stand alone mode or via Internet delivered services. These could be designed to suit the needs of not only the patients and their carers, but also those of the professional speech and language therapists (SLTs) who could tailor and monitor the treatment more regularly than presently possible. The process of creating and evaluating prototype applications with SLTs is described and recommendations are made for the direction of future research and development.

CONVENTIONAL SPEECH AND LANGUAGE THERAPY AIDS

Paper-based representations of lip and tongue positions for sounds are a venerable and common speech and language therapy aid. These are generally provided as line drawings, illustrating how specific sounds are made; the aids are intended for use both during therapy and in practice sessions outside of therapy. Figure 1 is typical of this type of therapy material.

While useful, this paper-based material has obvious limitations. The line drawings are static, and fail to accurately represent the movements necessary to make speech sounds (Harris, 2004). No spoken explanation or auditory reinforcement is possible—this would be provided by the speech therapist, during a session or by a carer. The latter would be untrained and may in certain circumstances reinforce errors. Aphasia frequently includes impairments in processing written language. Notice that these sheets include a relatively large amount of text—some of it possibly redundant ("Make the target sound as clear as possible"), some echoing in a less accessible form the line drawings ("Tongue tip raised behind teeth"), some vocabulary that is highly technical and likely to be unintelligible without training in speech therapy ("Quality: Approximant"). These issues can put off users from practising on their own, or can diminish the effectiveness of their practice.

The problem of confining effective therapy to formal, therapist-facilitated sessions is significant. A meta-analysis of evaluations of aphasia therapy (covering 864 individuals) concluded that concentrated therapy over a shorter period of time has a greater positive impact on recovery than less concentrated therapy over a longer period (Bhogal, Teasell, & Speechley, 2003). Clearly, face-to-face therapy with a trained therapist is the ideal, but availability is a bottleneck for aphasia treatment. Even in an economically well-developed country such as the UK, there are only, on average, 0.6 SLTs per 10 beds in rehabilitation units (Andrews & Turner-Stokes, 2005)

Early attempts at software support for SLT have had variable and limited success (Burton, Meeks, & Wright, 1991), in part because of hardware and

Figure 1. Sample paper-based therapy material

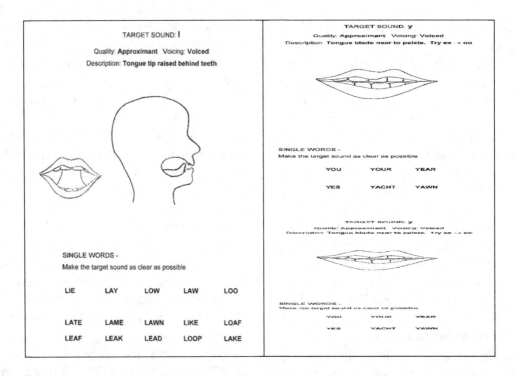

MOBILE DIGITAL COMMUNICATION AND THERAPY TOOLS

development environment limitations, and in part because there was a limited body of knowledge about the accessibility and usability for aphasic users. As will be discussed next, these barriers are less significant today.

Two broad categories of software speech and language therapy applications exist for use by people with aphasia:

- **Drill-and-practice software** that offer instruction and the opportunity to practise language skills. A typical application contains a variety of standard instructional material, and also allows the therapist to

record additional words, phrases, or other utterances for playback and practice by the patient.

- **Compensatory software** that provides alternative means for the user to communicate, for example by producing audio, image-based, or written messages for the user. The user selects an appropriate message for playback or display in situations where communication is required—for example, when dealing with commercial organizations or government departments.

Both types of software now commonly include multimedia facilities. Earlier applications were severely limited in scope by tiny (by today's standards) storage available with standard PCs. Larger hard drives, DVD storage, and high-bandwidth Internet connections now support applications that can include video and audio display of large practice sets, audio recording of user practice sessions (for both immediate feedback to the user, and for later evaluation by the therapist), icons for navigating the interface, and speech generation facilities to reduce the need for pre-recording messages used to communicate with others.

Both application categories also have potential to increase their effectiveness by moving to mobile devices. Compensatory communication devices, or AAC devices, would obviously be more useful if they were small and light enough to be easily carried along into conversational situations—so that the user does not have to struggle with a laptop while shopping. Similarly, portable drill-and-practice devices would allow users to practise frequently during the odd breaks that are inevitably sprinkled through the day, and would allow patients to continue therapy when away from home for more extended periods such as vacations (Glykas & Chytas, 2004). However, current commercially-available portable AAC devices are special-purpose pieces of equipment; as such, they are considerably more expensive than standard PDAs and mobiles, even with monochrome screens. As a consequence, portable AAC tools are not widely used at present.

Moving implementations of AAC tools from special-purpose hardware to standard, general purpose devices holds promise for supporting the development of cheaper, more readily available devices. Recent PDA models are now viable platforms for speech and language therapy applications. Screens, while small, now have a high enough resolution to provide a crisp display of images and line drawings. Memory remains relatively small but is sufficient to store a selection of drill exercises and common conversational phrases and sentences. Indeed, there is evidence that including smaller datasets makes these applications more, rather than less, usable. Experience with information display for non-literate (Deo, Nichols, Cunningham, Witten, & Trujillo, 2004) and communication impaired users (Dunlop,

Cunningham, & Jones, 2002) emphasizes that these users primarily depend on browsing rather than search to navigate the application. Browsing forces the user to rely on memory to find, and return to, desired documents or exercises, and a too-rich set of options to select from can quickly lead to frustration when the user cannot remember the location of a previously retrieved item or cannot efficiently navigate to a new item. For this type of application, it appears that the limitation is the capacity of the user's memory, rather than the device's storage. Some patients with aphasia have short-term memory problems so the retrieval issue is even more important.

ACCESSIBILITY ISSUES

Funding is one significant accessibility issue for speech and language therapy tools. The cost of a device, software, and training often puts these applications out of reach for many individuals, and local health authorities have extremely limited supplies, if any, for loan (Harris, 2004).

It is clear, however, that the provision of a mobile speech and communication aid is not sufficient in and of itself to ensure that it will be used, and useful, in everyday settings. One study in the UK and the Netherlands demonstrated that, for carefully selected patients, these applications can effectively support therapy—however, 11 of the 28 patients did not choose to use the device in non-clinical settings (van de Sandt-Koenderman, Wiegers, & Hardy, 2005).

Accessibility can also be limited by interface and interaction design. People with aphasia often experience difficulties with reading and understanding text, and may experience a related visual problem that makes reading tiring and error-prone. Investigations into accessibility and usability of software for people with aphasia have produced several recommendations for effective applications (FAST, 2004; Queensland Aphasia Groups, 2001a, 2001b):

- Interfaces should include few, and simple, words, in large print.
- White space should dominate—text should be as widely spaced as possible.
- When possible, include images and icons to explain the words, or to serve as substitutes/reminders of the words.
- Text, images, and functionality should be pertinent to the needs and interests of people with aphasia—not to therapists, caregivers, or members of the medical community. Note that a separate interface may be required for these other potential users of the application (for example, to allow therapists to create new drill exercises).
- Interaction should not be keyboard dependent, and should be designed for ease of use with alternative input/output devices such as screen readers and switches. Aphasia frequently makes construction of text via the keyboard difficult or impossible, and the condition causing the aphasia (for example, a stroke) may also create physical disabilities that hamper keyboarding.
- Applications should be interactive rather than static information displays, and if pos-sible should support the user in expressing his/her own thoughts (rather than literally putting words into the user's mouth with a standard set of utterances).

DEVELOPMENT OF MULTIMEDIA PROTOTYPES

Two prototypes of a speech therapy tool, Sound-Helper, were developed. The tool is intended to support drill and practice in phonetic sounds. In both versions, the user selects a sound for practice, and the sound is represented both through audio and through a demonstration of mouth placement for producing that sound. The design of these prototypes was informed by a speech and language therapist and by an expert in interaction design. On the advice of the therapist the prototypes focus on vowel sounds, as these are the first sounds which are usually addressed in therapy for a person newly aphasic after a stroke or other brain injury.

The SoundHelper interface is organized around the familiar and common "folder" metaphor. The top level of the system presents the folder "con-

Figure 2a. Top level of SoundHelper

Figure 2b. A second-level folder, organizing exercises based on vowel sounds

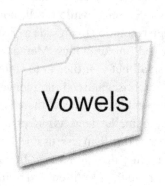

taining" all of the exercises (Figure 2a). At the next level, folders are also used to organize and group different classes of sound—for example, exercises pertaining to vowels (Figure 2b).

The two demonstration displays implemented are an animated line drawing (Figure 3a) and a short video clip (Figure 3b). The prototypes were developed using Adobe Premiere to capture the

Figure 3a. Prototype line drawing, pronouncing "ay"

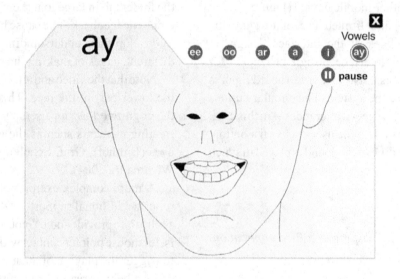

Figure 3b. Prototype video, pronouncing "ay"

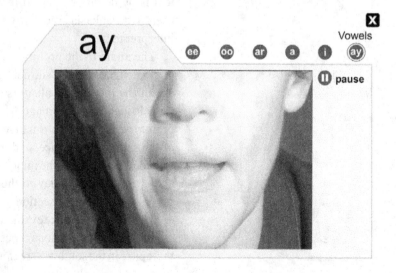

video and Macromedia Flash MX to compress the video clips. Macromedia Flash was then used to rotoscope or motion capture the lip and facial movements, to produce the animated line drawings. The vector graphics produced will automatically rescale to fit a browser window. On a PC monitor, this allows a life-size representation of the animation, creating a "mirror" effect for the user. Automatic scaling also allows the display to be easily ported to a smaller screen such as that of a PDA or other mobile device (Figure 4).

Simplicity and limited use of text are the guiding principles of the interface design, in conformity to interface design principles for users with aphasia. No text or keyboard input is required to use the system. The circular buttons are labelled with speech utterances—in this case, vowel sounds. When the user clicks on a button, an audio recording of the sound is played in synch

Figure 4. Sample display as a PDA application

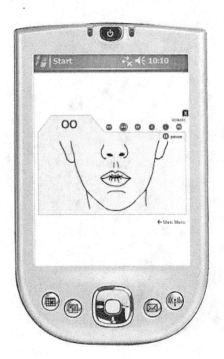

with the video or animation. If the user cannot read even this limited amount of text, then the small number of options helps the user to rely on memory to locate the desired sound. The feedback from selecting a sound is immediate—the audio and visual demonstration—and so this high degree of interactivity supports learning and exploring the application. A black circle around the selected button highlights the current sound, and the appropriate phonetic text is displayed on the folder tab in large font (here, "ay") to provide emphatic feedback. The pause button allows user to develop their understanding and practise the different stages of making the sound.

Note that the video and animation only include the lower part of the face. This further reduces the cognitive load for users by eliminating distracting elements such as the expression of the eyes (Bulthoff, Graf, Scholkopt, Simoncelli, & Wichmann, 2004).

A more complex exercise or interface would require additional support for the user. One possibility is to provide audio "mouseover" help—that is, for mouse pointer contact with screen elements to trigger the playing of an audio file containing spoken help or reading out button labels, rather than the conventional display of help text (Deo et al., 2004). Textual mouseover displays are unlikely to be useful for users with aphasia—the text is by default displayed in a small font, and users with aphasia frequently find it difficult to interpret text.

The audio in the prototypes was recorded by a female speaker of a similar age to the median age of the intended evaluators of the prototypes (female speech and language therapists). The video prototype is based on a native English speaker of the same gender, race, and median age as the intended evaluators. The importance of matching audio and visual display to the intended user is discussed in the next section.

The prototypes were evaluated in an empirical study with 20 professional speech and language therapists in the southwest region of England. The evaluators rated both prototypes as very easy to use and as potentially more helpful to their patients than conventional paper-based handouts. Eighteen of the evaluators (90%) could identify current patients who might benefit from using this type of application—but only if it were to be made available on a portable device (17 evaluators; 85%). This study was part of an action research project with the next stage being a refinement of the application to include 3D representations of the formation of sounds made inside the mouth and, provided ethical approval is granted, evaluation with a set of case studies is planned.

SOCIO-CULTURAL ISSUES IN SPEECH AND LANGUAGE THERAPY APPLICATION DESIGN

Voice characteristics of the speaker featured in speech and language therapy software should ideally match those of the user in significant factors such as age, cultural group, and sex, as it is commonplace for users of communication aids to adopt the intonation and accent of the recorded voice (Harris, 2004). Recording the voice of a family member or friend is therefore usually avoided, as this can cause confusion for the patient.

The user's cultural membership can also have a significant effect on the choice of voice for a speech and language therapy tool of this type—it is important that the user be able to understand and identify with the accent of the audio. The user's nationality or culture can also affect the example sounds and words to include in the application. New Zealand English provides extreme examples of the accommodations in the exercises that must be possible; for instance, for a New Zealand user the phrase "fish and chips" would be a poor choice for practice in the short i sound as the Kiwi accent usually renders that as "fush and chups", and vowel neutralisation causes word pairs such as "full" and "fill" to be indistinguishable when spoken. Alternative pronunciations are generally seen by individuals as incorrect, with their own pronunciation viewed as the only correct way to speak (Maclagan, 2000). Users are unlikely to accept software that uses pronunciation they regard as "not speaking properly".

The user should also be able to identify with the speaker images, or at the very least should not feel excluded or offended by them. Again, this points to the need to tailor this type of speech and language therapy software to groups of users. A system based on realistic videos, such as the prototype in Figure 3b, obviously would require a greater degree of customisation. The line drawings of Figure 3a are suitable for a broader intended target audience—the animation is less identifiable by gender, race, and age (though not entirely generic). The creation of icons and representations of humans that are completely culturally neutral is not possible, but guidelines exist to reduce the level of cultural bias (del Galdo & Nielsen, 1996).

Note that these problems with providing appropriate matches to the user in voice and appearance exist for face-to-face therapy as well. Again, New Zealand provides an extreme case in point: New Zealand speech is so distinctive that it can be difficult for New Zealand born therapists to work in other English-speaking countries, and conversely New Zealand born clients report difficulties in understanding non-New Zealand therapists (Maclagan, 2000). The creation of multimedia therapy tools offers the chance to provide tailored examples and exercises to an extent rarely possible in conventional therapy, where the number and age/sex/cultural distribution of therapists is rarely large enough to allow a patient to choose a therapist exactly matching his or her own background.

INTERNET-ENABLED THERAPY

The Internet has been suggested for use with speech and language applications in two ways:

for delivering therapy and for monitoring therapy. Internet delivery of exercises is highly appealing for the use of the prototypes on mobile devices, as this could address the problem of relatively small memory availability on PDAs and other mobiles: exercises could be streamed in on demand. This solution may be too financially costly for many users, however. Alternatively, selected exercises might be periodically uploaded via a "sync" with a PC or other larger storage device. Remote monitoring possibilities include storing logs or summaries of user sessions in central database, for therapists to later examine (Glykas & Chytas, 2004). This type of monitoring does not provide the capacity for the immediate feedback that is available in a face-to-face session, but does allow the therapist to maintain awareness of the user's progress between sessions. Therapist feedback could be delivered via the Internet, or the monitoring could be used to inform the next face-to-face session.

The potential of Internet-enabled therapy devices for supporting between-session monitoring or care raises several concerns—most notably, that the increased use of technology in therapy could raise barriers between the therapist and patient, and could lead to the dehumanisation of the professional-patient relationship. Issues of legal liability for the efficacy of this more attenuated version of treatment have yet to be fully resolved. There is also a perceived risk that other areas of care could suffer if scarce health funds are diverted to this potentially costly form of telemedicine (including cost of devices, development of multimedia therapy software, tailoring of software to individual patients, Internet transmission costs, therapist time devoted to remote monitoring, and so forth) (Rosenberg, 2004).

A more general anxiety voiced by the therapists evaluating the prototypes of the earlier section, *Development of Multimedia Prototypes*, is that of the ability of therapists to customize therapies as delivered via the Internet, or for that matter through any multimedia application or device. Therapists perceive that therapy as delivered

conventionally is tailored to an individual's needs, although it is difficult to see how the current generic paper-based materials are personalized to any great extent. Clearly these new technologies have raised expectations, and any system deployed would have to provide facilities for therapists to inspect, approve, and modify the therapy support being offered.

FURTHER DIRECTIONS FOR RESEARCH AND DEVELOPMENT

The preliminary evaluation with SLTs illustrated that although they were positive about the application, there are clear limitations to the prototypes already developed. In particular, SLTs identified the need for context words and illustrations for each phonetic sound and help with the production of sounds involving movements within the mouth. For example, the plosive sound "t" where the tongue is placed against the front upper palate. There is also a need for an SLT to be able to tailor the learning materials for a particular patient and monitor their progress.

It should be possible to create 3D models of the mouth rather than the vertical sectional drawings currently used. The model could then be rotated by the patient to see different viewpoints. The context words and images could be addressed relatively easily by building up a repository, or digital library, of learning objects which could be chosen by the SLT or added to by them in the same way that current mobile AAC devices can be tailored for a particular patient. A suitable system for this needs to be developed in a cooperative design project with SLT users which will allow the SLT to create drill and practice exercises, capture video and audio together with a further system to allow a patient to upload video of their progress for the SLT to monitor. The technical problems of suitable codecs to compress video need to be addressed although some of this work has already been done in the context of British

Sign Language learning (Andrews, 2005). The advent of full 3G services on mobile devices could be utilised to develop a fully mobile therapy service. However, the ergonomics of the devices used will need careful design as it is widely accepted that current mobile phones are unsuitable for use by elderly users due to the small size of their interaction buttons (Goodman, Dickinson, & Syme, 2004).

There are, therefore, a range of future directions in which research in mobile multimedia applications to enable people with aphasia to practise spoken language skills independently between sessions could progress. Blended approaches to therapy are likely to be positively received by SLTs to enable them to support patients with aphasia more flexibly than at present.

REFERENCES

Andrews, J. (2005). Using SignLab for formative and summative assessment. *The University of Bristol Learning Technology Support Service Fifth Annual National Conference*, Bristol, June 20. Retrieved April 24, 2006, from http://www.ltss.bris.ac.uk/vleconf05/Speakers/andrews.doc

Andrews, K., & Turner-Stokes, L. (2005). *Rehabilitation in the 21st century: Report of three surveys*. London: Royal Hospital for Neuro-disability.

Armstrong, L. (1993). Assessing the older communication-impaired person. In J. R. Beech, & L. Harding (Eds.), *Assessment in speech and language therapy* (pp. 163-166). London: Routledge.

Bhogal, S. J., Teasell, R. W., & Speechley, M. R. (2003). Intensity of Aphasia therapy, impact on recovery. *Stroke*, (34), 987-993.

Bulthoff, H. H., Graf, A. B. A., Scholkopt, B., Simoncelli, E. P., & Wichmann, F. A. (2004). Machine learning applied to perception: Decision-images for gender classification. *Advances in Neural Information Processing Systems, 17*. Retrieved April 24, 2006, from http://www.cns.nyu.edu/pub/eero/wichmann04a.pdf

Burton, E., Meeks, N., & Wright, K. (1991). Opportunities for using computers in speech and language therapy: A study of one unit. *British Journal of Disorders in Communication, 26*(2), 207-217.

Deo, S., Nichols, D. M., Cunningham, S. J., Witten, I. H., & Trujillo, M. F. (2004). Digital library access for illiterate users. *Proceedings of the 2004 International Research Conference on Innovations in Information Technology*, Dubai (UAE), October (pp. 506-516). United Arab Emirates: UAE University.

Dunlop, H., Cunningham, S. J., & Jones, M. (2002). A digital library of conversational expressions: Helping profoundly disabled users communicate. *Proceedings of the 2nd ACM/IEEE-CS Joint Conference on Digital Libraries* (JCDL), Portland (Oregon, USA), July 14-18 (pp. 273-274). New York: ACM Press.

Foundation for Assistive Technology (FAST). (2004, April). *Reporting on assistive technology in a rapidly changing world* (pp. 11-14). Retrieved April 24, 2006, from http://www.fastuk.org/RAPID.pdf

del Galdo, E. M., & Nielsen, J. (Eds.) (1996). *International user interfaces*. London: John Wiley & Sons.

Glykas, M., & Chytas, P. (2004). Technology assisted speech and language therapy. *International Journal of Medical Informatics, 73*, 529-541.

Goodman, J., Dickinson, A., & Syme, A. (2004). Gathering requirements for mobile devices using focus groups with older people. *Designing a More Inclusive World, Proceedings of the 2nd*

Cambridge Workshop on Universal Access and Assistive Technology (CWUAAT), Cambridge, UK, March. Retrieved April 24, 2006, from http://www.computing.dundee.ac.uk/projects/UTOPIA/publications/navigation_workshop.pdf

Harris, C. (2004). Progressing from paper towards technology. *Communication Matters, 18*(2), 33-37.

Maclagan, M. (2000). Where are we going in our language? New Zealand English today. *New Zealand Journal of Speech-Language Therapy, 53-54,* 14-20.

NHS. (2004). *Allied health professionals.* Retrieved April 24, 2006, from http://www.nhscareers.nhs.uk

Queensland Aphasia Groups. (2001a). *Web developer's guidelines.* Retrieved April 24, 2006, from http://www.shrs.uq.edu.au/cdaru/aphasiagroups/Web_Development_Guidelines.html

Queensland Aphasia Groups. (2001b). *What is aphasia-friendly?* Retrieved April 24, 2006, from http://www.shrs.uq.edu.au/cdaru/aphasiagroups/Aphasia_Friendly.html

Rosenberg, R. S. (2004). *The social impact of computers (3rd ed.).* USA: Elsevier Academic Press.

Royal College of Speech and Language Therapists (RCSLT). (2004). *What do speech and language therapists do?* Retrieved January 15, 2006, from http://www.rcslt.org/whatdo.shtml

van de Sandt-Koenderman, M., Wiegers, J., & Hardy, P. (2005, May). A computerised communication aid for people with aphasia. *Disability Rehabilitation, 27*(9), 529-533.

Chapter VI
A Proposed Framework for Mobile Services Adoption:
A Review of Existing Theories, Extensions, and Future Research Directions

Indrit Troshani
University of Adelaide, Australia

Sally Rao Hill
University of Adelaide, Australia

ABSTRACT

Mobile services are touted to create a significant spectrum of business opportunities. Acceptance of these services by users is, therefore, of paramount importance. Consequently, a deeper insight is required to better understand the underlying motivations leading users to adopting mobile services. Further, enhanced understanding would also help designing service improvements and appropriate adoption strategies. Most of the existing theoretical acceptance models available originate from organisational contexts. As mobile services bring additional functional dimensions, such as hedonic or experiential aspects, using extant models for predicting mobile services acceptance by individuals may be inadequate. The aim of this chapter is to explore and critically assess the use of existing acceptance theories in the light of evolving mobile services. Constructs affecting adoption behaviour are discussed and relevant extensions are made which culminate with a framework for mobile services adoption. Managerial implications are explored and future research directions are also identified.

INTRODUCTION

Mobile technologies and services are touted to create a significant spectrum of business opportunities. According to the International Telecommunications Union (ITU) mobile phone penetration rates have increased significantly in many countries in Northern Europe (e.g., Sweden—98.05%, Denmark—88.72%, Norway—90.89%) (Knutsen, Constantiou, & Damsgaard, 2005). Similarly, Japan and Korea have consistently experienced very high diffusion rates of mobile devices and services (Carlsson, Hyvonen, Repo, & Walden, 2005; Funk, 2005). While experts predict that by 2010 online access via mobile channels is expected to reach 24% of homes in North America, 27% in Eastern Europe, and 33% in North-Western Europe (Hammond, 2001), the current penetration rate in many countries in the Western hemisphere and Asia-Pacific, including the U.S. and Australia lags behind the forerunners (Funk, 2005; Ishii, 2004; Massey, Khatri, & Ramesh, 2005). Given the difference between rapid growth rates in the adoption of mobile technologies and associated services in some countries and the relatively slow growth rates in others (Bina & Giaglis, 2005; Knutsen et al., 2005), it is important to identify the factors and predictors of further adoption and integrate them into a consolidated framework.

Mobile technology is enabled by the collective use of various communication infrastructure technologies and portable battery-powered devices. Examples of mobile devices include notebook computers, personal digital assistants (PDAs) and PocketPCs, mobile, "smart" and Web-enabled phones, and global positioning system (GPS) devices (Elliot & Phillips, 2004). There is a variety of communication infrastructure technologies that can enable these devices. Data networking technologies, such as GSM, GPRS, and 3G, are typically used for connecting mobile phones. WiFi (wireless fidelity) is used for connecting devices in a local area network (LAN). Mobile devices can be connected wirelessly to peripherals

such as printers and headsets via the Bluetooth technology and virtual private networks (VPNs) enable secure access to private networks (Elliot & Phillips, 2004). Mobile devices are powered by mobile applications which deliver various services while enhancing flexibility, mobility, and efficiency for users within business and life domains. Despite the availability of technologically advanced mobile devices there is evidence that advanced mobile services which run on these have not been widely adopted (Carlsson et al., 2005; Khalifa & Cheng, 2002). The adoption of advanced mobile services is important for the mobile telecommunications industry because mobile services associated with technologically advanced devices constitute a massive source of potential revenue growth (Alahuhta, Ahola, & Hakala, 2005; Massey et al., 2005).

The adoption of advanced mobile technologies and services requires further research as most of the current technology acceptance models are based on research conducted in organisational contexts (Carlsson et al., 2005), and there has been only limited research from consumers' perspective (Lee, McGoldrick, Keeling, & Doherty, 2003). The features of mobile technologies and services, such as short message service (SMS), multimedia messaging service (MMS), e-mail, map, and location services, allow for single wireless devices, such as mobile phones, to be used seamlessly and pervasively across traditionally distinct spheres of life, such as work, home, or leisure, and with various levels of time commitment and self-ascribed roles (Dholakia & Dholakia, 2004). The interactions of these aspects are more intense than ever before (Knutsen et al., 2005). As mobile technologies and services add other functional dimensions, such as hedonic and/or experiential aspects (Kleijen, Wetzels, & de Ruyter, 2004; Mathwick, Malhotra, & Rigdon, 2001), applying extant theories outright to determine the acceptance and adoption by individual users may be questionable and inadequate (Knutsen et al., 2005).

Moreover, more research is called for in the adoption of mobile technologies because of the levels of complexity and diversity that may be encountered during their adoption. A number of factors contribute to this level of complexity and diversity. First, there is a strong relationship between the mobile devices and their users because the former always carries the identity of the latter (Chae & Kim, 2003). As a result, spatial positioning and identification of users is easier in the mobile context than in the traditional innovation adoption (Figge, 2004). Second, most mobile devices have limited available resources including memory, processing power, and user interface, which have the potential to offset ubiquity benefits (Chae & Kim, 2003; Figge, 2004). Third, the lifecycle of mobile technologies is usually short, which increases adoption risks because new technologies become rapidly obsolete and may, therefore, need to be replaced by newer ones. During this process, a certain amount of consumer learning might be required before adopters can be confident and satisfied in using the mobile devices and services (Saaksjarvi, 2003). Again, this supports the argument that current models of technology acceptance may not be applied directly in predicting mobile adoption behaviour because they do not reflect the levels of complexity and diversity in the adoption of mobile technologies.

This chapter focuses on mobile phones and the associated services. Examples of mobile services include mobile e-mail, commercial SMS, and MMS services, downloads to portable devices, access to news through a mobile phone, mobile ticket reservations, mobile stock trading, as well other customised services which may be made available by mobile phone operators (Bina & Giaglis, 2005). Research shows that ownership of technologically advanced mobile phones is a main driver for advanced mobile services (Carlsson et al., 2005). Therefore, the adoption of mobile services should also be considered in the context and the technologies which enable them.

The aim of this chapter is to extend the existing models and to propose an integrated conceptual and parsimonious framework which explains adoption behaviour of users of mobile technologies and services. To accomplish this, we first provide an overview of recent developments of mobile technologies and services. Then, a critical assessment of existing acceptance models is made. Next, acceptance constructs and their relevance to mobile technologies and services are discussed. These constructs are then integrated into a new framework about mobile services adoption. In the last section, the implications of this model and future research directions are also discussed.

OVERVIEW OF MOBILE TECHNOLOGY EVOLUTION

In this section, an overview of the evolution of mobile phone technologies is provided. The recognition the evolution of these technologies is important because it puts the adoption constructs discussed later in the appropriate context. The diagram in Figure 1 summarises the evolution of the technologies.

Second Generation Wireless Devices

The second generation of wireless devices (2G) introduced the digitisation of mobile communication and encompasses several standards which incrementally introduced new services and improved existing ones. It was a big leap forward from the first generation wireless communication (1G) which used analog standards and was characterised by poor quality and narrow bandwidth which resulted in limited adoption by both businesses and individuals (Elliot & Phillips, 2004). The commonly used standards by 2G are the Global System for Mobile Communications, the Wireless Applications Protocol, and the General Packet Radio Service. These are explained in more detail in the following sections.

Figure 1. Evolution of wireless technologies (Source: Carlsson et al., 2005)

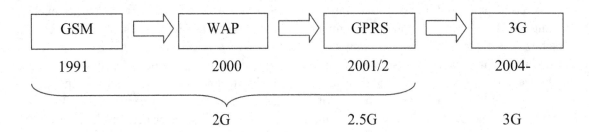

Global System for Mobile Communication

Launched in the early 1990s, the Global System for Mobile Communications (GSM) constitutes the world's fastest growing and most popular mobile telephony. Available through over 500 networks and serving almost a billion customers in 195 countries, GSM has been expanding exponentially (UMTS, 2003). Most countries, including underdeveloped and developing or even countries with a very low population density have at least two GSM network operators (Rossotto, Kerf, & Rohlfs, 2000). This has increased product and service offerings and competitiveness which has boosted GSM popularity even further.

One of the key advantages of the GSM technology is that it unified a range of different mobile communication standards into a single standard which constitutes a complete and open network architecture. This allows GSM-compatible mobile devices to be connected to any GSM network, therefore, enhancing interoperability. Further, GSM uses digital encoding which encrypts communications between a mobile phone and its base station, which makes interception more difficult. This results in improved security (Elliot & Phillips, 2004).

Another feature of the GSM technology is the automatic country-to-country communication, also known as global roaming. Because international travel for both business and pleasure has increased in recent years, roaming between mobile networks has become valuable as it generates as much as 15% of mobile operator's average revenue per user (ARPU) (UMTS, 2003). The subscriber identity module (SIM) card is another aspect of the GSM technology which is central to its popularity. The SIM card allows operators to manage information about their customers, including customer profile and billing, security access and authentication, virus intrusion and downloading capabilities (UMTS, 2003).

In addition to features such as caller ID, call forwarding, and call waiting, SMS emerged as the first unique mobile service and became the most popular mobile service after 1995 when adopters began using mobiles to send and receive limited amount of data in the form of short messages (Carlsson et al., 2005). While SMS later became the foundational platform for a variety of other services, it is considered to be cumbersome by many users because in addition to memorise service codes, users are also required to type text using the keypad of the mobile device (Carlsson et al., 2005).

Wireless Application Protocol

The Wireless Application Protocol (WAP) was introduced with the aim of providing advanced telephony and data access from the Internet using mobile terminals such as mobile phones, PDAs, smart phone, and other portable handheld devices (van Steenderen, 2002). With WAP, mobile devices can access Web sites specifically designed and built for them. WAP was, therefore, expected to provide the opportunity for connecting two of the fastest growing sectors of the telecommunications industry, namely, the Internet and mobile communications. As a matter of fact, the hype generated by WAP reached such dizzying heights that from January to August 2000, the number of WAP-compatible Web pages increased from almost zero to 4.4 million (Teo & Pok, 2003).

The benefits of WAP were supposed to extend to several industries ranging from mobile operators, developers of WAP applications, manufacturers of mobile devices, and consumers in terms of various services, including banking, ticket reservations, entertainments, voice and fax mail notifications (Klasen, 2002; van Steenderen, 2002). Nevertheless, except for NTT DoCoMo's i-mode successful demonstration of mobile Internet, WAP has turned out to be a major disappointment with early adopters and other enthusiasts experiencing cognitive dissonance due to the relative oversell (Carlsson et al., 2005; Ratliff, 2002; Teo & Pok, 2003; Xylomenos & Polyzos, 2001).

Other challenges have also adversely affected widespread diffusion of WAP. Narrow bandwidth, low storage memory, and small screen limitations have resulted in slow communications; abridged Internet access has resulted in mediocre interfaces and almost no graphics. This has considerably limited Web site effectiveness (Klasen, 2002). By the end of 2000, only 12 million Europeans had WAP-compatible mobile devices, and of these, only 6% regularly used WAP functionality (Robins, 2003). Worldwide, only 10-15% of whom own WAP-compatible mobile devices would ever use WAP services, suggesting that "WAP had no future" (Klasen, 2002, p. 196). Nonetheless, the introduction of WAP constitutes a major step forward as it showed that Internet browsing is possible in mobile devices in general and phones in particular (Carlsson et al., 2005).

General Packet Radio Service

Simply known as the GPRS, the General Packet Radio Service constitutes an improvement over the GSM technology. GPRS uses packet-based data transfer mechanisms to provide continuous Internet accessibility (Elliot & Phillips, 2004; Hart & Hannan, 2004). With GPRS, users are not required to stay connected all the time in order to use a service. As a result, they are not charged on the basis of the connection time, rather, on the basis of the amount of downloaded data (Carlsson et al., 2005). Overall, GPRS is more efficient and cheaper than GSM, and yet, less widespread among users. Advancements associated with GPRS include the introduction of cameras, colour screens, multimedia messaging service (MMS), and video streaming (Carlsson et al., 2005). Because GPRS enhances 2G services, it is often referred to as the 2.5G technology (Elliot & Phillips, 2004).

Third Generation Wireless Devices

3G represents the next generation of mobile communication technologies, and it makes considerable improvements over its predecessors. These improvements include broad bandwidth which results in higher connection speeds, variety of multimedia capabilities and improved screen display, enhanced security features, and increased storage capacity (Elliot & Phillips, 2004). These enhancements enable users to receive digital photographs, moving video images, high quality sound in their mobile devices, and full unabridged e-mail and Internet access (Elliot & Phillips, 2004). Corporate users are also able to connect remotely to office computers and networks in order to access and download files quickly and easily (Robins, 2003). Because 3G technology mainly improves and enhances many existing services, it is considered to be an evolution rather than a revolution over the previous generation (Carlsson et al., 2005).

3G ensures that anybody, anywhere can access the same services (Grundström & Wilkinson, 2004). Further, 3G aims at integrating both the business and the social domains of the user's life which is the reason why 3G terminals are also referred to as "lifestyle portals" (Elliot & Phillips, 2004, p. 7). Another feature of the 3G technology is its capability to provide location-based services (LBS) which could support health, transport, entertainment, data mining, and so forth (Casal, Burgelman, & Bohlin, 2004; UMTS, 2003). There is evidence that there is demand for such services which constitutes the main economic incentive for the development of the 3G technology (Alahuhta et al., 2005; Repo, Hyvonen, Pantzar, & Timonen, 2004).

A CRITICAL REVIEW OF THEORETICAL MODELS OF TECHNOLOGY ACCEPTANCE

A review of technology acceptance literature revealed many competing theoretical models, each with different focus and tested in different contexts. A significant amount of research effort has been put into building theories to examine how and why individuals adopt new information technologies and predict their level of adoption and acceptance. While one stream of research focuses on individual acceptance of technology (Compeau & Higgins, 1995; Davis, Bagozzi, & Warshaw, 1989), other streams have focused on implementation success at the organizational level (Leonard-Barton & Deschamps, 1988).

Many of the previously empirically researched models have been drawn from social psychology,

for example, theory of reasoned action (TRA), motivational model, theory of planned behaviour (TPB), and sociology, for example, social cognitive theory (SCT) and innovation diffusion theory (IDT). Others specifically apply to technology adoption, for example, technology acceptance model (TAM). While each of these models made unique contributions to the literature on technology acceptance and adoption, most of these theoretical models theorise behaviour intention and/or usage as the key dependent variable in explaining acceptance of information technology because behavioural intentions are motivational factors that capture how hard people are willing to try to perform a behaviour (Ajzen, 1991). For example, TPB suggests that behavioural intention is the most influential predictor of behaviour; after all, a person does what s/he intends to do. In a meta-analysis of 87 studies, an average correlation of 0.53 was reported between intentions and behaviour (Sheppard, Hartwick, & Warshaw, 1988). As mobile services and underlying technologies are emerging information technologies, it is appropriate to consider this as the point of departure and use it to form the basis of a theoretical framework in mobile services and technology acceptance and adoption. The models that have been most frequently quoted in the technology acceptance and adoption literature are discussed next.

Theory of Reasoned Action (TRA)

Theory of reasoned action models are considered to be the most systematic and extensively applied approaches to attitude and behaviour research. According to TRA, the proximal determinant of a behaviour is a behavioural intention, which, in turn, is determined by attitude. These models propose that an individual's actual behaviour is determined by the person's intention to perform the behaviour, and this intention is influenced jointly by the individual's attitude and subjective norm. Attitude is defined as "a learned predisposition to respond in a consistently favourable or unfavourable manner with respect to a given object" (Fishbein & Ajzen, 1975, p. 6). A person's attitude towards a behaviour is largely determined by salient beliefs about the consequences of that behaviour and the evaluation of the desirability of the consequences (Fishbein & Ajzen, 1975). Subjective norm is defined as "the person's perception that most people who are important to him think he should or should not perform the behaviour in question" (Dillon & Morris, 1996). In brief, TRA asserts that attitude and subjective norm and their relative weights directly influence behavioural intention.

Theory of Planned Behaviour (TPB) and Decomposed Theory of Planned Behaviour

TPB, which generalizes TRA by adding a third construct—perceived behavioural control (Ajzen, 1991)—has been one of the most influential theories in explaining and predicting behaviour, and it has been shown to predict a wide range of behaviours (Sheppard et al., 1988). TPB asserts that the actual behaviour is determined directly both by behavioural intention and perceived behavioural control. Behavioural intention is formed by one's attitude, subjective norm, and perceived behavioural control (Ajzen, 1991). Further, a decomposed TPB includes constructs such as relative advantage, compatibility, influence of significant others, and risk from the innovation diffusion literature, and decomposing the three perceptions in TPB into a variety of specific belief dimensions. This model offers several advantages over TPB and is considered more complete and management-relevant by focusing on specific factors that may influence adoption and usage (Teo & Pok, 2003).

Technology Acceptance Model (TAM)

TAM can be seen as an adaptation of the theory of reasoned action (TRA) and was developed to predict and explain individual system use in the workplace (Davis, 1989). This model further suggests that two beliefs—perceived usefulness and perceived ease of use—are instrumental in explaining the user's intentions of using a system. Perceived usefulness refers to the degree to which "a person believes that use of the system will enhance his or her performance" whereas perceived ease of use is the degree to which "a person believes that using the system will be free of effort". Simply put, a technology that is easy to use and is useful will lead to a positive attitude and intention towards using the technology.

The main advantage of this model over others is that the two related beliefs can generalize across different settings. Thus, some argue that it is the most robust, parsimonious, and influential model in explaining information technology adoption behaviour (Elliot & Loebbecke, 2000; Teo & Pok, 2003; Venkatesh, Morris, Davis, & Davis, 2003). Indeed, since its development, it has received extensive empirical support through validations, applications, and replications for its prediction power (Taylor & Todd, 1995, 1995a; Venkatesh & Morris, 2000a). A number of modified TAM models were proposed to suit new technologies including Internet and intranet (Agarwal & Prasad, 1998; Chau, 1996; Chau & Hu, 2001; Horton, Buck, Waterson, & Clegg, 2001). For example, TAM has been used to predict Internet purchasing behaviour (Gefen, Karahanna, & Straub, 2003; Kaufaris, 2002).

A major theoretical limitation of TAM is the "exclusion of the possibility of influence from institutional, social, and personal control factors" (Elliot & Loebbecke, 2000, p. 49). Thus the suitability of the model for predicting general individual acceptance needs to be re-assessed as the main TAM constructs do not fully reflect the specific influences of technological and usage-context factors that may alter user acceptance (King, Gurbaxani, Kraemer, McFarlan, Raman, & Yap, 1994; Taylor & Todd, 1995). In response to this, a number of modifications and changes to the original TAM models have been made. The most prominent of these is the unified theory of acceptance and use of technology (UTAUT), a unified model that integrates constructs across eight models (Venkatesh et al., 2003). UTAUT provides a refined view of how the determinants of intention and behaviour evolve over time and assumes that there are three direct determinants of intention to use (performance expectancy, effort expectancy, and social influence) and two direct determinants of usage behaviour (intention and facilitating conditions). However, both TAM and UTAUT have received criticisms with the fundamental one being about the problems in applying these beyond the workplace and/or organisation for which originally created (Carlsson et al., 2005).

Motivational Theories

Motivation theories are rooted in psychological research to understand individuals' acceptance of information technology (Davis, Bagozzi, & Warshaw, 1992; Igbaria, Parasuraman, & Baroudi, 1996). These theories often distinguished extrinsic and intrinsic motivation. While extrinsic motivation refers to the performance of an activity in helping achieve valued outcomes, intrinsic motivation puts emphasis on the process of performing an activity (Calder & Staw, 1975; Deci & Ryan, 1985). For example, perceived usefulness is an extrinsic source of motivation (Davis et al., 1992) while perceived enjoyment (Davis et al., 1992), perceived fun (Igbaria et al., 1996), and perceived playfulness (Moon & Kim, 2001) can be considered intrinsic sources of motivation. Both sources of motivation affect usage intention and actual usage. Therefore, in addition to ease of use and usefulness, intrinsic motivators, such as playfulness, will also play an important role in increasing usability in a usage environment in which information technology applications are both used for work and play (Moon & Kim, 2001).

Innovation Diffusion Theory

The innovation diffusion theory is concerned with how innovations spread and consists of two closely related processes: the diffusion process and adoption process (Rogers, 1995). Diffusion is a macro process concerned with the spread of an innovation from its source to the public whereas the adoption process is a micro process that is focused on the stages individuals go through when deciding to accept or reject an innovation. Key elements in the entire process are the innovation's perceived characteristics, the individual's attitude and beliefs, and the communication received by individuals from their social environment. In relation to the factors pertaining to innovation, factors such as, relative advantage, complexity, trialability, observability, and compatibility, were considered important in influencing individual's acceptance of the innovation (Rogers, 1995).

TOWARDS AN ACCEPTANCE MODEL FOR MOBILE SERVICES

This section develops an acceptance model for mobile technology and services that may be empirically tested. This development begins with identifying the latent constructs in extant technology adoption literature. However, mobile services differ from traditional systems in that mobile services are ubiquitous, portable, and can be used to receive and disseminate personalised and localised information (Siau, Lim, & Shen, 2001; Teo & Pok, 2003). Thus, the models examined in the previous section and the constructs included in these models may not be applicable to mobile services adoption. In particular, we discuss the various antecedents of attitude towards mobile services and develop a new model based on the widely used TAM model to predict adoption of new mobile services.

User Predisposition

User predisposition refers to the internal factors of an individual user of mobile services. Personal differences strongly influence adoption. There is evidence that successful acceptance of innovations depends as much on individual adopter differences as on the innovation itself. Indeed, individual differences help identify segments of adopters who are more likely to adopt technology innovations than others, which in turn, helps providers address adopter needs more closely (Massey et al., 2005). Diffusion resources can also be used more effectively and efficiently (Agarwal & Prasad, 1998). Early adopters, for example, can then act as opinion leaders or change agents to facilitate the diffusion of the technology further (Rogers, 1995). There are several dimensions used to capture individual differences, including personal innovativeness, perceived costs, demographic factors, psychographic profiles, and personality traits (Dabholkar & Bagozzi, 2002). In this chapter, we define user predisposition as the collection of factors including the individual's prior knowledge and experience of existing mobile services, compatibility, behavioural control, personal innovativeness, perceived enjoyment, and price sensitivity.

First, *prior knowledge* is essential for the comprehension of the technology and related services. According to Rogers (1995), knowledge occurs when a potential adopter learns about the existence of an innovation and gains some understanding concerning its functionality. Like other technologies, the mobile technology is comprised of both the hardware (i.e., the physical mobile device) and software domains (i.e., the applications consisting of the instructions to use the hardware as well as other information aspects) (Rogers, 1995). Thus knowledge from both hardware and software domains might be required for complete comprehension (Moreau, Lehmann, & Markman, 2001; Saaksjarvi, 2003). Prior knowledge consists of two components, namely, familiarity and ex-

pertise. For instance, the former constitutes the number of mobile services-related experiences accumulated by consumers over time, which includes exposure to advertising, information search, interaction with salespersons, and so on. The latter represents the ability to use the mobile services, and it includes beliefs about service attributes (i.e., cognitive structures) as well as decision rules for acting on those beliefs (i.e., cognitive processes) (Alba & Hutchinson, 1987). In any case, familiarity alone cannot capture the complexity of consumer knowledge (Alba & Hutchinson, 1987), which suggests the learning is required (Saaksjarvi, 2003).

With learning, consumers use the "familiar" component of existing knowledge as a means to understand and comprehend new phenomena in the innovation which is being adopted (Roehm & Sternthal, 2001). Specifically, existing knowledge in general and analogical learning in particular have been shown to be powerful and highly persuasive communication devices in acquiring in-depth understanding of innovation benefits and functionality (Moreau et al., 2001; Roehm & Sternthal, 2001; Yamauchi & Markman, 2000). An analogy compares and contrasts a known base innovation to an unknown target innovation. The base and the target share structural attributes, but are different in terms of surface attributes. A cellular phone versus a personal digital assistant (PDA) versus a "smart phone" are good examples. Research shows that "a message containing an analogy is better comprehended and is more persuasive when the recipient has expertise with regard to the base product [innovation]." (Roehm & Sternthal, 2001, p. 269). However, expertise alone is insufficient to ensure analogy persuasiveness. Substantial resources, training/usage instructions, and a positive mood are also required to facilitate learning (Roehm & Sternthal, 2001). However, while knowledge is important, by itself, it has limited usefulness, and therefore, "knowledge alone cannot determine the basis for adoption" (Rogers, 1995, p. 167) of a technology or service.

Adopters' previous positive or negative *experiences* with a technology or service can have a significant impact on their perceptions and attitudes towards that technology (Lee et al., 2003; Taylor & Todd, 1995a). Specifically, experience may influence adopters in forming positive or negative evaluations concerning innovations, which can boost or impair adoption of mobile technologies and services. Because of their greater clarity and certainty, direct prior experiences are likely to have a stronger impact on perceptions and attitudes towards usage than indirect or incomplete evidence (i.e., pre-trial) (Knutsen et al., 2005; Lee et al., 2003).

The second variable within the user predisposition construct is *compatibility*. Rogers (1995) defines compatibility as the degree to which an innovation is perceived to be consistent with existing values of potential adopters. In general, high incompatibility will adversely affect potential adopters of an innovation, which decreases the likelihood of adoption (Saaksjarvi, 2003). In contrast, high compatibility is likely to increase adoption propensity. In the context of wireless devices, lifestyle compatibility is the extent to which adopters believe mobile devices and services can be integrated into their daily lives. For example, adopters' lifestyle in terms of degree of mobility is likely to have a strong impact on their decision to adopt the technology (Pagani, 2004; Teo & Pok, 2003). For example, a person who leads a busy lifestyle, and is employed in an information-intensive job, and is always on the move is more likely to adopt a wireless device and its associated services compared to a person who leads a sedentary lifestyle.

Third, perceived *behavioural control*, a dynamic and socio-cognitive concept, has attracted a lot of attention in adoption literature. Earlier work by Ajzen (1991) considered it as a uni-dimensional variable. More recent empirical findings suggest that perceived behavioural control has two distinct components: self-efficacy, which is an individual's judgement of their capability to perform a behaviour, and controllability, which constitutes an individual's beliefs if they have the necessary resources and opportunities to adopt the innovation. It denotes a subjective judgment of the degree of control over the performance of a behaviour not the perceived likelihood that performing the behaviour will produce a given outcome (Ajzen, 1991). In the context of mobile service adoption, perceived behavioural control refers to the individual perception of how easy or difficult it is to get mobile services.

Fourth, *personal innovativeness* is the willingness of an individual to try out and embrace new technologies and their related services for accomplishing specific goals. Also known as technology readiness, personal innovativeness embodies the risk-taking propensity which exists in certain individuals and not in others (Agarwal & Prasad, 1998; Massey et al., 2005; Parasuraman, 2000). This definition helps segment potential adopters into what Rogers (1995) characterises as innovators, early adopters, early and late majority adopters, and laggards. Personal innovativeness represents a confluence of technology-related beliefs which jointly determine an individual's predisposition to adopt mobile devices and related services. The adoption of any innovation in general, and of innovative mobile phones and services in particular is inherently associated with greater risk (Kirton, 1976). Therefore, given the same level of beliefs and perceptions about an innovation, individuals with higher personal innovativeness are more likely to develop positive attitudes towards adopting it than less innovative individuals (Agarwal & Prasad, 1998).

Fifth, *perceived enjoyment* refers to the degree to which using an innovation is perceived to be enjoyable in its own right and is considered to be an intrinsic source of motivation (Al-Gahtani & King, 1999). Because the market for mobile innovations and services is comprised of both corporate users and consumers, factors focusing on perceived enjoyment constitute an important consideration (Carlsson et al., 2005; Pagani, 2004). That is, adopters use an innovation for the pleasure or enjoyment its adoption might bring and, therefore, serve as an end unto itself. Further, intrinsic enjoyment operates outside valued outcomes or immediate material needs (i.e., extrinsic motivations), such as enhanced job performance, increased pay, and so forth (Mathwick et al., 2001; Moon & Kim, 2001). Most research on enjoyment is based on the "flow theory" according to which flow represents "the holistic sensation that people feel when they act with total involvement" (Csikszentmihalyi, 1975). In a "flow state" individuals interact more voluntarily with innovations within their specific context, which determines their subjective experiences (Csikszentmihalyi, 1975). Consequently, individuals who have a more positive enjoyment experience with an innovation are likely to have stronger adoption intentions than those who do not (Moon & Kim, 2001).

That is, intrinsic enjoyment can positively affect the adoption and use of innovative mobile services, and is therefore, a significant determinant of intention and attitude towards adoption (Kaufaris, 2002; Novak, Hoffman, & Yung, 2000). Further, upon adoption, individuals are more likely to use the mobile services that offer enjoyment more extensively than those which do not. As a consequence, perceived enjoyment is also seen to have a significant effect beyond perceived usefulness (Davis et al., 1989a). However, the complexity of a mobile innovation or service has a negative effect on perceived enjoyment, suggesting that the potential impact of enjoyment may not be fully realised (Igbaria et al., 1996).

The final variable that needs to be added to the existing technology adoption models is *price sensitivity*. In the original technology acceptance models, the costs of adopting an innovation were not considered to be a relevant construct because the actual users did not have to pay for the technology. In an organisational setting, the cost would be incurred by the organisation. However, in the context of individual private adoption, cost becomes a relevant factor. There is evidence showing that perceived financial resources required to adopt mobile technologies and services constitute a significant determinant of behavioural intention (Kleijen et al., 2004; Lin & Wang, 2005). However, evidence also shows that adopters of mobile devices and services also attempt to assess the value of adoption by comparing perceived costs against the benefits (Pagani, 2004). Perceived costs are directly related to income and socioeconomic status of potential adopters which are recognised to have a strong impact on technology adoption and diffusion (Lu, Yu, Liu, & Yao, 2003). For example, in Europe individuals earning income beyond certain levels were found to have a high propensity to embrace mobile technologies, such as WAP mobile phones, handheld computers, and so forth (Crawford, 2002). Similarly, there's evidence that in fast growing economies, individuals with higher income spend more on mobile devices (Lu et al., 2003).

Perceived Usefulness

Perceived usefulness is "the degree to which a person believes that using a particular system would enhance his or her job performance" (Davis, 1989, p. 320). Perceived usefulness is also known as performance expectancy (Venkatesh et al., 2003). An innovation is believed to be of high usefulness when a potential adopter believes that there is a direct relationship between use on the one hand and productivity, performance, effectiveness, or satisfaction on the other (Lu et al., 2003).

Usefulness recognition is important because it has been found to have a strong direct effect on the intention of adopters to use the innovation (Adams, Nelson, & Todd, 1992; Davis, 1989). In addition, potential adopters assess the consequences of their adoption behaviour and innovation usage in terms of the ongoing desirability of usefulness (Chau, 1996; Venkatesh & Davis, 2000). Although an innovation might provide at least some degree of usefulness, a potential reason not to adopt exists when adopters fail to see the "need" to adopt (Zeithaml & Gilly, 1987). Adopters may not be able to recognise their needs until they become aware of the innovation or its consequences (Rogers, 1995). Need recognition is, therefore, likely to drive potential adopters to educate themselves in order to be able to utilise the innovation fully before being able to recognise its usefulness. This in turn is likely to lead to a faster rate of adoption (Rogers, 1995; Saaksjarvi, 2003).

Perceived usefulness can be split into two parts. Near-term usefulness is perceived to have an impact on the near-term job fit, such as job performance or satisfaction (Thompson, Higgins, & Howell, 1994). Long-term usefulness is perceived to enhance the future consequences of adoption including career prospects, opportunity for preferred job assignments, or social status of adopters (Chau, 1996; Thompson et al., 1994). Evidence shows that even though perceived near-term usefulness has the most significant impact on the behavioural intention to adopt an innovation, perceived long-term usefulness also exerts a positive, yet lesser impact (Chau, 1996; Jiang, Hsu, Klein, & Lin, 2000).

In the case of mobile technology and services, perceived usefulness is defined as the degree to which the mobile technology and services provide benefits to individuals in every day situations (Knutsen et al., 2005). The range and type of service offerings as well as the compatibility of the user's existing computing devices influence perceived usefulness (Pagani, 2004). In addition, Pagani (2004) also finds that usefulness emerges as the strongest determinant in the adoption of three generation mobile services which is consistent finds of research concerning the adoption of other innovations (Venkatesh et al., 2003).

Perceived Ease of Use

Perceived ease of use is the "degree to which a person believes that using a particular system would be free of effort" (Davis, 1989, p. 320). Other constructs that capture the notion of perceived ease of use are complexity and effort expectancy (Rogers, 1995; Venkatesh et al., 2003). Perceived ease of use may contribute towards performance, and therefore, near-term perceived usefulness. In addition, lack of it can cause frustration, and therefore, impair adoption of innovations. Nevertheless, "no amount of EOU [ease of use] will compensate for low usefulness" (Keil, Beranek, & Konsynski, 1995, p. 89).

In the mobile setting, perceived ease of use represents the degree to which individuals associate freedom of difficulty with the use of mobile technology and services in everyday usage (Knutsen et al., 2005). For example, there is evidence in the media that using certain services on a mobile device can be quite tedious, especially when browsing Internet-like interfaces on mobile devices is required (Teo & Pok, 2003). Together with relatively small screen sizes and associated miniaturized keypads, the overall usage experience may be adversely affected. This suggests that input and output devices are likely to influence perceived ease of use (Pagani, 2004). In addition, user-friendly and usable intuitive man-machine interfaces, including clear and visible steps, suitable content and graphical layouts, help functions, clear commands, symbols, and meaningful error messages are likely to influence adoption as well (Condos, James, Every, & Simpson, 2002). Further, Pagani (2004) argues the mobile system response time affects perceived ease of use suggesting that mobile bandwidth is important as well.

Social Influences

Social influence constitutes the degree to which individuals perceive that important or significant others believe they should use an innovation (Venkatesh et al., 2003). Venkatesh et al. (2003) believe that the social influence constructs may only become significant drivers on intention to adopt when users adopt an innovation in order to comply mandatory requirements. In these circumstances, social influence seems to be significant in the early phases of adoption and its effect decreases with sustained usage (Venkatesh & Davis, 2000). Conversely, in voluntary settings, social influence appears to have an impact on perceptions about the innovation (Venkatesh et al., 2003). Social influence is related to three similar constructs, namely, subjective and social norms, and image.

In Taylor and Todd's study (1995), subjective norms are defined to include the influence of other people's opinions otherwise known as reference groups. These include peers, friends, superiors, computer, and technology experts. Subjective norms have a greater impact during the initial adoption phase when potential adopters have little or no experience or when the adoption behaviour is new (Thompson et al., 1994). Research shows that pressure from reference groups to adopt an innovation is effective because it contributes to reducing perceived risk associated with adoption (Teo & Pok, 2003).

Social factors constitute another construct of social influence. Social factors represent cues individuals receive from members of their social structure which prompt them to behave in certain ways (Thompson, Higgins, & Howell, 1991). For example, in Japan, teenagers regard smart phones as fashion items (Lu et al., 2003). Further, there is evidence that unique communications patterns determined by key social and cultural factors, such as group-oriented nationality, have positively affected adoption practices of using the Internet via mobile phones in East Asia (Ishii, 2004).

A third critical construct related to social influence is image. The adoption of an innovation can be seen to enhance one's status or image in their social system. For certain adopters a mobile device may be more of a lifestyle than a necessity (Bina & Giaglis, 2005; Teo & Pok, 2003). For example, early adopters of mobile computing devices might be image-conscious users who wish to be seen as trend-setters or technology savvy enthusiasts.

Facilitating Conditions

Facilitating conditions refer to external controls and catalysts in the adoption environment which aim at facilitating adoption and diffusion of new technologies (Terry, 1993). Facilitating conditions are important because they are considered to be direct usage antecedents, and are therefore, likely to make adoption behaviour less difficult by removing any obstacles to adoption and sustained usage (Thompson et al., 1994; Venkatesh et al., 2003). These conditions can be provided by both governments and mobile operators. For example, governments or the representative agencies can act as facilitators by bringing together the telecommunication industry, academia, and research community. Government agencies can also set up protocol standardization policies and regulations favouring the future growth of mobile communication systems (Lu et al., 2003). Likewise, mobile operators can encourage adoption by mass advertising campaigns and active promotion aimed at increasing awareness about mobile devices and related services (Teo & Pok, 2003). Further, promoting and enforcing appropriate interconnection agreements and adequate regulatory mechanisms among mobile operators help adopters of mobile devices take advantage of roaming services and consequently be conducive to adoption (Rossotto et al., 2000).

Facilitating conditions also capture the existence of a trusting environment that is external to the mobile operator's control. A trusting environment constitutes an important factor in the adoption of mobile technologies and services. It determines the user's expectations from the relationship with their service providers, and it increases their perceived certainty concerning the provider's expected behaviour. Generally, trust is essential in all economic activities where undesirable opportunistic behaviour is likely to occur (Gefen et al., 2003). However, trust becomes vital in a mobile environment, where situational factors such as uncertainty or risk and information asymmetry are present (Ba & Pavlou, 2002). On the one hand, adopters of mobile technology are unable to judge the trustworthiness of service providers, and on the other, the latter can also easily take advantage of the former by engaging in harmful opportunistic behaviours. For example, service providers can sell or share the transactional information of its users or their personal information.

There are two key elements in a trusting environment, namely, security and privacy (Lu et al., 2003). In a wireless environment, security encompasses confidentiality, authentication, and message integrity. Because mobile devices have limited computing resources and wireless transmissions are more susceptible to hacker attacks, security vulnerabilities can have serious consequences (Galanxhi-Janaqi & Nah, 2004; Lu et al., 2003). There are several remedies against the dangers of insecurity, for example, public key infrastructure and certificate authority which use public key cryptography to encrypt and decrypt mobile transmissions and authenticate users.

Ironically, the same information practices which provide value to both users and providers of mobile technology and services also cause privacy concerns. Some of these concerns include: the type of information that can be collected about users and the ways in which it will be protected; the entities that can access this information and their accountability; and the ways in which the information will be used (Galanxhi-Janaqi & Nah, 2004). In mobile adoption research the trust environment has been encapsulated in a construct called perceived credibility (Lin & Wang, 2005; Wang, Wang, Lin, & Tang, 2003). Evidence shows that there is a "significant direct relationship between perceived credibility and behavioural intention" (Lin & Wang, 2005, p. 410) to use mobile services.

Moderating Variables

Evidence shows that gender and age might influence the adoption of technology and related services due to their moderating effects on other constructs (Venkatesh et al., 2003). In general, men tend to exhibit task-oriented attitudes suggesting that usefulness expectations might be more accentuated in men than women (Minton & Scheneider, 1980). This is particularly the case for younger men (Venkatesh & Morris, 2000a). On the other hand, ease of use expectations are more salient for women and older adopters (Bozionelos, 1996). Further, women are predisposed to be more sensitive to the opinions of members of their social structure. As a result women are more likely to be affected by social influence factors when deciding to adopt new mobile technologies and services (Venkatesh & Morris, 2000a). Similarly, because affiliation needs increase with age (Rhodes, 1983), older adopters are more likely to be affected by social influence.

Quality of life of potential adopters is another moderating variable which is likely to affect the adoption of mobile devices and services. "Quality of life" is an established social sciences notion which represents "a global assessment of a person's life satisfaction according to his chosen criteria" (Diener & Suh, 1997; Shin & Johnson, 1978). There is evidence which indicates that mobile technology and services have enhanced the perceived quality of social and work life of adopters (Jarvenpaa, Lang, Takeda, & Tuunainen, 2003). Bina and Giaglis (2005) present evidence that the reverse is also true. They indicate that adopters who are satisfied with specific life domains exhibit favourable attitudes towards the adoption of specific mobile services (Bina & Giaglis, 2005).

Further, evidence also shows that stage of use and voluntariness of usage have moderating effects on adoption attitudes through various constructs. For example, perceived ease of use is significant during the initial period of usage when process-

Figure 2. Proposed model of acceptance of mobile services

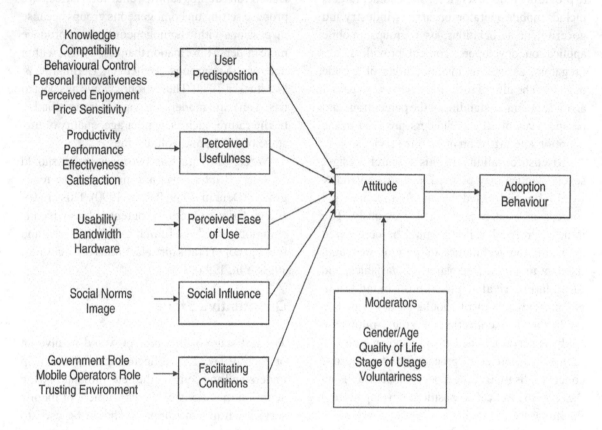

related issues constitute obstacles to be overcome (Venkatesh et al., 2003). Perceived ease of use, however, becomes insignificant during periods of extended usage (Agarwal & Prasad, 1998; Davis et al., 1989a; Thompson et al., 1994). Similarly, ease of use has a significant positive effect on attitude toward use in both voluntary and mandatory usage contexts (Al-Gahtani & King, 1999; Keil et al., 1995; Venkatesh & Davis, 2000).

To summarise the constructs discussed in this section, Figure 2 portrays a proposed model of acceptance of mobile services. The implications of this model are discussed next.

PROPOSED METHOD

In order to validate the model discussed in the previous section, we propose a two-stage research design, consisting of both qualitative and quantitative approaches.

Qualitative Stage

This stage is the first round of the fieldwork data collection. Data collection at this stage would involve conducting face-to-face in-depth interviews in order to study the perceptions of all stakeholders who contribute directly or indirectly to providing mobile services. Stakeholder targets include mobile operators or carriers, industry and government associations, user groups, mobile application developers, content providers, aggregators, as well as manufacturers of mobile devices. The aim of these interviews is to gather an in-depth understanding of the perceptions and perspectives of all stakeholders involved in the adoption and diffusion of mobile services.

To ensure validity in this research design, several tactics may be used. Construct validity would be addressed by using the multiple sources of evidence as noted previously. The issue of internal validity would be considered by using the techniques of pattern matching the data to a predicted pattern of variables, and formulating rival explanations. In addition, an interview protocol should be developed to guide the data collection. A single pilot case study is recommended to be used in order to refine data collection procedures and improve conceptualisation of the model prior to finalising the set of theoretical propositions developed from the literature.

The second round of the qualitative stage would involve focus groups with mobile users. Because the research phenomenon is contemporary and little prior research has been conducted, focus groups would be appropriate for generating ideas and obtaining insights from existing mobile service users and potential users (Carson, Gilmore, Gronhaug, & Perry, 2001). Focus groups are useful when investigating complex behaviour and motivations. By comparing the different points of view that participants exchange during the interactions in focus groups, researchers can examine motivation with a degree of complexity that is usually not available with other methods (Morgan & Krueger, 1993). The use of a focus group is more valuable many times over compared with any representative sample for situations requiring the investigation of complex decision-

making processes, as is the case for this research. Based on demographic characteristics, we propose setting up homogeneous groups because discussions within homogeneous groups produce more in-depth information than discussions within heterogenous groups (Bellenger, Bernhardt, & Goldtucker, 1989). These groups would be selected based on main moderating variables identified in the literature, such as, gender/age, quality of life, stage of usage, and voluntariness.

We believe that at least two investigators should conduct all interviews and moderate the focus groups (Denzin, 1989; Patton, 1990). This kind of triangulation reduces the potential bias which is commonly cited as a limitation of interviews and focus gruops (Frankfort-Nachmias & Nachmias, 1996; Yin, 1994).

Quantitative Stage

The last stage of this project would involve an online survey. The collected data would help understand and confirm the determinants and the adoption intentions of the consumers of mobile services. Random sampling should be used to select the sample. We propose that two types of data analysis should be performed on the survey data: descriptive analysis and inferential analysis. Descriptive analysis should be carried out for transformation of raw data into a form that would provide information to describe a set of factors in a situation (Sekaran, 2000). For the inferential analysis, a structural equation model (SEM) should be used to test the refined model.

MANAGERIAL IMPLICATIONS

Mobile technologies and the associated services integrate both the business and social domains of the user's life (Elliot & Phillips, 2004; Knutsen et al., 2005). 3G services in general and location-based services in particular can provide anytime-anyplace tracking of adopters (UMTS, 2003). This creates the opportunity for developing accurate adopter profiles both in their work- and leisure-related domains. In addition, live video and location-based information can also be gathered (Robins, 2003). While such information can help address the needs of adopters better, it can also be misused by businesses for unethical direct business-to-consumer marketing (Casal et al., 2004), raising privacy concerns, overcontrol and overwork of individual adopters (Yen & Chou, 2000). For example, by reducing space and time constraints, mobile communications provide an immensely flexible work environment for some individuals while bringing about overwork or intrusion problems for others (Gerstheimer & Lupp, 2004). As a result, existing privacy protection policies and regulations about employees and consumers should reflect these new conditions. These policies should also account for overcontrol prevention that is likely to result from organisations' attempts to monitor individual performance (Yen & Chou, 2000).

Designing content suitable for mobile phones constitutes an important issue that affects the adoption and diffusion of mobile technology and associated services. This has implications for service providers, developers, policymakers, and academics. Content providers must design content "for value-contexts specific for mobile use which provide users freedom from complicated configuration procedures, and ubiquitously serve and support current day-to-day individual social practices" (Knutsen et al., 2005a, p. 7). Developers of mobile applications need to recognise that mobile applications are quite different from PC applications (Funk, 2005). Developers should use established standards, such as HTML and Java. More importantly, the usage contexts, the adopters, and their evolving behaviour should be important considerations. Further, because "made-for-the-medium" content type and design may be required (Massey et al., 2005), the avail-

able technologies which determine screen size, display quality and processing speeds should be taken into consideration as well (Funk, 2005). The combined effect of these factors on navigation patterns, adopters' cognitive overload, and subjective perceptions about the usability and ease of use of mobile applications can have a critical impact on uptake (Chae & Kim, 2004).

Segmentation of mobile service adopters must not only be based on adopter type (e.g., pioneers, early adopters, majority adopters, and laggards) but also on individual differences. The basis of segmentation should constitute the foundation in developing marketing strategies. For example, individuals with high personal innovativeness or novelty seekers are likely to be willing to experiment with new mobile devices and services, in which case these should be marketed as technological innovations. For individuals who are reluctant to use the same devices and services and are likely to feel discomfort and insecurity while using them, lifestyle promotions may be more appropriate (Dabholkar & Bagozzi, 2002; Teo & Pok, 2003). In addition, endorsements by peers, famous celebrities, or other referent groups may be adequate if these individuals appreciate social norms and image (Hung, Ku, & Chang, 2003; Teo & Pok, 2003). Marketing mobile applications for adopters in one category is likely to frustrate adopters in the other. Therefore, developers and marketers should be prudent in recognising that the confluence of various individual characteristics with varying levels of prior experience, perceptions, and learning predispositions are all likely to influence adoption and retention patterns (Card, Moran, & Newell, 1983; Hung et al., 2003; Massey et al., 2005).

Further, the interface design of mobile applications should encompass both intrinsic and extrinsic motivation dimensions (Moon & Kim, 2001). Based on the proposed model, marketers should promote attributes such as usefulness, ease of use, and enjoyment as important aspects when attempting to persuade potential users in adopting specific mobile phones and services as well as to increase their loyalty and retention (Dabholkar & Bagozzi, 2002; Hung et al., 2003; Lin & Wang, 2005). In particular, personalisation is a well-suited and an achievable goal as mobile phones are identifiable. 3G phones also enable identification of the location of individual handsets, making location-specific marketing possible. Messages promoting the services of businesses, such as restaurants, hotels, grocery stores, and so forth, can be transmitted when users are detected within range (Robins, 2003). Evidence shows that despite privacy concerns, many users of mobile devices are happy to receive unsolicited promotional messages provided that such messages are relevant and personalised (Robins, 2003).

Governments and mobile operators should design appropriate and dedicated strategies to promote the relative advantages of mobile phones and services. Such promotion strategies are important because of their impact on the perceptions of potential adopters (Knutsen et al., 2005). Moreover, the development of wireless communication infrastructures and the provision of incentives are likely to contribute towards the minimisation of the digital divide which results from demographic factors such as varying income levels, education and experience, gender, and age (Lin & Wang, 2005). The digital divide not only prevents the exploitation of the full market potential, but it also adversely impacts the maximization of benefits for current adopters due to limited network externalities effects (Katz & Shapiro, 1986).

CONCLUSIONS AND FUTURE RESEARCH

User acceptance of mobile technology and related services is of paramount importance. Consequently, a deeper insight into theory-based research is required to better understand the underlying motivations and barriers that will lead users to inhibit them from adopting these technologies and services. This in turn will also help designing technology and service improvements as well as appropriate adoption and diffusion strategies. There are several theoretical models in the literature which attempt to determine acceptance and adoption of new technologies. However, most of these models originate from organisational contexts. As mobile technologies and services add other functional dimensions such as hedonic or experiential aspects, applying extant theories outright to determine the acceptance and adoption of mobile services may be questionable and inadequate.

In this chapter, we have explored and critically reviewed existing technology acceptance theories. Relevant constructs of extant models were discussed in the light of evolving mobile technologies and services and then incorporated into a synthesised acceptance model of mobile services. The proposed model attempts to view acceptance of mobile services beyond traditional organisational borders and permeate everyday social life practices. The proposed model which can be tested empirically provides the foundation to guide further validation and future research in the area of mobile services adoption.

In addition, a plethora of mobile services have become available recently (Alahuhta et al., 2005). Because all services would be available to adopters through a single user interface of the current technology, the appropriation of these services by users may be interconnected and at different stages of maturity (Knutsen et al., 2005). These interconnections are temporal and are also likely to have mutually enhancing, suppressing, or compensating effects on each other (Black & Boal, 1994). This adds dynamism and complexity to acceptance and, therefore, cannot be explained by simply considering factors impacting individual or aggregate adoption at single points in time (Knutsen et al., 2005; Pagani, 2004).

Consequently, future research should develop and test dynamism-compatible acceptance models because these models may provide a deeper understanding and help in explaining how and why technology acceptance perceptions change as the appropriation process progresses. Further, with a wide variety of mobile devices and services available and their applicability in distinct spheres of life, the definition of a unit of analysis in mobile services adoption has become a challenging task (Knutsen et al., 2005). Additional research in this aspect is also needed.

REFERENCES

Adams, D. A., Nelson, R. R., & Todd, P. A. (1992). Perceived usefulness, ease of use, and usage of information technology: A replication. *MIS Quarterly, 16*(2), 227-247.

Agarwal, R., & Prasad, J. (1998). A conceptual and operational definition of personal innovativeness in the domain of information technology. *Information Systems Research, 9*(2), 204-215.

Ajzen, I. (1991). The theory of planned behavior. *Organisational Behavior and Human Decision Process, 52*(2), 179-211.

Alahuhta, P., Ahola, J., & Hakala, H. (2005). *Mobilising business applications: a survey about the opportunities and challenges of mobile business applications and services in Finland* (Technology Review No. 167/2005). Helsinki: Tekes.

Alba, J. W., & Hutchinson, J. W. (1987). Dimensions of consumer expertise. *Journal of Consumer Research, 13*(3), 411-454.

Al-Gahtani, S. S., & King, M. (1999). Attitudes, satisfaction and usage: Factors contributing to each in the acceptance of information technology. *Behaviour & Information Technology, 18*(4), 277-297.

Ba, S., & Pavlou, P. A. (2002). Evidence of the effect of trust building technology in electronic markets: price premiums and buyer behavior. *MIS Quarterly, 26*(3), 243-268.

Bellenger, D. N., Bernhardt, K. L., & Goldtucker, J. L. (1989). Qualitative research techniques: Focus group interviews. In T. J. Hayes, & C. B. Tathum (Eds.), *Focus group interviews: A reader* (pp. 7-28). Chicago: American Marketing Association.

Bina, M., & Giaglis, G. M. (2005). *Exploring early usage patterns of mobile data services.* Paper presented at the International Conference on Mobile Business, Sydney, Australia, July 11-13.

Black, J. A., & Boal, K. B. (1994). Strategic resources: Traits, configurations, and paths to sustainable competitive advantage. *Strategic Management Journal, 15,* 131-148.

Bozionelos, N. (1996). Psychology of computer use: Prevalence of computer anxiety in British managers and professionals. *Psychological Reports, 78*(3), 995-1002.

Calder, B. J., & Staw, B. M. (1975). Self-perception of intrinsic and extrinsic motivation. *Journal of Personality and Social Psychology, 31*(4), 599-605.

Card, S. K., Moran, T. P., & Newell, A. (1983). *The psychology of human-computer interaction.* Hillsdale, NJ: Lawrence Earlbaum Associates.

Carlsson, C., Hyvonen, K., Repo, P., & Walden, P. (2005). *Adoption of mobile services across different platforms.* Paper presented at the 18th Bled eCommerce Conference, Bled, Slovenia, June 6-8.

Carson, D., Gilmore, A., Gronhaug, K., & Perry, C. (2001). *Qualitative research in marketing.* London: Sage.

Casal, C. R., Burgelman, J. C., & Bohlin, E. (2004). Propects beyond 3G. *Info, 6*(6), 359-362.

Chae, M., & Kim, J. (2003). What's so different about the mobile Internet? *Communications of the ACM, 46*(12), 240-247.

Chae, M., & Kim, J. (2004). Do size and structure matter to mobile users? An empirical study of the effects of screen size, information structure, and task complexity on user activities with standard web phones. *Behaviour & Information Technology, 23*(3), 165-181.

Chau, P. Y. K. (1996). An empirical assessment of a modified technology acceptance model. *Journal of Management Information Systems, 13*(2), 185-204.

Chau, P. Y. K., & Hu, P. J.-H. (2001). Information technology acceptance by individual professionals: a model comparison approach. *Decision Science, 32*(4), 699-719.

Compeau, D. R., & Higgins, C. A. (1995). Computer self-efficacy: Development of a measure and initial test. *MIS Quarterly, 23*(2), 189-211.

Condos, C., James, A., Every, P., & Simpson, T. (2002). Ten usability principles for the development of effective WAP and m-commerce services. *Aslib Proceedings, 54*(6), 345-355.

Crawford, A. M. (2002). International media habits on the rise. *Ad Age Global, 2*(11). Retrieved from http://web.ebscohost.com/ehost/detail?vid=3&hid=101&sid=ff86c2ae-e7f7-4388-96b4-7da9c1bc4eb3%40sessionmgr106

Csikszentmihalyi, M. (1975). *Beyond boredom and anxiety.* San Francisco: Jossey-Bass.

Dabholkar, P. A., & Bagozzi, R. P. (2002). An attitudinal model of technology-based self-service: Moderating effects of consumer traits and situational factors. *Journal of Academy of Marketing Science, 30*(3), 184-201.

Davis, F. D. (1989). Perceived usefulness, perceived ease of use, and user acceptance in information technology. *MIS Quarterly, 13*(3), 319-340.

Davis, F. D., Bagozzi, R. P., & Warshaw, P. R. (1989). User acceptance of computer technology: A comparison of two theoretical models. *Management Science, 35*(8), 982-1002.

Davis, F. D., Bagozzi, R. P., & Warshaw, P. R. (1992). Extrinsic and intrinsic motivation to use computers in the workplace. *Journal of Applied Social Psychology, 22*, 1111-1132.

Deci, E. L., & Ryan, R. M. (1985). *Intrinsic motivation and self-determination in human behavior.* New York: Plenum Press.

Denzin, N. K. (1989). *The research act: A theoretical introduction to sociological methods (3rd ed.).* Englewood Cliffs, N. J.: Prentice Hall.

Dholakia, R. R., & Dholakia, N. (2004). Mobility and markets: Emerging outlines for m-commerce. *Journal of Business Research, 57*(12), 1391-1396.

Diener, E., & Suh, E. (1997). Measuring quality of life: Economic, social and subjective indicators. *Social Indicators Research, 40*(1-2), 189-216.

Dillon, A., & Morris, M. (1996). User acceptance of information technology: theories and models. *Journal of American Society for Information Science, 31*, 3-32.

Elliot, G., & Phillips, N. (2004). *Mobile commerce and wireless computing systems.* Harlow: Pearson Education Limited.

Elliot, S., & Loebbecke, C. (2000). Interactive, inter-organizational innovations in electronic commerce. *Information Technology & People, 13*(1), 46-66.

Figge, S. (2004). Situation-dependent services: A challenge for mobile operators. *Journal of Business Research, 57*(12), 1416-1422.

Fishbein, M., & Ajzen, I. (1975). *Belief, attitude, intention and behaviour: An introduction to theory and research.* Reading, MA: Addison-Wesley.

Frankfort-Nachmias, C., & Nachmias, D. (1996). *Research methods in the social sciences (5th ed.).* New York: St. Martin's Press.

Funk, J. L. (2005). The future of the mobile phone Internet: An analysis of technological trajectories and lead users in the Japanese market. *Technology in Society, 27*(1), 69-83.

Galanxhi-Janaqi, H., & Nah, F. F.-H. (2004). U-commerce: Emerging trends and research issues. *Industrial Management & Data Systems, 104*(9), 744-755.

Gefen, D., Karahanna, E., & Straub, D. W. (2003). Trust and TAM in online shopping: an integrated model. *MIS Quarterly, 27*(1), 51-90.

Gerstheimer, O., & Lupp, C. (2004). Needs versus technology: The challenge to design third-generation mobile applications. *Journal of Business Research, 57*(12), 1409-1415.

Grundström, C., & Wilkinson, I. F. (2004). The role of personal networks in the development of industry standards: A case study of 3G mobile telephony. *Journal of Business and Industrial Marketing, 19*(4), 283-293.

Hammond, K. (2001). B2C e-commerce 2000-2010: What experts predict. *Business Strategy Review, 12*(1), 43-50.

Hart, J., & Hannan, M. (2004). The future of mobile technology and mobile wireless computing. *Campus-Wide Information Systems, 21*(5), 201-204.

Horton, R. P., Buck, T., Waterson, P. E., & Clegg, C. W. (2001). Explaining intranet use with the technology acceptance model. *Journal of Information Technology, 16,* 237-249.

Hung, S.-Y., Ku, C.-Y., & Chang, C.-M. (2003). Critical factors of WAP services adoption: An empirical study. *Electronic Commerce Research and Applications, 2*(1), 42-60.

Igbaria, M., Parasuraman, S., & Baroudi, J. J. (1996). A motivational model of microcomputer usage. *Journal of Management Information Systems, 13*(1), 127-143.

Ishii, K. (2004). Internet use via mobile phone in Japan. *Telecommunications Policy, 28*(1), 43-58.

Jarvenpaa, S. L., Lang, K. R., Takeda, Y., & Tuunainen, V. K. (2003). Mobile commerce at crossroads. *Communications of the ACM, 46*(12), 41-44.

Jiang, J. J., Hsu, M. K., Klein, G., & Lin, B. (2000). E-commerce user behaviour model: An empirical study. *Human Systems Management, 19*(4), 265-276.

Katz, M. L., & Shapiro, C. (1986). Technology adoption in the presence of network externalities. *Journal of Political Economy, 94*(4), 822-841.

Kaufaris, M. (2002). Applying the technology acceptance model and flow theory to online consumer behaviour. *Information Systems Research, 13*(2), 205-223.

Keil, M., Beranek, P. M., & Konsynski, B. R. (1995). Usefulness and ease of use: Field study evidence regarding task considerations. *Decision Support Systems, 13*(1), 75-91.

Khalifa, M., & Cheng, S. K. N. (2002). *Adoption of mobile commerce: Role of exposure.* Paper presented at the 35th Hawaii International Conference on System Sciences, Hilton Waikoloa Village, Hawaii, January 7-10 (pp. 46-52). IEEE Computer Society.

King, J. L., Gurbaxani, V., Kraemer, K. L., McFarlan, F. W., Raman, K. S., & Yap, C. S. (1994).

Institutional factors in information technology innovation. *Information Systems Research, 5*(2), 139-169.

Kirton, M. (1976). Adopters and innovators: a description and measure. *Journal of Applied Psychology, 61*(5), 622-629.

Klasen, L. (2002). Migrating an online service to WAP: Case study. *The Electronic Library, 20*(3), 195-201.

Kleijen, M., Wetzels, M., & de Ruyter, K. (2004). Consumer acceptance of wireless finance. *Journal of Financial Services Marketing, 8*(3), 206-217.

Knutsen, L., Constantiou, I. D., & Damsgaard, J. (2005). *Acceptance and perceptions of advanced mobile services: Alterations during a field study.* Paper presented at the International Conference on Mobile Business, Sydney, Australia, July 11-13.

Lee, M. S. Y., McGoldrick, P. J., Keeling, K. A., & Doherty, J. (2003). Using ZMET to explore barriers to the adoption of 3G mobile banking services. *International Journal of Retail & Distribution Management, 31*(6), 340-348.

Leonard-Barton, D., & Deschamps, I. (1988). Managerial influence in the implementation of new technology. *Management Science, 34*(10), 1252-1265.

Lin, H., & Wang, Y. (2005). *Predicting consumer intention to use mobile commerce in Taiwan.* Paper presented at the International Conference on Mobile Business, Sydney, Australia, July 11-13.

Lu, J., Yu, C., Liu, C., & Yao, J. E. (2003). Technology acceptance model for wireless Internet. *Internet Research: Electronic Networking Applications and Policy, 13*(3), 206-222.

Massey, A. P., Khatri, V., & Ramesh, V. (2005). *From the Web to the wireless Web: Technology readiness and usability.* Paper presented at the 38th Hawaii International Conference on System Sciences, Hilton Waikoloa Village, Hawaii, January 3-6 (p. 32b). IEEE Computer Society.

Mathwick, C., Malhotra, N., & Rigdon, E. (2001). Experiental value: Conceptualization, measurement and application in the catalog and Internet shopping environment. *Journal of Retailing, 77*(1), 39-56.

Minton, G. C., & Scheneider, F. W. (1980). *Differential psychology.* Prospect Heights, IL: Waveland Press.

Moon, J.-W., & Kim, Y.-G. (2001). Extending the TAM for a World-Wide-Web context. *Information & Management, 38*(4), 217-230.

Moreau, C. P., Lehmann, D. R., & Markman, A. B. (2001). Entrenched knowledge structures and consumer response to new products. *Journal of Marketing Research, 38*(1), 14-29.

Morgan, D. L., & Krueger, R. A. (1993). When to use focus groups and why. In D. L. Morgan (Ed.), *Successful focus groups* (pp. 1-19). London: Sage Publications.

Novak, T. P., Hoffman, D. L., & Yung, Y. (2000). Measuring the customer experience in online environments: A structural modeling approach. *Marketing Science, 19*(1), 22-42.

Pagani, M. (2004). Determinants of adoption of third generation mobile multimedia services. *Journal of Interactive Marketing, 18*(3), 46-59.

Parasuraman, A. (2000). Technology readiness index: A multiple item scale to measure readiness to embrace new technologies. *Journal of Service Research, 2*(4), 307-320.

Patton, M. Q. (1990). *Qualitative evaluation and research methods (2nd ed.).* London: Sage Publications.

Ratliff, J. M. (2002). NTT DoCoMo and its i-mode success: Origins and implications. *California Management Review, 44*(3), 55-71.

Repo, P., Hyvonen, K., Pantzar, M., & Timonen, P. (2004). *Users intenting ways to enjoy new mobile services: The case of watching mobile videos.* Paper presented at the 37th Hawaii International Conference on System Sciences, Hawaii, January 5-8 (p. 40096.3). IEEE Computer Society.

Rhodes, S. R. (1983). Age-related differences in work attitudes and behavior: A review of conceptual analysis. *Psychological Bulletin, 93*(2), 328-367.

Robins, F. (2003). The marketing of 3G. *Marketing Intelligence & Planning, 21*(6), 370-378.

Roehm, M. L., & Sternthal, B. (2001). The moderating effect of knowledge and resources on the persuasive impact of analogies. *Journal of Consumer Research, 28*(2), 257-272.

Rogers, E. M. (1995). *Diffusion of innovations.* New York: Free Press.

Rossotto, C. M., Kerf, M., & Rohlfs, J. (2000). Competition in mobile telecommunications: Sector growth, benefits for the incumbent and policy trends. *Info, 2*(1), 67-73.

Saaksjarvi, M. (2003). Consumer adoption of technological innovations. *European Journal of Innovation Management, 6*(2), 90-100.

Sekaran, U. (2000). *Research methods for business: A skill building approach.* New York: John Wiley and Sons.

Sheppard, B. H., Hartwick, J., & Warshaw, P. R. (1988). The theory of reasoned action: A meta-analysis of past research with recommendations for modifications and future research. *Journal of Consumer Research, 15*(3), 325-343.

Shin, C. C., & Johnson, D. M. (1978). Avowed happiness as an overall assessment of quality of life. *Social Indicators Research, 5,* 475-492.

Siau, K., Lim, E. P., & Shen, Z. (2001). Mobile commerce: Promises, challenges, and research agenda. *Journal of Databases Management, 12*(2), 4-13.

Taylor, S., & Todd, P. A. (1995). Understanding information technology usage: A test of competing models. *Information Systems Research, 6*(2), 144-176.

Taylor, S., & Todd, P. A. (1995a). Assessing IT usage: The role of prior experience. *MIS Quarterly, 19*(4), 561-570.

Teo, T. S. H., & Pok, S. H. (2003). Adoption of WAP-enabled mobile phones among Internet users. *Omega: The International Journal of Management Science, 31*(6), 483-498.

Terry, D. J. (1993). Self-efficacy expectancies and the theory of reasoned action. In D. C. Terry, C. Gallois, & M. McCamish (Eds.), *The theory of reasoned action: Its application to AIDS-preventive behaviour* (pp. 135-152). Oxford: Pergamon.

Thompson, R., Higgins, C., & Howell, J. (1994). Influence of experience on personal computer utilization: Testing a conceptual model. *Journal of Management Information Systems, 11*(1), 167-187.

Thompson, R. L., Higgins, C. A., & Howell, J. M. (1991). Personal computing: Toward a conceptual model of utilization. *MIS Quarterly, 15*(1), 125-143.

UMTS. (2003). *Mobile evolution: Shaping the future.* Retrieved August 28, 2005, from http://www.umts-forum.org/servlet/dycon/ztumts/umts/Live/en/umts/Resources_Papers_index

van Steenderen, M. (2002). Business applications of WAP. *The Electronic Library, 20*(3), 215-223.

Venkatesh, V., & Davis, F. D. (2000). A theoretical extension of the technology acceptance model: four longitudinal field studies. *Management Science, 46*(2), 186-204.

Venkatesh, V., & Morris, M. G. (2000a). Why don't men ever stop to ask for directions? Gender, social influence, and their role in technology acceptance and usage behavior. *MIS Quarterly, 24*(1), 115-139.

Venkatesh, V., Morris, M. G., Davis, G. B., & Davis, F. D. (2003). User acceptance of information technology: Toward a unified view. *MIS Quarterly, 27*(3), 425-478.

Wang, Y.-S., Wang, Y.-M., Lin, H.-H., & Tang, T.-I. (2003). Determinants of user acceptance of Internet banking: An empirical study. *International Journal of Service Industry Management, 14*(5), 501-519.

Xylomenos, G., & Polyzos, G. C. (2001). Quality and service support over multi-service wireless Internet links. *Computer Networks, 37*(5), 601-615.

Yamauchi, T., & Markman, A. B. (2000). Inference using categories. *Journal of Experimental Psychology: Learning, Memory, and Cognition, 26*(3), 776-795.

Yen, D. C., & Chou, D. C. (2000). Wireless communications: Applications and managerial issues. *Industrial Management & Data Systems, 100*(9), 436-443.

Yin, R. K. (1994). *Case study research: Design and methods.* Beverley Hills: Sage.

Zeithaml, V. A., & Gilly, M. C. (1987). Characteristics affecting the acceptance of retailing technologies: A comparison of elderly and nonelderly consumers. *Journal of Retailing, 63*(1), 49-68.

Chapter VII
Video Coding for Mobile Communications

Ferdous Ahmed Sohel
Monash University, Australia

Gour C. Karmakar
Monash University, Australia

Laurence S. Dooley
Monash University, Australia

ABSTRACT

With the significant influence and increasing requirements of visual mobile communications in our everyday lives, low bit-rate video coding to handle the stringent bandwidth limitations of mobile networks has become a major research topic. With both processing power and battery resources being inherently constrained, and signals having to be transmitted over error-prone mobile channels, this has mandated the design requirement for coders to be both low complexity and robust error resilient. To support multi-level users, any encoded bit-stream should also be both scalable and embedded. This chapter presents a review of appropriate image and video coding techniques for mobile communication applications and aims to provide an appreciation of the rich and far-reaching advancements taking place in this exciting field, while concomitantly outlining both the physical significance of popular quality image and video coding metrics and some of the research challenges that remain to be resolved.

INTRODUCTION

While the old adage is that *a picture is worth thousands of words*, in the *digital era* a colour image typically corresponds to more like a million words (double bytes). While an image is a two-dimensional spatial representation of intensity that remains invariant with respect to time (Tekalp,

1995), video is a three-dimensional time-varying image sequence (Al-Mualla, Canagarajah, & Bull, 2002) and as a consequence represents far more information than a single image. Mobile technologies are becoming omnipresent in our lives with the common mantra to communicate *with anybody, anytime, anywhere*. This has fueled consumer demand for richer and more diverse mobile-based applications, products, and services, and given the *human visual system* (HVS) is the most powerful perceptual sensing mechanism, it has inevitably meant that image and latterly video technologies are the drivers for many of these new mobile solutions.

Second generation (2G) mobile communication systems, such as the *Global System for Mobile* (GSM) started by supporting a number of basic multimedia data services including voice, fax, *short message services* (SMS) and information-on-demand (news headlines, sports scores and weather). *General Packet Radio Service* (GPRS), which has often been referred to as 2.5G, extends GSM to provide packet switching services and afford the user facilities including e-mail, still-image communication, and basic Internet access. By sharing the available bandwidth, GPRS offers efficiency gains in applications where data transfer is intermittent like Web-browsing, e-mail, and instant messaging. The popularity of GSM and GPRS led to the introduction of *third generation* (3G) mobile technologies which address live video applications, with real-time video telephony being advertised as the flagship application for this particular technology, offering a maximum theoretical data rate of 2Mbps, though in practice this is more likely to be 384Kbps. Multimedia communications along with bandwidth allocation for video and Web applications remains one of the primary focuses of 3G as well as the proposed *fourth generation* (4G) mobile technologies, which will provide such functionality as broadband wireless access and interactivity capability, though it is not due to be launched until 2010 at the earliest. Many technological challenges remain including

the need for greater coding efficiency, higher data rates, lower computational complexity, enhanced error resilience and superior bandwidth allocation, and reservation strategies to ensure maximal channel utilisation. When these are resolved, mobile users will benefit from a rich range of advanced services and enhanced applications including video-on-demand, interactive games, video telephony, video conferencing and tele-presence, tele-surveillance, and monitoring.

As video is a temporal sequence of still frames, coding in fact involves both single (intra) and multiple (inter) frame coding algorithms, with the former being merely still image compression. Since, for mobile applications only low bit-rate video sequences are suitable, this chapter analyses both high image and video compression techniques. Approaches to achieving high image compression are primarily based upon either the *discrete cosine transform* (DCT), as in the widely adopted *Joint Picture Expert Group* (JPEG) standard or the *discrete wavelet transform* (DWT) which affords scalable and embedded sub-band coding in the most recent interactive JPEG2000 standard. In contrast, a plethora of different inter-frame coding techniques have evolved within the generic block-based coding framework, which is the kernel of most current video compression standards such as the *Moving Picture Expert Group* family of MPEG-1, MPEG-2, and MPEG-4, together with the symmetrical video-conferencing H.261 and H.263 coders and their variants. MPEG-4, which is the latest audio/video coding family member, offers object-based functionality and is primarily intended for Internet-based applications. It will be examined later in the chapter, together with the main features of the newest video coding standard, somewhat prosaically known as H.264 or *advanced video coding* (AVC), which is now formally incorporated into MPEG-4.

All these various compression algorithms remove information content from the original video sequence in order to gain compression efficiency, and without loss of generality the quality of the encoded video will be compromised to some extent. As a consequence, the issue of quality assessment arises, which can be subjective, objective, or both, and this chapter will explore both the definition and physical significance of some of the more popular quality metrics. In addition, the computational complexity of both the encoder and decoder directly impacts upon the limited power resources available for any mobile unit. Moreover, in this consumer driven age, the insatiable desire for choice means some people will pay more to get a higher quality of service product, while others will be more than happy with basic functionality and reasonable signal quality. In order to ensure the availability of different consumer levels within the same framework, it is essential to ensure that signal coding is both scalable and embedded.

PERFORMANCE AND QUALITY METRICS OF VIDEO CODING ALGORITHMS

The performance of all contemporary video coding systems is normally assessed using a series of well-accepted metrics mentioned by Bull, Canagarajah, and Nix (1999), including:

- Coding efficiency,
- Picture reconstruction quality,
- Scalable and embedded representations,
- Error resilience,
- Computational complexity, and
- Interactivity.

Coding Efficiency

This is one of the prime metrics for low bit-rate coding in mobile communications, with the inherent bandwidth limitations of mobile networks propelling research to explore highly efficient video coding algorithms. Compression is achieved by reducing the amount of data required to represent

the video signals by minimising inherent redundancies in both the spatial and temporal domains, as well as to some degree, dropping insignificant or imperceptible information at the pyrrhic cost of a loss in quality. In addition, higher coding gains can be achieved using lower spatial and temporal (frame rate) resolution video formats, such as the *common interchange format* (CIF) and sacrificing the colour depth of each pixel, though again this impacts on perceived quality. *Compression ratio* (CR) is the classical metric for measuring coding efficiency in terms of the information content of the video and can be evaluated in numerous ways (Al-Mualla et al., 2002). For example,

$$CR = \frac{\text{number of bits in original video}}{\text{number of bits in compressed video}} \quad (1)$$

From a purely compression perspective, an encoder generating a higher CR is regarded as superior to one with a lower CR, as the clear advantage secured is that without loss of generality, a smaller bit-stream incurs a lower transmission time. An alternative representation is to use *compression* (C) which is quantified in *bits per pixel* (*bpp*), where the best encoder generates the lowest *bpp* value. This is formally defined as:

$$C = \frac{\text{size of the compressed video (bits)}}{\text{number of pels in original video}} \quad bpp \quad (2)$$

Picture Reconstruction Quality

Video coding for mobile communications is by its very nature lossy, so it is essential to be able to quantitatively represent the loss and reflect the compression achieved. To specify, evaluate, compare, and analyse video coding and communication systems, it is necessary to determine the level of picture quality of the decoded images displayed to the viewer. Visual quality is inherently *subjective* and is influenced by many factors that make it difficult to obtain a completely accurate measure for perceived quality. For example, a viewer's opinion of visual quality can depend very much on their psycho-physical state or the task at hand such as passively watching a movie, keenly watching the last few overs of a tense cricket match, actively participating in a video conference session, or trying to identify a person in a video surveillance scene. Measuring visual quality using objective criteria can give both accurate and repeatable results, but as yet there is no unified quantitative measurement system that entirely reproduces the perceptual experience of a human observer (VQEG, 1998) or no single metric that consistently outperforms other (objective) techniques from a subjective viewpoint (Wu & Rao, 2006). In the following section, both subjective and objective quality measuring techniques are examined.

Subjective Quality Measurement

Human perception of a visual scene is formed by a complex interaction between the components of the HVS, particularly through the eye to the brain. Perceived visual quality is affected by many different factors, including:

- Spatial fidelity (how clear parts of a scene are to the viewer, whether there is obvious distortion, whether the objects retain their geometric or structural shape, what the objects look like, and other fine detail concerning colour, lighting, and shading effects);
- Temporal fidelity (whether the motion appears natural, continuous and smooth);
- Viewing conditions: Distance, lighting, and colour;
- Viewing environment: a comfortable, non-distracting environment usually leads to the perception of higher quality, regardless of the actual quality of the scene;

- Viewer's state of mind, domain knowledge, training and expertise, interest, and the extent to which the observer interacts;
- The *recency effect*: the psychological opinion upon a visual sequence is more heavily influenced by recently-viewed rather than older video material (Wade & Swanston, 2001);
- Viewer's psycho-physical condition.

All these factors combine to make it expensive, time consuming, and extremely difficult to accurately measure visual quality, though a number of subjective assessment methodologies do exist, such as the *double stimulus impairment scale* (DSIS), *double stimulus continuous quality scale* (DSCQS), and *single stimulus continuous quality scale* (SSCQS) as of Ghanbari (1999). Moreover, the *Video Quality Expert Group* (VQEG), which was established in 1997, is currently working on the establishment a unified quality measurement standard (Wu et al., 2006) to enable subjective testing of both image and video data. The current status of the VQEG will be discussed shortly.

Objective Quality Measurement

Since subjective quality measurement is so sensitive to a large number of factors and may not be repeatable, measuring the visual quality using objective criteria becomes of paramount importance. Objective measurements give accurate and repeatable results at low cost and so are widely employed in video compression systems. A number of objective quality measuring techniques have been adopted by researchers and these will now be briefly investigated:

PSNR: Among all the objective measurements the logarithmic *peak-signal-to-noise-ratio* (PSNR) metric is most widely used in the literature and is defined as:

$$PSNR_{DB} = 10\log_{10}\frac{(2^n-1)^2}{MSE} \qquad (3)$$

where MSE is the mean squared error between the original and approximating video and (2^n-1) is the maximum possible signal value in an n-bit data representation. The MSE for each frame is given by:

$$MSE = \frac{1}{H \times V}\sum_{x=0}^{H}\sum_{y=0}^{V}\left[f(x,y)-\tilde{f}(x,y)\right]^2 \qquad (4)$$

where H and V are respectively the horizontal and vertical frame dimensions, while $f(x,y)$ and $\tilde{f}(x,y)$ are the original and approximated pixel values at location (x,y). Using this definition, a video with a higher PSNR is therefore rated better than one with a lower value. PSNR is commonly applied for three basic reasons as summarised by Topiwala (1998): (1) it is a first order analysis and treats data samples as independent events using a sum of squares of the error measure; (2) it is straightforward to compute and leads to easily tractable optimisation approaches; and (3) it has a reasonable correspondence with perceived image quality as interpreted by either humans or machine interpreters. It does, however, have some limitations (Richardson, 2003) most notably that it requires an unimpaired original video as a reference, though this may not be always available and also not easy to verify the original video had perfect fidelity and does not necessarily equate to an absolute subjective quality.

L_p *Metrics:* In addition to the MSE metric, various other weightings derived from the L_p norms can be used as quality measures. While closed-form solutions are feasible for minimising MSE, they are virtually impossible to obtain for normalised L_p metrics, which are formally defined as:

$$\varepsilon_p = \frac{1}{H \times V}\sum_{x=0}^{H}\sum_{y=0}^{V}\left|f(x,y)-\tilde{f}(x,y)\right|^p, p \neq 2 \qquad (5)$$

Fast and efficient search algorithms make L_p norms ideally applicable, especially if they correlate more precisely with subjective quality. Two p–norms, namely $p = \infty$ (L_∞) and $p = 1$ (L_1), correspond to the peak absolute error and sum-of-error magnitudes respectively and are widely referred to as the *class one* and *class two* distortion metrics, in the literature (Katsaggelos et al., 1998; Kondi et al., 2004; Meier, Schuster, & Katsaggelos, 2000; Schuster & Katsaggelos, 1997; Sohel, Dooley, & Karmakar, 2006a).

In the vertex-based video-object shape coding algorithms (Katsaggelos, Kondi, Meier, Ostermann, & Schuster, 1998; Kondi, Melnikov, & Katsaggelos, 2004; Schuster et al., 1997), distortion is measured from a slightly different perspective. Instead of considering the entire object, only the geometric distortion at the object boundary points is considered. The *shortest absolute distance* (SAD) between the shape boundary points and the corresponding approximating shape is then applied as the measurement strategy, though this can lead to erroneous distortion measures, especially at shape corners and sharp edges. In fact, the SAD guarantees every point on the approximated shape is within the correct geometric distortion but does not ensure all points on the shape boundary produce the distortion accurately. To overcome this anomaly, a new distortion measurement strategy has been developed (Sohel, Dooley, & Karmakar, 2006a) that accurately measures the Euclidean distance between the reference shape and its approximation for generic shape coding.

In the MPEG-4 standard, the *relative area error* (RAE) D_n measure is used to represent shape distortion (Brady, 1999):

$$D_n = \frac{\text{number of mismatched pixels in the approximated shape}}{\text{number of pixels in the original shape}}$$

$$(6)$$

Though it should be noted that since different shapes can have different ratios of boundary pels to interior pels, D_n only provides physical meaning when it is used to measure different approximations of the same shape (Katsaggelos et al., 1998).

As there are numerous quality metrics for both subjective and objective evaluation, it has become essential to attempt to formally standardise them. As alluded earlier, the VQEG has the objective of unifying objective picture quality assessment methods to reflect the subjective perception of the HVS. Despite their best efforts and rigorous testing in two phases, Phase I (1997-1999) and Phase II (2001-2003), an overall decision has yet to be made, though Phase I concluded that no objective measurement system was able to replace subjective testing and no single objective model outperformed all others in all cases (VQEG, 1999; Wu et al., 2006). Details of the various quality measurement techniques and vision model-based digital impairment metrics, together with perceptual coding techniques are described and analysed by Wu et al. (2006).

Scalability and Embedded Representations

Scalable compression refers to the generation of a coded bit-stream that contains embedded subsets, each of which represents an efficient compression of the original signal. The one major advantage of scalable compression is that either the target bit-rate or reconstruction quality does not need to be known at the time of compression (Taubman, 2000). A related advantage of practical significance is that the video does not have to be compressed multiple times in order to achieve a target bit-rate, so scalable encoding enables a decoder to only selectively decode portions of the bit-stream. It is very common in multicast/broadcast systems that there are different receivers having different capacities and different users supposed to be receiving different *quality-of-service* (QoS) levels. In scalable encoding, the bit-stream comprises one *base layer* and either one or more associated *enhancement layers*. The base layer can be independently decoded and the various enhancement layers conjointly decoded with the base layer to progressively increase the perceived picture quality. For mobile applications such as video-on-demand and TV access for mobile terminals, the server can transmit embedded bit-streams while the receiver processes the incoming data at its capacity and eligibility. In recent times, for example in Taubman (2000), Taubman and Marcellin (2002), Taubman and Zakhor (1994), and Atta and Ghanbari (2006), scalable video coding has assumed greater priority than other related issues including optimality and compression ratio.

As the example in Figure 1 shows, *Decoder 1* only processes the base layer while *Decoder N* handles the base and all enhancement layers, thereby generating a range of possible picture qualities from basic through to the very highest possible quality, all from a single video bit-stream. An intermediate decoder, such as *Decoder i* then utilises the base layer and enhancement layers up to the ith inclusive layer to generate a commensurate picture quality. Video coding systems are typically required to support a range of scalable coding modes, with the following three being particularly important: (1) *spatial scalability,* which involves increasing or decreasing picture resolution, (2) *temporal scalability* which provides varying picture (frame) rates, and (3) *quality (amplitude) scalability* which varies the picture quality by changing the PSNR or L_p metric for instance.

Figure 1. Generic concept of scalable and embedded encoding

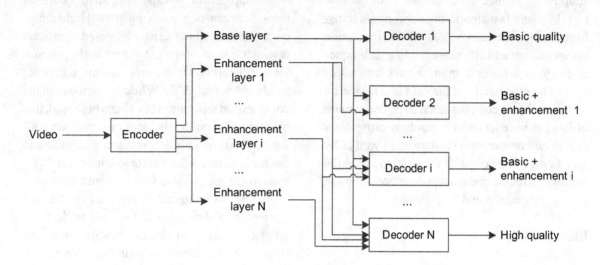

Error Resilience

Mobile communication channels are notorious hostile environments with high error rates caused by many different loss mechanisms ranging from multi-path fading, carrier signal strengths, co-channel interference, network congestion, misrouting through to channel noise (Wu et al., 2006). For coded video, the impact of these errors is magnified due to the fact that the bit-stream is highly compressed. Indeed, the greater the compression, the more sensitive the bit-stream is to errors, since each bit represents a larger portion of the original video and crucially, the bit-stream synchronization may become disturbed. The effect of errors on video is also exacerbated by the use of predictive and *variable-length coding* (VLC), which can lead to both temporal and spatial error propagation, so it is clear that transmitting compressed video over mobile channels may be hazardous and prone to degradation. Error-resilience techniques are characterised by their ability to tolerate errors introduced into the compressed bit-stream while maintaining an acceptable video

quality, with such strategies occurring at the encoder and/or decoder (Redmill, 1994; Salama, Shroff, Coyle, & Delp, 1995). The overall objective of error resilient coding is to reduce the effect of data loss by taking remedial action and displaying a quality video or image representation at the decoder, despite the fact the encoded signal may have been corrupted in transmission.

Computational Complexity

In mobile terminals, both processing power and battery life are scarce resources and given the high computational overheads necessitated to process video signals, employing computationally efficient algorithms is mandated. Moreover, for real-time video applications over mobile channels, the transmission delay of the signal must be kept as low as possible. Both symmetrical (video conferencing and telephony) and asymmetric (TV broadcast access and on-demand video) mobile applications mean that video coding-decoding algorithms should be designed so that mobile terminals incur minimal computational overheads. There are various steps that can be adopted to reduce computational complexity. First by using fast algorithms, appropriate transformations, fast search procedures for motion compensation and efficient encoding techniques at every step. Second, minimise the amount of data to be processed—if the data size is small it requires a lower computational cost. The amount of data can be reduced for example by using lower spatial and/or temporal resolutions as well as by sacrificing pixel colour depth so attenuating the bandwidth requirements and power consumption in both processing and transmission.

Interactivity

In many mobile applications involving Web browsing, video downloading and enjoying on-demand video, playing online games, interactivity has become a key element and with it also the user's expectation over the degree of interactivity available. For instance, users normally expect to have control over the standard suite of video recorder functions like play, pause, stop, rewind/forward, and record, but may in addition, also wish to be able to select a portion of video and to edit it or insert into another application similar to a multimedia-authoring tool (Bull et al., 1999). MPEG-4 functionality includes object-based manipulation so providing much greater flexibility for interactive indexing, retrieval, and editing of video content to the mobile users. Moreover, the H.264 standard enables switching between multiple bit-rates while browsing and downloading, together with interactive bandwidth allocation and reservation.

HIGH COMPRESSION INTRA-FRAME VIDEO CODING AND IMAGE CODING TECHNIQUES

As a single video frame is in fact a still image, *intra-frame video coding* and *image coding* have

exactly the same objective, namely to achieve the best compression by exploiting spatial correlations between neighbouring pels. High image compression techniques have attracted significant research interest in recent years, as they permit visible distortions to the original image in order to obtain a high CR. While numerous image compression techniques have been proposed, this chapter will specifically focus on waveform coding methods including transform and sub-band coding together with vector quantisation (VQ) since these are suitable for and commonly used in mobile communication applications. Second generation techniques that attempt to describe an image in terms of visually meaningful primitives including shape contour and texture will then be analysed.

Waveform-Based Coding

Waveform-based coding schemes typically comprise the following three principal steps:

- Decomposition/transformation of the image data,
- Quantisation of the transform co-efficients, and
- Rearrange and entropy coding of the quantised co-efficients.

Figure 2 shows the various constituent processing blocks of a waveform coder, each of which will now be considered.

Transform Coding

The first step of a waveform coder is *transformation* that maps the image data into an alternative representation so most of the energy is compacted into a limited number of transform co-efficients with the remainder either being very small or zero. This de-correlates the data so low energy co-efficients may be discarded with minimal impact upon the reconstruction image quality. In addition,

Figure 2. A generic waveform-based image coder

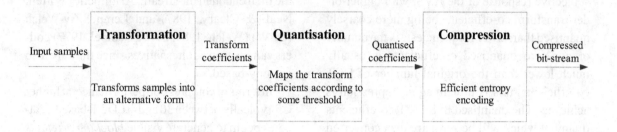

Figure 3. JPEG coding principles: (a) partition the image into 8×8 macroblocks, (b) the 64 pels in each block, (c) the conventional zigzag scan ordering for the quantised DCT co-efficients which are represented by the respective cells of the matrix

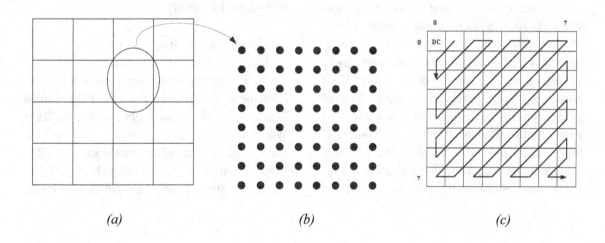

(a)	*(b)*	*(c)*

the HVS exhibits varying sensitivity to different frequencies with normally a greater sensitivity towards the lower than the high frequencies. There are many possible waveform transforms, though the most popular is the *discrete cosine transform* (DCT) which has as its basis the discrete fourier transform (DFT). Indeed, the DCT can be viewed as a special case of the DFT as it decomposes images using only cosines or even-symmetrical functions and for this reason, it is the fundamental building block of both the JPEG image and MPEG video compression standards. With the DCT, the image is first subdivided into blocks known as *macroblocks* (MB), which are normally of fixed size and usually 8×8 pels, with the DCT applied to all pels in that block (see Figures 3(a) and 3(b). The next major step after applying the DCT is quantisation, which maps the resulting 64 DCT co-efficients into a much smaller number of output values. The *Q-Table* is the formal mechanism whereby the quantised DCT co-efficients are then adapted to reflect the subjective response of the HVS, with higher-order transform co-efficients being more coarsely quantised than lower frequencies. As the number of non-zero quantised co-efficients is usually much lower than the original number of DCT co-efficients, this is how image compression is achieved. The quantised 8×8 DCT co-efficients (many of which will be zero) are then converted into a single sequence, starting with the lowest DCT frequency and then progressively increasing the spatial frequency. As all horizontal, vertical, and diagonal components must be considered, rather than read the co-efficients either laterally or longitudinally from the matrix, the distinctive *zigzag* pattern shown in Figure 3(c) is used to scan the quantised co-efficients in ascending frequency which tends to cluster low-frequency, non-zero co-efficients together. The first frequency component (top left of the DCT matrix) is the average (DC) value of all the DCT frequencies and so does not form part of the *zigzag* bit-stream, but is instead differentially coded using DPCM, with other DC components from adjacent MB. The resulting sequence of AC frequency co-efficients, will after quantisation, contain many zeros so the final step

is to employ lossless entropy coding such as the VLC *Huffman Code* to minimise redundancy in the final encoded bit-stream. Arithmetic (Witten, Neal, & Cleary, 1987) and Lempel-Ziv-Welch (LZW) (Welch, 1984; Ziv & Lempel, 1977) coders can be used as alternatives since they are also entropy-based.

In terms of compression performance, at higher CR typically between 30 and 40, DCT-based strategies begin to generate visible *blocking artefacts* and as all block-based transforms suffer from this distortion to which the HVS is especially sensitive, DCT-based coders are generally considered inappropriate for very low bit-rate image and video coding applications.

Sub-Band Coding

Sub-band image coding which does not produce the aforementioned blocking artefacts, has been the subject of intensive research in recent years (Crochiere, Webber, & Flanagan, 1976; Said & Pearlman, 1996; Shapiro, 1993; Taubman, 2000; Woods & O'Neil, 1986). It is observed from the mathematical form of the rate-distortion (RD) function that an efficient encoder splits the original signal into spectral components of infinitesimally

Figure 4. Sub-band decomposition: (a) first level DWT based on Daubechies (1990), (b) second level, (c) parent-child dependencies in the three level sub-band, and (d) the overall scanning order of the decomposed levels

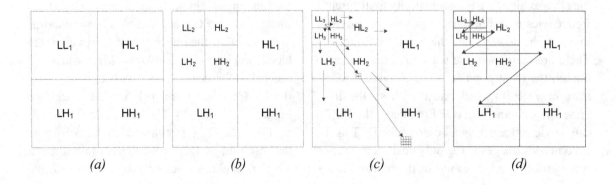

small bandwidth and then independently encodes each component (Nanda & Pearlman, 1992). In sub-band coding, the input image is passed through a set of band-pass filters to decompose it into a set of sub-band images prior to critical sub-sampling (Johnston,1980). For example, as shown in Figure 4(a), following the first decomposition level, the image is divided into 4 sub-bands where *L* and *H* respectively represent the *low* and *high* pass filtered outputs (for the horizontal and vertical directions), while the numerical subscript denotes the decomposition level. Subsequently, the lowest resolution sub-image (LL_1) is further decomposed at the 2nd level (see Figure 4(b)) because, as mentioned in the previous section, most signal energy tends to be concentrated in this sub-band. As each resulting sub-image has a lower spatial resolution (bandwidth) they are down-sampled before each is independently quantised and coded. It is worth noting that like the DCT, sub-band decomposition does not in itself lead to compression as the number of the sub-bands remains equal to the number of samples in the original image. However, the elegance of this approach is that each sub-band can be coded efficiently according to its statistics and visual prominence, leading to an inherent embeddedness and scalability in the sub-band coding process. As the example in Figure 4(c) illustrates, sub-bands at lower resolution levels contain more coarse information about their dependent levels in the hierarchy. For instance, LL_3 contains information about HL_3, LH_3, and HH_3 and these four sub-bands form LL_2 which contains information about HL_2, LH_2, and HH_2. LL_3 therefore contains the coarse and basic information about the image and the dependent levels contain some more hierarchical information so the sub-band process inherently affords both embedded and scalable coding. The *discrete wavelet transform* (DWT) (Daubechies, 1990) is most commonly used for decomposition as it has the capability to operate at various scales and resolution levels. As with the DCT, DWT co-efficients are quantised before encoded and then

a number of strategies can be used to code the resulting sub-bands using various scanning processes analogous to the *zigzag* pattern employed by JPEG. Note, due to the sub-band decomposition, scanning DWT co-efficients in ascending order of frequency is far more complex, though alternative techniques exist, with one of the most popular and efficient being Shapiro's *embedded zero tree* (EZW) wavelet compression (Shapiro, 1993), which introduced the concept of *zero trees* to derive a rate-efficient embedded coder. Essentially, the correlation and self-similarity across decomposed wavelet sub-bands is exploited to reorder the DWT co-efficient in terms of *significance* for embedded coding. Said and Pearlman (1996) presented a further advancement with the *spatial partitioning of images into hierarchical trees* (SPIHT) which at the time, was recognised as the best compression technique. Inspired by the EZW coder, they developed a set-theoretic data structure to achieve very efficient embedded coding that improved upon EZW in terms of both complexity and performance, though the major limitation of SPIHT is that it does not provide quality (SNR) scalability. Taubman subsequently introduced the *embedded block coding with optimised truncation* (EBCOT) (Taubman, 2000) method which affords higher performance as well as both SNR and spatial scalable image coding. EBCOT outperforms both EZW and SPIHT and is generally considered by the research community as the best DWT-based image compression technique to the extent that it has now been incorporated into the new JPEG2000 still image coding standard (Taubman, 2002). Each sub-band is partitioned into small blocks of samples called *code-blocks* so EBCOT generates a separate highly scalable bit-stream for each *code-block* which can be independently truncated to any of a collection of different lengths. The EBCOT block coding algorithm is built around the concept of fractional bit-planes (Li & Lei, 1997; Ordentlich, Weinberger, & Seroussi, 1998) which ensures efficient and finely embedded coding.

Figure 5. Encoder/decoder in a VQ arrangement. Given an input vector, the best matched codeword is found and its index in the codebook is transmitted. The decoder uses the index and outputs the codeword using the same VQ codebook.

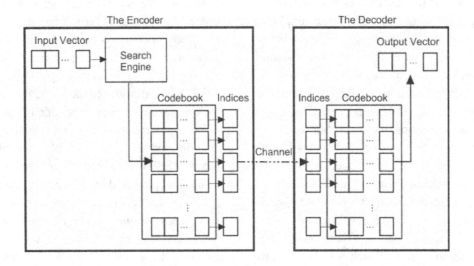

From a mobile communication perspective, the scalability, high compression, and low complexity performance of JPEG2000 make it an increasingly attractive coding option for low bit-rate applications. The one drawback of sub-band image coding is however, that since higher frequency co-efficients are discarded from the encoded image, blurring effects occur at high CR levels, though perceptually this is preferable to the inherently blocky effects of transform coding.

Vector Quantisation (VQ)

In VQ, the input image data is first decomposed into a *k*-dimensional input *vector* and a corresponding *code-vector* which is stored in a predefined lookup table known as a *codebook*, which is searched so that it provides the best match for each input vector. The corresponding index of the code-vector in the codebook is then transmitted to

the decoder, where it is used to retrieve the relevant code-vector using exactly the same codebook, so enabling the image to be efficiently reconstructed. Figure 5 shows the schematic diagram of a typical VQ-based system. There are many VQ variants (Rabbani & Jones, 1991) including, for example, adaptive VQ, tree structured VQ, classified VQ, product VQ, pyramid VQ, while the challenging matter of how best to design the codebook is another well-researched area (Linde, Buzo, & Gary, 1980; Chou, Lookabaugh, & Gary, 1989).

Second Generation Techniques

Waveform-based image coding techniques operate either on individual pels or blocks of pel using a statistical model, which can lead to some disadvantages including: (1) Greater emphasis being given to a codeword assignment that statistically reduces the bit-requirement, rather than the extraction of representative messages from the image; (2) The encoded entities are consequences of the technical constraints in transforming images into digital data, that is, from the spatial to frequency domains or RD constraints, rather than being real entities; (3) They do not fully exploit the properties of the HVS. This led to a new coding class collectively known as *second generation* methods (Kunt, Ikonomopoloulos, & Kocher, 1985) that decompose the image data into visual primitives such as contours and textures. There are many approaches for this type of coding such as, for example, dividing an image into directional primitives using segmentation-based techniques to extract regions from the image which are represented by their shape and texture content. Sketch-based coding also uses a similar segmentation based approach, and details on these and other *second generation* techniques may be found in Kunt et al. (1985).

Second generation methods provide higher compression than waveform-coding methods for the same reconstruction quality level (Al-Mualla et al., 2002) and do not possess the problems of blocking and blurring artefacts at very low bit-rates. They are particularly suitable for encoding images and video sequences from regular domains known *a prior* such as, for example, animations. The extraction of real objects however, is both intractable and computationally expensive, and in addition these methods suffer from unnatural contouring effects like the loss of continuity and smoothness which can make image detail look artificial.

Other intra-frame coding techniques include *iterated function* systems (IFS) (Distasi, Nappi, & Riccio, 2006; Øien, 1993), fractal geometry-based coding (Barnsley, 1988; Jacquin, 1992), prediction coding, block-truncation coding, quad-tree coding, recursive coding, and multi-resolution coding. IFS expresses an image as the attractor of a *contractive function* system which can be retrieved simply by progressively iterating the set of functions starting from any initial

arbitrary shape. IFS-based compression affords good performance at high CR in the range of 70-80 (Barthel, Voye, & Noll, 1993; Jacobs, Fisher, & Boss, 1992), though this is counterbalanced by the fact that such techniques are computationally complex and hence time consuming. A comprehensive review of second generation techniques can be found in Clarke (1995).

INTER-FRAME VIDEO CODING

As video is a sequence of still frames, a naïve yet simple approach to video coding is to employ any of the still image (intra-frame) coding methods previously discussed on a frame-by-frame basis. *Motion JPEG* (M-JPEG) is one such approach that contiguously applies JPEG intra-frame coding (Wallace, 1991) to each individual frame, and while it has never been standardised, unlike the new M-JPEG2000 which is formally defined as part of the JPEG2000 compression standard, the drawback in both approaches is that they do not exploit the obvious temporal correlations that exist between many consecutive video frames, so limiting their coding efficiency. As the example in Figure 6 highlights, there is considerable similarity between the two frames of the popular *Miss America* test video sequence, so if the first frame is encoded in intra-mode and the difference between the current and the next frame is coded instead, a large bit-rate saving can be achieved. Inter-frame video coding refers to coding techniques that achieve compression by reducing the temporal redundancies within multiple frames. In addition, to reduce spatial redundancy, existing intra-frame coding techniques can serve as the basis for the development of inter-frame coding. This can be done either by generalising them for 3D signals, viewing the temporal as the third dimension, or by predicting the motion of the video in the current frame from some already encoded frame(s) as the reference to reduce the temporal redundancy. Inter-frame coding alone however, is

Figure 6. Temporal redundancy between successive frames: (a) and (b) are respectively the 29th and 30th frames of the Miss America video sequence, (c) pixel-wise difference between them

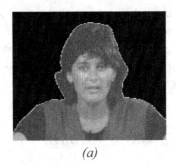
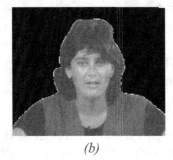
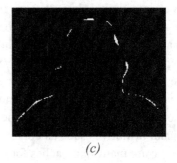

(a) (b) (c)

inappropriate for many video applications which for instance require random access within the frames, so all reference frames have to be intra-coded. In practice a combination of intra- and inter-frame coding is usually applied whereby certain frames are intra-frame coded (so called *I-frames*) at specific intervals within the sequence and the remaining frames are inter-frame coded (*P-frames*) with reference to the I-frames. Some frames known as *B-frames*, may also have both forward and backward reference frames. There are also some video coding systems, such as the latest H.264 standard, which have a provision for switching between intra- and inter-frame coding modes within the same frame, and introduce new picture types known as *Switching-P* (SP) and *Switching-I* (SI) frames which enable drift-free switching between different bit-streams.

Three categories of inter-frame video coding suitable for mobile communications will now be discussed.

Waveform-Based Techniques

The easiest way to extend the 2D (spatial) image coding to inter-frame video coding is to consider 3D (spatial and temporal) waveform coding. The basic framework will be similar to that in Figure 2, with the notable exception that 3D transformations are used followed by quantisation and entropy coding rather than the 2D transformation. The main advantage of this approach is that the computationally intensive process of motion compensation is not required, though it suffers from a number of major shortcomings including the requirement for a large frame memory which renders it inappropriate for real-time applications like video telephony, while blocking artefacts (as in the DCT) also make it unsuitable for low bit-rate coding. One other limitation, especially for the 3D sub-band based approaches is that the temporal filtering is not performed in the direction of the motion, and so temporal redundancies are not fully utilised to gain the highest compression efficiency. A solution of these problems is to combine the temporal components with motion compensation as proposed in Dufaux and Moscheni (1995).

Motion Compensation

The generic framework for motion compensated video coding techniques is given in Figure 7, with the primary difference with Figure 2 being the additional motion compensation block, where

Figure 7. A generic waveform-based inter-frame video coder

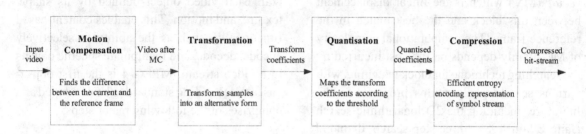

the difference between the current and reference frame is predicted. To appreciate the development of motion compensation strategies, it is worth reviewing the *conditional replenishment* (Haskell, Mounts, & Candy, 1972) technique, which represents one of the earliest approaches to inter-frame coding, with the input frame separated into "changed" and "unchanged" regions with respect to a previously coded (reference) frame. Only the changed regions needed to be encoded while the unchanged regions were simply copied from the reference frame and for this purpose only the relative addresses of these regions were required to be transmitted. Coding of the changed regions can, in principle, be performed using any intra-frame coding technique, though improved performance can be achieved by predicting the changed regions using well established *motion estimation* (ME) and *motion compensation* (MC) processes. In fact, changes in a video are primarily due to the movement of objects in the sequence, so therefore by using an object motion model between frames, the encoder can estimate the motion that has occurred between the current and reference frames, in a process commonly referred to as ME. The encoder then uses this motion model and estimated motion information to move the content of the reference frame to provide a better prediction of the current frame, which is MC, and collectively the complete prediction process

is known as *motion compensated prediction*. The reference frame used for ME may appear temporarily either before or after the current frame in the video sequence, with the two cases respectively being known as *forward* and *backward prediction*. *Bidirectional prediction* employs two frames (one each for forward and backward prediction) as the reference. As mentioned earlier, there are three different frame types used in the motion prediction process: I-frames which are intra-coded; P-frames which use either the previous or next I-frame as the reference frame; and B-frames which use the previous and next P-frames as the reference frames. The ME and MC-based coder is the most commonly used inter-frame coding method and is the bedrock for a range of popular video coding standards including MPEG-1 and MPEG-2 as well as the tele-conferencing coding H.261 and H.263 family.

In all these video coding standards, each frame is divided into regularly sized pixel blocks for ME (though the most recent H.264 standard also supports variable-sized MB), before block-by-block matching is performed. This block-matching motion estimation (BMME) strategy (Jain & Jain, 1981) is in fact the most commonly used ME algorithm, with the current frame first divided into blocks and then the motion of each block estimated by finding the best matching block in the reference frame. The motion of the current block is then represented by a *motion vector* (MV) which is the linear displacement between this block and the best match in the reference frame. The computational complexity of MC mainly depends on the cost incurred by the searching technique for block matching, with various searching algorithms proposed in the literature, including the 2D logarithmic search (Jain & Jain, 1981), three-step search, diamond search (Tham, Ranganath, Ranganath, & Kassim, 1998), minimised maximum-error (Chen, Chen, Chiueh, & Lee, 1995), fast full search algorithms (Toivonen & Heikkilä, 2004), successive elimination algorithm (Li & Salari, 1995), and the simplex minimisation search (Al-Mualla, Canagarajah, & Bull, 2001). Following the MC step, all remaining steps are similar to those delineated for intra-frame coding, that is, transformation/sub-band formation, quantisation, and compression using entropy coding.

Amongst the waveform inter-frame coding, block DCT-based methods are the most widely employed in the various standards, though the increasing requirement for scalability and higher compression ratios to enable very low bit-rate coding, has been the catalyst for wavelet-based image coders to be increasingly popular with considerable research being undertaken for instance into 3D sub-band coding (Ghanbari, 1991; Karlsson & Vetterli, 1988; Man, de Queiroz, & Smith, 2002; Ngan & Chooi, 1994; Podilchuk, Jayant, & Farvardin, 1995; Taubman & Zakhor, 1994) and motion compensated sub-band coding (Choi & Woods, 1999; Katto, Ohki, Nogaki, & Ohta, 1994).

Object-Based Video Coding

Object-based coding techniques can be viewed as an extension of second generation image coding techniques in the sense that a video object is defined in terms of visual primitives such as shape, colour, and texture. These techniques achieve very efficient compression by separating coherently moving objects from a stationary background, with each video object defined by its shape, texture, and motion. This enables content-based functionality such as the ability to selectively encode, decode, and manipulate specific objects in a video stream. MPEG-4 is the first object-based video coding standard to be developed and comprises of the following major steps:

- Moving object detection and segmentation,
- Shape coding,
- Texture coding,
- Motion estimation and compensation,
- Motion failure region detection, and
- Residual error encoding.

Figure 8 shows the overall block diagram for a generic object-based encoder. The first stage involves separating moving objects in the video sequence from the stationary background using fast and effective motion segmentation techniques, where the aforementioned ME and MC techniques are equally applicable. After the segmentation, shape coding is performed followed by ME and motion-compensated texture coding. The object segmentation, shape coding, motion estimation, and texture coding techniques are discussed in the MPEG-4 section.

The overall performance of the encoder is highly dependent on the performance of the segmentation approach employed. Following MC there may still be a significant amount of residual energy in certain areas of the image where MC alone was insufficient. The object segmentation is then re-applied to the compensated and original image to isolate these motion failure regions with high prediction errors (Bull et al., 1999). The residual information in the motion failure regions are then encoded using a block-based DCT scheme similar to the still image compression standard, for example, JPEG. The primary advantage of object-based coding is that it provides content-based flexibility and interactivity in video

Figure 8. The overall block diagram for a generic object-based encoder

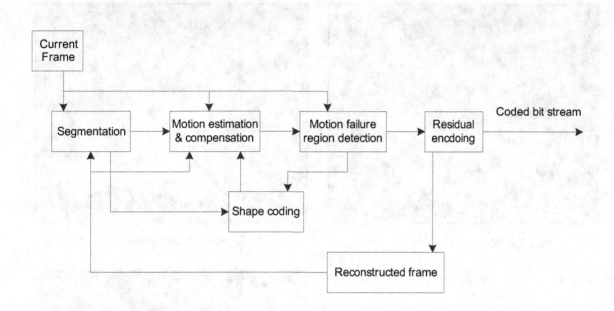

Figure 8. The overall block diagram for a generic object-based encoder

processing: encoding, decoding, manipulation, scalability, and also the interactive editing and error concealment.

Model-Based Video Coding

All compression techniques are based to some extent on an underlying model. The term model-based coding, however, refers specifically to an approach that seeks to represent the projected 2D image of a 3D scene using a semantic model. The aim is then to find an appropriate model together with the corresponding parameters, a task which can be divided into two main steps: *analysis* and *synthesis*. Model parameters are obtained by analysing the object's appearance and motion in the video scene, and these are then transmitted to a remote location, where a video display of the object is synthesised using pre-stored models at the receiver. In principle, only a small number of parameters are required to communicate the

changes in complex objects, thus enabling a very high CR. Analysis is by far the more challenging task due to the complexity of most natural scenes, such as a head and shoulder sequence (Aizawa, Harashima, & Saito, 1989; Li & Forchheimer, 1994) or face model (Kampmann, 2002). The synthesis block is easier to realise as it can build on techniques already developed for image synthesis in the field of computer graphics. For more detail about model-based video coding, the interested reader is referred to the comprehensive tutorial provided in Pearson (1995).

THE MPEG-4 VIDEO STANDARD

MPEG-4 is officially termed the *generic coding of audio-visual objects*, and the philosophy underpinning this latest video compression standard has shifted from the traditional perspective of considering a video sequence as simply being

Figure 9. Video object concepts, access, and manipulation (images from IMSI[1])

(a) Original video scene   (b) Segmented background object

(c) Segmented foreground object

(d) A different scene

(e) Edited scene, with the segmented object of (c) is inserted into scene (d)

a collection of rectangular video frames in the temporal dimension. Instead, MPEG-4 treats a video sequence as a collection of one or more video objects (VO) which are defined as a flexible entity that a user is allowed to access and manipulate. A VO may be arbitrarily shaped and exist for an arbitrary length of time, with a video scene made up of a background object and separate foreground objects. Consider the example in Figure 9, where the scene in Figure 9(a) is separated into two elements namely, the background (Figure 9(b)) and a single foreground object (Figure 9(c)), with both objects able to be independently used in other scene creations, so for instance, the VO in Figure 9(c) can be scaled and inserted into the new scene in Figure 9(d) giving the composite image in Figure 9(e). Clearly, for object manipulation, the object area is required to be defined and this leads to the challenging research area of object segmentation.

Object Segmentation

This has been the focus of considerable research and based upon the user interaction requirements, object segmentation methods usually fall into three distinct categories:

Manual Segmentation

This requires human intervention to manually identify the contour of each object in every source video frame, so it is very time-consuming and obviously only suitable for off-line video content. This approach however, can be appropriate for segmenting important visual objects that may be viewed by many users and/or re-used many times in differently composed sequences, such as cartoon animations.

Semi-Automatic Segmentation

Examples of this approach include ISO/IEC (2001) and Sun, Haynor, and Kim (2003), where a human operator either inputs some rough initial objects that resemble the original objects or identifies the objects and even the objects' contour in a single frame. The segmentation algorithm then refines the object contours and tracks the objects through successive frames of the video sequence. Semi-automatic techniques are useful in applications where domain knowledge concerning the intended object is known as *a priori*. Semi-automatic segmentation algorithms may also require some different types of information, for instance, the number of objects that the user intends for as the input. One very good example of this type is the fuzzy clustering-based image segmentation algorithm (Ali, Dooley, & Karmakar, 2006), where the algorithm comes up with a number of segmented objects equal to the user input.

Fully-Automatic Segmentation

Paradoxically, semi-automatic segmentation has the potential to provide better results than the fully-automatic counterpart, since in the semi-automatic segmentation some relevant and domain specific information or an outline of the object is provided as an input, while in case of automatic segmentation, all of the information about an object is developed by the application itself which is both computationally expensive and sometimes can lead to erroneous results as the perceptual notion of what exactly is an object is not well defined. However, the main problem with semi-automatic segmentation is that it requires user inputs. Fully-automatic segmentation algorithms, such as those in Karmakar (2002) and Kim and Hwang (2002), attempt to perform a complete segmentation of the visual scene without any user intervention, based on for instance, spatial characteristics such as edges, colour, and distance, together with temporal characteristics such as the object motion between frames.

Again, as video is a sequence of still frames, a naïve approach, much similar to that of the image coding, would be employing the image segmentation methods for video segmentation on a frame-by-frame basis. The image segmentation approaches can also be extended for video segmentation exploiting the temporal correlations. This can be done either by generalising the image segmentation algorithms for 3D signals, viewing the temporal as the third dimension or by utilising the motion of the objects in the consecutive frames. In recent times, detection and tracking of video objects has become an important and increasingly popular research area (Goldberger & Greenspan, 2006; Greenspan, Goldberger, & Mayer, 2004; Tao, Sawhney, & Kumar, 2002). As already mentioned a VO is primarily defined by its shape, texture, and motion; in MPEG-4 the VO shape and motion compensated texture are independently encoded. The following sections briefly discuss shape and texture coding paradigms for video objects.

Shape Coding

In computer graphics, the shape of an object is defined by means of an α-map (plane) M_j of size $H \cdot V$ pels where $M_j = \{m_j(x, y) \mid 0 \leq x < H, 0 \leq y < V\}$ $0 \leq m_j \leq 255$ where H and V are respectively the horizontal and vertical frame dimensions. The grey scale shape M_j defines for each pixel whether it belongs to a particular video object or not, so if $m_j(x, y) = 0$, then pixel (x, y) does not belong to the shape. In the literature, for binary shapes $m_j(x, y) = 0$ refers to the background, while $m_j(x, y) = 255$ is a foreground object. Binary shape coders can be classified into two major classes: *bitmap*-based which encode every pixel as to whether it belongs to the object, and *contour*-based, which encodes the outline of the shape (Katsaggelos et al., 1998). The former are used in the fax standards G4 (Group 4) (CCITT, 1994) and JBIG (Joint Bi-level Image Experts Group) (ISO, 1992) and within the MPEG-4 coding standard, two bitmap-based shape coders have been developed: the non-adaptive *context-based arithmetic encoder* (CAE)

(Brady, Bossen, & Murphy, 1997), the *adaptive modified modified-read* (MMR) (Yamaguchi, Ida, & Watanabe, 1997) shape coder, and the newly developed digital straight line-based shape coding (DSLSC: *digital straight line based shape coding*) technique (Aghito & Forchhammer, 2006). Conversely, many different applications have fueled research into contour-based shape coding, including chain coders (Eden & Kocher, 1985; Freeman, 1961), parametric Bezier curve-based shape descriptors (Sohel, Karmakar, Dooley, & Arkinstall, 2005, 2007), polygon (H'otter, 1990; Katsaggelos et al., 1998; Kondi et al., 2004; Meier et al., 2000; O'Connell, 1997; Schuster et al., 1997; Sohel, Dooley, & Karmakar, 2006b) and B-spline based approximations (Jain, 1989; Katsaggelos et al., 1998; Meier et al., 2000; Kondi et al., 2004; Schuster et al., 1997; Schuster & Katsagellos, 1998). Within the MPEG-4 framework, two interesting contour-based shape coding strategies have been developed: (1) the vertex-based polynomial shape approximation based upon (Katsaggelos et al., 1998); and (2) the baseline-based shape coder (Lee et al., 1999). CAE is embedded in the MPEG-4 shape coder and so in the next section, both CAE and vertex-based operational rate-distortion optimal shape coding framework (Katsaggelos et al., 1998) will be outlined.

Context-based arithmetic coder: MPEG-4 has adopted a non-adaptive context-based arithmetic coder for shape information, since it allows regular memory access to the shape information and as consequence affords easier hardware implementation (Katsaggelos et al., 1998), and resourceful use of the existing block based motion compensation to exploit temporal redundancies. The binary α–planes are encoded by the CAE, while the grey scale α–planes are encoded by motion compensated DCT coding, which is similar to texture coding. For binary shape coding, a rectangular box enclosing the arbitrarily shaped *Video Object Plane* (VOP) is formed and the bounded box is divided into 16×16 *macro-blocks*, which are called binary-alpha-blocks (BAB). As

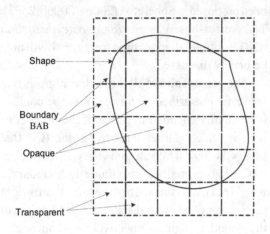

Figure 10. Binary α–block and its classification

illustrated in Figure 10, BAB are classified into three categories: transparent, opaque, and alpha or shape block. The transparent block does not contain any information about the object. The opaque block is located entirely inside an object, while the shape block is partially located in the object boundary, that is, part in the object and part background; thus these alpha-blocks are required to be processed by the encoder both for intra- and inter-coding modes.

CAE is a binary arithmetic encoder where the symbol probability is determined from the context of the neighbouring pixels based on templates, with Figure 11(a) and Figure 11(b) showing the templates for intra- and inter-modes respectively. In CAE, pixels are coded in scan-line order and in a three stage process:

- Compute a context number based on the template and encoding mode.
- Index a probability table using the context number.
- Use the indexed probability to drive an arithmetic encoder.

In the case of the inter-mode, the alignment is performed after MC. For further details on CAE, the interested reader is referred to (Brady et al., 1997).

Vertex-Based Shape Coding

Vertex-based shape coding algorithms can be efficiently used in the high compression mobile communication applications and involve encoding the outline of an object's shape using either a polygon or B-spline based approximation for

Figure 11. Templates for defining those pels (x) to be encoded, c_i are pels in the neighbourhood of (x) within the templates: (a) Intra-mode, (b) Inter-mode (Note: Alignment is performed after MC)

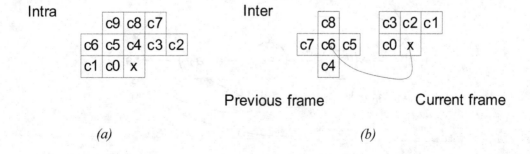

(a)   (b)

lossy shape coding. The placement of vertices allows easy control of local variations in the shape approximation error. For lossless (zero geometric distortion) shape coding, the polygon approximation simply becomes that of a chain code (Lynn, Aram, Reddy, & Ostermann, 1997; Sikora, Bauer, & Makai, 1995). A series of vertex-based rate-distortion optimal shape coding algorithms has been proposed in Schuster et al. (1997), Katsaggelos et al. (1998), Meier et al. (2000), Kondi et al. (2004), and Sohel et al. (2006b), which employ weighted *directed acyclic graph* (DAG) based dynamic programming using polygons or parametric B-spline curves. The aim of these algorithms (Katsaggelos et al., 1998), is that for some prescribed admissible distortion, a shape contour is optimally encoded in terms of the number of bits, by selecting the set of control points that requires the lowest bit-rate and vice versa. These algorithms select the vertex on the shape-contour having the highest curvature as the starting vertex, and formulate the shape coding problem as finding the shortest path from the starting vertex to the last vertex of the shape-contour. The edge-weights are determined based on the admissible distortion and the bit requirement for the differential coding (Schuster et

al., 1997) of the vertices. A number of performance enhancement techniques for these algorithms have been proposed in Sohel et al. (2006a, 2006b, 2007). The vertex-based *operational rate distortion* (ORD) optimal shape coding framework will now be briefly discussed.

The general aim of all these algorithms is that for some prescribed distortion, a shape contour is optimally encoded in terms of the number of bits, by selecting a set of *control points* (CP) that incurs the lowest bit-rate and vice versa. To select all CP that optimally approximate the boundary, a weighted DAG is formed and the minimum weight path is searched, with the start and end points of the boundary being respectively the source and destination vertices in the DAG. Both polygonal and quadratic B-spline based frameworks have been developed in Katsaggelos et al. (1998), with the *admissible control points* being considered as the vertices of DAG in the former case, and a trellis of *admissible control points* pairs are considered as the DAG vertices in the latter case. In this chapter only polygonal encoding will be discussed, while for B-spline based encoding the interested reader is referred to Katsaggelos et al. (1998), Kondi et al. (2004), and Sohel, Dooley, and Karmakar (2007).

Figure 12. DAG of five ordered admissible control points for polygonal encoding. There is a path in the DAG from a_i to a_j provided $i < j$.

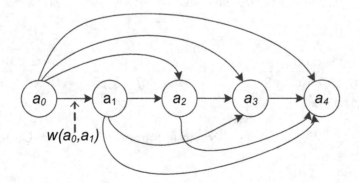

Figure 12 illustrates the DAG formation for polygonal encoding for five admissible CP namely a_0, a_1, a_2, a_3, a_4 with a_0 and a_4 being the start and end vertices respectively. Initially, admissible CP are restricted to be selected from only the boundary points, however this is subsequently relaxed by forming a fixed width band known as the *admissible control point band* (ACB) around the boundary, so points lying within this band can be admissible CP. This means that a point, though not on the boundary of an object, can still be selected as the CP and thereby further reduce the bit-rate. The framework presented in Katsaggelos et al. (1998) uses a single admissible distortion (D_{max}), which is also used as the width of the ACB around the boundary as shown in Figure 13, so any point lying inside the ACB can be a CP for the shape approximation within the prescribed

Figure 13. Admissible control point band around a shape boundary

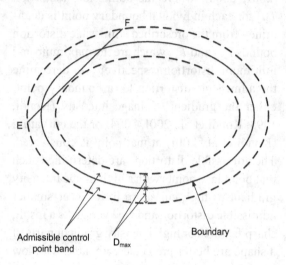

Algorithm 1. The polygonal ORD optimal shape coding algorithm

Inputs: B – the boundary; T_{max} and T_{min} – the admissible distortion bounds.
Variables: $MinRate(a_{i,m})$ – current minimum bit-rate to encode up to vertex $a_{i,m}$ from b_0; $pred(a_{i,m})$ – preceding CP of $a_{i,m}$ (double subscripts is used to denote the ACB); $N[i]$ – the number of vertices in A associated to b_i.
Output: P – the ordered set of CP approximating B.

Determine the admissible distortion $T[i]$ for $0 < i < N_B - 1$;
Determine the *sliding window* width $L[i]$ for $0 < i < N_B - 1$ according to Sohel (2007);
Form the ACB A using $T[i]$ for $0 < i < N_B - 1$ according to Sohel (2007);
Initialise $MinRate(a_{0,0})$ with the total bits required to encode the first boundary point b_0;
Set $MinRate(a_{k,n})$, $0 < k < N_B$, $0 \leq n < N[k]$ to infinity;
FOR each vertex $a_{i,m}$, $0 \leq i < N_B - 1$, $0 \leq m < N[i]$
  FOR each vertex $a_{j,n}$, $i < j \leq \min\{(i + L[i])(N_B - 1)\}$, $0 \leq n < N[j]$
    Check the edge-distortion $dist(a_{i,m}, a_{j,n})$;
    IF $dist(a_{i,m}, a_{j,n})$ maintains the admissible distortion THEN
      Determine bit-rate $r(a_{i,m}, a_{j,n})$ and edge-weight $w(a_{i,m}, a_{j,n})$;
      IF $((MinRate(a_{i,m}) + w(a_{i,m}, a_{j,n})) < MinRate(a_{j,n}))$ THEN
        $MinRate(a_{j,n}) = MinRate(a_{i,m}) + w(a_{i,m}, a_{j,n})$;
        $pred(a_{j,n}) = a_{i,m}$;
Obtain P with properly indexed values from *pred*.

admissible distortion. The notion of fixed admissible distortion has been generalised by Kondi et al. (1998) and Kondi, Melnikov, and Katsaggelos (2001, 2004), where the admissible distortion $T[i]$ for each individual boundary point is determined from the prescribed admissible distortion bounds T_{max} and T_{min} which are the maximum and minimum distortion respectively. To determine the admissible distortion for a boundary point, either the gradient of image intensity (Kondi, 1998; Kondi et al., 2001, 2004) or the curvature (Kondi et al., 2001) at that point is considered. The admissible distortions are determined such that boundary points with a high image intensity gradient or high curvature have lower smaller admissible distortion and vice versa. As a result, sharp features or high intensity gradient parts of a shape are better protected compared with low image gradient or flatter shape portions from an approximation perspective. Within the variable admissible distortion framework, the philosophy of ACB has also been generalised in Sohel, Dooley, and Karmakar (2006b, 2007) to support variable ACB so it can fully exploit the variable admissible distortion in reducing the bit-rate for a prescribed admissible distortion pair, with the width of the ACB for each boundary point being set equal to the admissible distortion. These works also defined the ACB width for individual boundary points for the B-spline based framework.

Each edge in the DAG is considered in the optimisation process for approximating the shape, though for a particular edge it is required to check whether all boundary points in between the end points of the edge maintain the admissible distortion, so in the example in Figure 13, edge *EF* does maintain the admissible distortion. If the admissible distortion is maintained, this edge is further considered in the rate-distortion optimisation process, so it becomes crucial to determine the level of distortion of each boundary point from the candidate DAG edge. The ORD framework in Katsaggelos et al. (1998) employs either the *shortest absolute distance* or alternatively the *distor-*

tion band approach; while in Kondi et al. (2004) the *tolerance band* which is the generalisation of the *distortion band* is used. The performance of these various distortion measurement techniques can be further enhanced by adopting the recently introduced *accurate distortion* metric in Sohel et al. (2006b) and the computationally efficient *chord-length-parameterisation* based approach in Sohel, Karmakar, and Dooley (2007b). Moreover, these algorithms use a *sliding window* (SW) which enforces the encoder to follow the shape boundary and also limits the search space for the next CP within the SW-width (Katsaggelos et al., 1998). The SW provides three fold benefit to the encoder: (1) avoid the trivial solution problems, (2) preserve the sharp features of the shape, and (3) computationally speed up the process. However, since the SW constricts the search space for the next CP within SW-width, the optimality of the algorithms is compromised in a bit-rate sense (Sohel, Karmakar, & Dooley, 2006). The techniques (Sohel, Dooley, & Karmakar, 2007; Sohel, Karmakar, & Dooley, 2006) formally define the most appropriate and suitable SW-width for the rate-distortion constrained algorithms.

After the distortion checking process, the edge-weight is determined which is *infinite* if the edge fails to maintain the admissible distortion for all the relevant boundary points. If the edge passes the distortion check, the edge weight is the bit required to encode the edge differentially, so for example, the edge weight $w(a_i, a_j)$ is equal to the edge bit-rate $r(a_i, a_j)$ which is the total number of bits required to differentially encode the vertex a_j given that vertex a_i is already encoded. For vertex encoding purposes, a combination of orientation dependent chain code and logarithmic run-length code are used in these algorithms (Schuster et al., 1997).

To summarise, the vertex-based ORD optimal shape coding algorithms seek to determine and encode a set of CP to represent a particular shape within prescribed RD constraints. Assume boundary $B = \{b_0, b_1, ..., b_{N_B-1}\}$ is an ordered set

Figure 14. Polygonal approximation results for the 1ˢᵗ frame of the Kids sequence with T_{max} = 3 and T_{min} = 1pel (Legends: Solid line—Approximated boundary; Dashed line—Original boundary; Asterisk—CP)

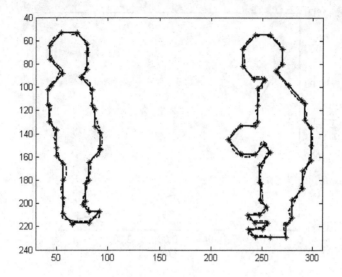

of shape points, where N_B is the total number of points and $b_0 = b_{N_B-1}$ for a closed boundary. $P = \{p_0, p_1, ..., p_{N_P-1}\}$ is an ordered set of CP used to approximate B, where N_P is the total number of CP and $P \subseteq A$, where A is the ordered set of vertices in ACB. For a representative example, the ORD polygonal shape coding algorithm for determining the optimal P for boundary B within RD constraints is formalised in Algorithm 1, with the detailed analysis provided in Schuster et al. (1997), Katsaggelos et al. (1998), Meier et al. (2000), Kondi et al. (2004), and Sohel, Dooley, and Karmakar (2007).

Some experimental results from this ORD shape coding framework are now presented.

Figure 14 shows the subjective results upon the 1ˢᵗ frame of the popular multiple-object *Kids* test video test sequence with L_∞ distortion bounds of T_{max} = 3 and T_{min} = 1*pel* respectively. In the experiments, the curvature-based approach of Kondi et al. (2001) was adopted from which it is visually apparent that those shape regions having high curvature are well preserved in the approximation with lower admissible distortion, while in the smoother shape regions, the higher admissible distortion is fully utilised to ensure that the bit-rate requirement is minimised, while upholding the prescribed distortion bounds.

Figure 15 shows the corresponding rate-distortion (RD) results for the 1ˢᵗ frame Kids sequence. The bit-rate is plotted along the ordinate in bit units, while the MPEG-4 relative area error (D_n) is shown along the abscissa in percentiles. The curve reveals that according to the ORD theory as the distortion decreases the required bit-rate increases and vice versa, however as anticipated, a *diminishing rate of return* trend is observed at higher distortion values. At lower D_n values, a much higher bit-rate reduction is achieved for only a small increase in the distortion, while at higher distortion values, a change in distortion generates only a comparatively moderate improvement in the bit-rate.

Figure 15. Rate-distortion results upon the 1ˢᵗ frame of Kids sequence

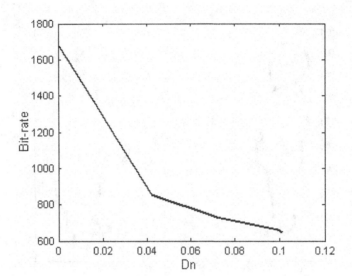

Motion Compensation

Motion estimation (ME) and compensation (MC) methods in MPEG-4 are very similar to those employed in the other standards, though the primary difference is that block-based ME and MC are adapted to the arbitrary-shape VOP structure. Since the size, shape, and location of a VOP can change from one instance to another, the absolute (frame) coordinate system is applied to reference each VOP. For opaque blocks, motion is estimated using the usual block matching method, however for the BAB, motion is estimated using a modified block matching algorithm, namely polygon matching where the distortion is measured using only those pixels in the current block. Padding techniques are used to define the values of pels where the ME and MC may require to access pels from outside the VOP. For the BAB in intra-mode, it is padded with horizontal and vertical

repetition. For the inter-alpha blocks, not only are alpha blocks repeatedly padded, but also the region outside the VOP within the block is padded with zeros.

Texture Coding

Texture is an essential part of a video object and which is reflected by it being assigned more bits than the shape in the coded bit-stream (Bandyopadhyay & Kondi, 2005; Kaup, 1998; Kondi et al., 2001). Each intra VOP and MC inter VOP is coded using a 8×8 block DCT, with the DCT performed separately on each of the luminance and chrominance planes. The opaque alpha blocks are encoded with block-based DCT, with BAB padding techniques used as outlined in the previous section, while all transparent blocks are skipped and so not encoded.

Padding removes any abrupt transitions within a block and hence reduces the number of significant DCT co-efficients. Since the number of opaque pixels in the 8×8 blocks of some of the boundary alpha blocks is usually less than 64 pixels, it is more efficient if these opaque pixels are DCT coded without padding in a technique known as *shape adaptive DCT* (Sikora & Makai, 1995). In Kondi et al. (2004), a joint optimal texture and shape encoding strategy was proposed based on a combination of the shape adaptive DCT and vertex-based ORD optimal shape coding framework. While block transforms such as DCT are widely considered to be the best practical solution for MC video coding, the DWT is particularly effective in coding still images. Recent research findings including, the MPEG-4 visual (MPEG-4: Part 2) use the DWT (Daubechies, 1990) as the core basis texture compression tools, moreover shape adaptive DWT (Li & Li, 1995) has been employed in the texture coding algorithms as, for example, in the *joint contour-based shape and texture coding* strategy proposed in Bandyopadhyay et al. (2005).

THE H.264 STANDARD

H.261 (ITU-T, 1993) was the first widely-used standard for videoconferencing and was primarily developed to support video telephony and conferencing applications over ISDN circuit-switched networks, hence the constraint that H.261 could only operate at multiples of 64Kbps, though it was specifically designed to offer computationally simple video coding at these bit-rates. H.261 employed a DCT model with integer-accuracy MC, while the next version known as H.263 (ITU-T, 1998) provides improved compression performance with half-pel MC accuracy and is able to provide high video quality at bit-rates lower than 30 kbps, as well as operating over both circuit- and packet-switched networks. The MPEG and the *Video Coding Experts Group*

(VCEG) subsequently developed the *advanced video coding* (AVC) standard H.264 (ISO/IEC, 2003) that aims to provide better video compression. H.264 does not explicitly define a CODEC as was the trend in the earlier standards, but rather defines the syntax of an encoded video bit-stream together with the method of decoding the bit-stream. The main features of H.264 as mentioned in Richardson (2003).

It supports multi-frame MC using previously-encoded frames as references in a more flexible way than other standards. H.264 permits up to 32 reference frames to be used in some cases, while in prior standards this limit was typically one or two only in the case of B-frames. This particular feature allows modest improvements in bit rate and quality in most video sequences, though for certain types of scenes, particularly rapidly repetitive flashing, back-and-forth scene cuts[2] and newly revealed background areas, significant bit-rate reductions are achievable. The computational cost of MC however, is increased with the increase in the search space for the best matched block.

It introduces the *tree-structured motion compensation*. While using the same basic principle of block-based motion compensation that has been employed since the original H.261 standard was established, a major departure is the support for a range of different sized blocks from the usual fixed 8×8 DCT-based block size used in MPEG-1, MPEG-2, and H.263, through to the smaller 4×4 and larger 16×16 block sizes, with various intermediate combinations including 16×8 and 4×8. The tree structure comes from the actual method of partitioning the MB into motion compensated sub-blocks. Choosing a large block size, such as 16×16 or 8×16, means a smaller number of bits are required to represent the MV and partition choice, however the corresponding motion compensated residual signal may be large, especially in areas of high detail. Conversely, choosing a small block size, that is, 4×4 or 4×8, results in a much lower energy in the motion compensated residual signal, but a larger number of bits will be required to represent the MV and partition choice. The variable block-size notion of H.264 has been illustrated in with the example in Figure 16. The choice of the MB partition size is therefore crucial to the efficiency of the compression. H.264 adopts what may be thought of as very much the intuitive approach, with a larger-sized partition being appropriate for predominantly smooth or homogeneous regions, while a smaller size is used in areas of high detail. The "best" partition size decision is made during encoding such that the residual energy and MV are minimised.

It also uses fractional (one-quarter) pel accuracy for MC and incorporates weighted prediction that allows an encoder to specify the use of scaling and offset when performing MC. This provides a significant benefit in performance in special cases, as for example, in fade-to-black, fade-in, and cross-fade transitions.

To reduce the blocking artefacts of DCT-based coding techniques, an in-loop de-blocking filters are employed. Moreover, the filtered MB is used in subsequent motion-compensated prediction of future frames, resulting in a lower residual error after prediction. Figure 17 presents an example of the effect of the de-blocking filter in the decoding loop in reducing the visual blocky artefacts.

It incorporates either a *context-adaptive binary arithmetic coder* (CABAC) or a *context-adaptive variable-length coder* (CAVLC). CABAC is an intelligent technique that compresses in a lossless manner the syntax elements in a video stream knowing their probabilities in a given context. CAVLC is a low-complexity alternative to CABAC for the coding of quantised transform co-efficient values. It is more elaborate and more efficient than the methods typically employed to code the quantised transform co-efficients in previous designs.

Figure 16. Tree structured variable block motion compensation in H.264

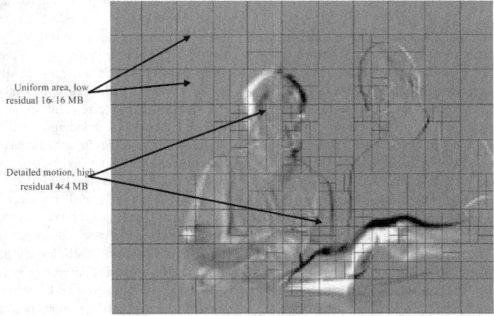

Uniform area, low residual 16×16 MB

Detailed motion, high residual 4×4 MB

Figure 17. Effect of de-blocking filter: (a) reference frame, (b) reconstructed frame without a filter, (c) reconstructed frame with filter

(a)

(b)

(c)

A *network abstraction layer* (NAL) is defined that allows the same video syntax to be used in many network environments, including features such as *sequence parameter sets* (SPS) and *picture parameter sets* (PPS) providing greater robustness and flexibility than previous standards.

Switching slices (known as SP and SI slices) features facilitate an encoder with efficient switching between different video bit-streams. Consider for instance a video decoder receiving multiple bit-rate streams across the Internet. The decoder attempts to decode the highest rate stream, but

may need, if data throughput falls, to switch automatically to decoding a lower bit-rate stream. The example in Figure 18 explains the switching using I-slice. Having already decoded frames A_0 and A_1 the decoder wishes to switch across to the other bit-stream at B_2. This however is a P-frame (synonymously P-slice) with no reference at all to the previous P-frames in video stream A. This means a solution is to code B_2 as an I-frame so it does not involve prediction and can therefore be coding independently. However, this results in an increase in the overall bit-rate, since the coding

Figure 18. Switching between video streams using SI-slices

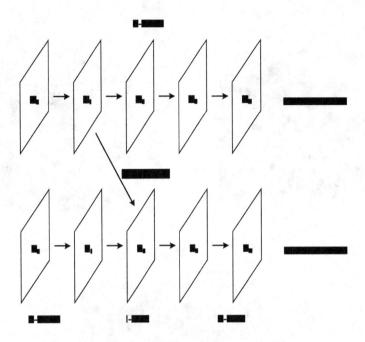

efficiency of an I-frame is much lower than that of a P-frame. H.264 provides an elegant solution to this problem through the use of SP-slices. Figure 19 illustrates an example of switching using SP-slices. At the switching points (frame 2 in both Streams A and B) which would be at regular intervals in the coded sequence, there are now three SP-slices involved (highlighted), which are all encoded using motion compensation prediction, so they will be more efficient than I-frame coding. SP-slice A_2 is decoded using reference frame A_1 and SP-slice B_2 is decoded using reference frame B_1; however the key to this technique is SP-slice AB_2—the switching slice. This is generated in such a manner that it is able to decode using motion-compensated prediction frame A_1 to produce slice B_2. This means the decoder output frame B_2 is the same regardless of

whether it is directly decoding B_1 or A_1 followed by AB_2. A reciprocal arrangement means an extra SP-slice BA_2 will also be included to facilitate switching from bit-stream B to A, though this is not shown in the diagram. While an extra SP-slice will be required at every switching point, the additional overhead this incurs is more than offset by not requiring the decoding of I-frames at these switching points.

All H.264 frames are numbered which allows the creation of *sub-sequences* (enabling temporal scalability by the optional inclusion of extra pictures between other pictures), and the detection and concealment of losses of even entire pictures (which can occur due to network packet losses or channel errors).

It ranks picture count which keeps the ordering of the pictures and the values of samples in the

Figure 19. Switching between video streams using SP-slices

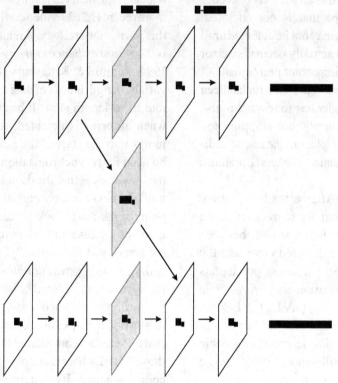

decoded pictures isolated from timing information, allowing timing information to be carried and controlled/changed separately by a system without affecting decoded picture content.

ERROR RESILIENT VIDEO CODING

There are many causes in mobile communication systems whereby the encoded data may experience error, which readily become magnified because in the compressed data, one single bit usually represents much more information than it did in the original video. Error resilient techniques therefore have become very important and have

attracted the attention of researchers. This section provides the reader with a lucid insight into some of the requirements, challenges, and various methods that exist for error resilient video coding in mobile communications.

Requirements of an Error Resilient CODEC

An error-resilient video coding system should be able to provide the following functionality:

- **Error Detection:** This is the most fundamental requirement. The system decoder can encounter a syntactic error, that is, an illegal code word of variable or fixed length, or a semantic error such as an MPEG-4 decoder generating a shape that is not enclosed. However, the error may not be detected until some point after it actually occurs, so error localisation is an important prerequisite.

- **Error Localisation:** When an error has been detected, the decoder has to resynchronise with the bit-stream without skipping too many bits, for example, via the use of additional resynchronisation markers (Ebrahimi, 1997).

- **Data Recovery:** After error localisation, data recovery attempts to recover some information from the bit-stream between the location of the detected error and the determined resynchronisation point, thus minimising information loss. *Reversible variable length coding* (RVLC) (VLC that can be decoded both in a forward and backward directions, a double-ended decodable code) (Wen & Villasenor, 1997) can be explicitly used for this purpose.

- **Error Concealment:** Finally, error concealment tries to hide the effects of the erroneous bit-stream by replacing lost information by meaningful data, that is, copying data from the previous frame into the current frame. The smaller the spatial and temporal extent of the error, the more accurate the concealment strategies can be.

Error Resilience at the Encoder

There are both pre- and post-processing techniques available for error correction. In the former, the encoder plays a pivotal role by introducing a controlled level of redundancy in the video bit-stream to enhance the error resilience, by sacrificing some coding efficiency. Resynchronisation of the code-words inserts unique markers into the encoded bit-stream so enabling the decoder to localise the detected error in the received bit-stream. When an error is detected at the decoder, it can skip the remaining bits until it locates the next resynchronisation marker. The more recent H.263+ video coding standard adopts this particular strategy. An alternative approach is the *error resilience entropy encoder* (Redmill, 1994; Redmill & Kingsbury, 1996), which takes variable length blocks of data and rearranges them into fixed-length slots. It has the advantage that when an error is detected, the decoder simply jumps into the start of the next block so there is no need for resynchronisation keywords, though the drawback is that the decoder discards all data until the next resynchronisation code or starting point of the next block is reached, even though much of the discarded data may have been correctly received. Reversible-VLC (Wen et al., 1997) coding also known as *double-ended*, decodes the received bits in reverse order instead of blindly discarding them when a resynchronisation code or start of the next block has been received, so that the decoder can attempt to recover and utilise those bits which were simply discarded with other coding schemes. It is noteworthy to mention that RVLC also keeps on proceeding with the incoming bit-streams. There are some other common forward techniques such as layered coding with prioritisation (Ghanbari, 1989), multiple-description coding (Kondi, 2005; Vaishampayan, 1993), and interleaved coding (Zhu, Wang, & Shaw, 1993) which are all designed for very low bit-rate video coding and so are well suited for video communications over the mobile networks.

Error Resilience at the Decoder

For either the post-processing or concealment techniques, the decoder plays the primary role in attempting to mask the effects of errors by providing a subjectively acceptable approximation of the original data using the received data. Error concealment is an ill-posed problem since there is no unique solution for a particular problem. Depending on the information used for concealment, these are divided into three major categories:

spatial, temporal, and hybrid techniques. Spatial approaches use the inherently high *spatial correlation* of video signals to conceal erroneous pels in a frame, using information from correctly received and/or previously concealed neighbouring pels within the same frame (Ghanbari & Seferides, 1993; Salama et al., 1995). Temporal methods exploit the latent inter-frame correlation of video signals and conceal damaged pels in a frame again using the information from correctly received and/or previously concealed pels within the reference frame (Narula & Lim, 1993; Wang & Zhu, 1998), while *hybrid* techniques seek to concomitantly exploit both spatial and temporal correlations in the concealment strategy (Shirani, Kossentini, & Ward, 2000).

There also exist some *interactive* approaches to error concealment, including the *automatic repeat request* (ARQ), sliding window, refreshment based on feedback, and selective repeat. A comprehensive review of these techniques can be found in Girod and Farber (1999). In all these cases, the encoder and decoder cooperate to minimise the effects of transmission errors, with the decoder using a feedback channel to inform the encoder about the erroneous data. Based on this information the encoder adjusts its operation to combat the effects of the errors. For shape coding techniques, there are number of efficient error concealment techniques (Schuster & Katsaggelos, 2006; Schuster, Katsaggelos, & Xiaohuan, 2004; Soares & Pereira, 2004, 2006), some of them use the parametric curves, such as Bezier curves (Soares et al., 2004) and Hermite splines (Schuster et al., 2004). While the techniques in Schuster et al. (2004) and Soares et al. (2004) are designed to conceal the errors by exploiting only the spatial information within intra-mode, the techniques by Schuster et al. (2006) and Soares et al. (2006) work in inter-mode and also utilises the temporal information.

Moreover, Bezier curve theory has been extended, by incorporating the localised control point information so that it reduces the gap be-

tween the curve and the control polygon, as the *half-way shifting Bezier curve,* the *dynamic Bezier curve,* and the *enhanced Bezier curve*s respectively in Sohel, Dooley, and Karmakar (2005a, 2005b), Sohel et al. (2005), and Sohel, Karmakar, and Dooley (2007). To improve the respective performance, these new curves can be seamlessly embedded into algorithms, for instance, the ORD optimal shape coding frameworks and the shape error concealment techniques, where currently the B-splines and the Bezier curves are respectively used. While these techniques conceal the shape error independent of the underlying image information, an image dependent shape error concealment technique has been proposed by Sohel, Karmakar, and Dooley (2007b), which utilises the underlying image information in the concealment process and obtains a more robust performance.

DISTRIBUTED VIDEO CODING

This is a novel paradigm in the video coding applications where instead of doing the compression at the encoder, it is either partially or wholly performed at the decoder. It is not, however, a new concept as the origins of this interesting idea can be traced back to the 1970s and the information-theoretic bounds established by Slepian and Wolf (1973) for distributed lossless coding, and also by Wyner and Ziv (1976) for lossy coding with decoder side information.

Distributed coding exploits the source statistics in the decoder, so the encoder can be very simple, at the expense of a more complex decoder, so the traditional balance of a complex encoder and simple decoder is essentially reversed. A high-level schematic visualization of distributed video coding is provided in Figure 20, where Figure 20(a) and (b) respectively contrast the conventional and distributed video coding paradigms. In a conventional video coder, all compression is undertaken at the encoder which must therefore

Figure 20. High-level view of: (a) conventional and (b) distributed video coding

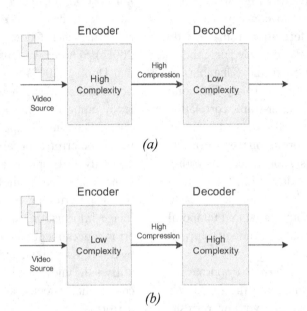

(a)

(b)

be sufficiently powerful to cope with this requirement. Many applications however, may require a dual system, that is, lower-complexity encoder at the possible expense of a higher-complexity decoder. Examples of such systems include: wireless video sensors for surveillance, wireless personal-computer cameras, mobile camera-phones, disposable video cameras, and networked camcorders. In all these cases, with conventional coding the compression must be implemented at the camera where memory and computation are scarce, so a framework is mandated where the decoder performs all the high complexity tasks, which is the essence of distributed video coding, namely to distribute the computational workload incurred in the compression between the encoder and decoder. This approach is actually more robust than conventional coding techniques in the sense of handling packet loss or frame dropping which are fairly common events in hostile mobile channels. Girod, Aaron, Rane, and Rebollo-Monendero

(2005) and Puri and Ramchandran (2002) have pioneered this research field and provide a good starting point for the interested reader on this contemporary topic.

FUTURE TREND

It was mentioned early that the VQEG have been striving for sometime to establish a single quality metric that truly represents the best of what is perceived by the HVS within the image and video coding domains. When eventually this is formally devised, it will command considerable attention from the research community as they rapidly endeavour to ensure that new findings and outcomes are fully compliant with this quality metric and that the performance of algorithm and system is superior from this new perspective (Wu et al., 2006).

Bandwidth allocation and *reservation* will inevitably, as it has in recent times, remain a very challenging research topic especially for mobile technologies (Bandyopadhyay et al., 2005; Kamaci, Altunbasak, & Mersereau, 2005; Sun, Ahmad, Li, & Zhang, 2006; Tang, Chen, Yu, & Tsai, 2006; Wang, Schuster, & Katsaggelos, 2005), and this can only be expected to burgeon in the future as the next series of mobile generation technologies, namely 3G and 4G mature.

Finally, as discussed previously, distributed video coding has gained increasing popularity among researchers as it affords a number of potential advantages for mobile operation over traditional and well-established compression strategies. Much work, however, remains in both revisiting and innovating new compression techniques for this distributed coding framework (Girod et al., 2005).

CONCLUSION

This chapter has presented an overview of video coding techniques for mobile communications where low bit-rate and computationally efficient coding are mandated in order to cope with the stringent bandwidth and processing power limitations. It has provided the reader with a comprehensive review of contemporary research work and developments in this rapidly burgeoning field. The evolution of high compression, intra-frame coding strategies from JPEG to JPEG2000 (version 2) and very low bit-rate inter-frame coding from block-based motion compensated MPEG-2 to the flexible object-based MPEG-4 coding have been outlined. Moreover, the main features of the AVC/H.264 have also been outlined together with a discussion on emerging distributed video coding techniques. It also has provided a functional discussion on the physical significance of the various video coding quality metrics that are considered essential for mobile communications

in conjunction with the aims and interests of the Video Quality Expert Group.

REFERENCES

Aghito, S. M., & Forchhammer, S. (2006). Context based coding of bi-level images enhanced by digital straight line analysis. *IEEE Transactions on Image Processing, 15*(8), 2120-2130.

Aizawa, K., Harashima, H., & Saito, T. (1989). Model-based analysis synthesis image coding (MBASIC) system for a person's face. *Signal Processing: Image Communication, 1*(2), 139-152.

Al-Mualla, M. E., Canagarajah, C. N., & Bull, D. R. (2001). Simplex minimization for single- and multiple-reference motion estimation. *IEEE Transactions on Circuits and Systems for Video Technology, 11*(12), 1209-1220.

Al-Mualla, M. E., Canagarajah, C. N., & Bull, D. R. (2002). *Video coding for mobile communications: Efficiency, complexity, and resilience.* Amsterdam: Academic Press.

Ali, M. A., Dooley, L. S., & Karmakar, G. C. (2006). *Object based segmentation using fuzzy clustering.* IEEE International Conference on Acoustics, Speech, and Signal Processing (ICASSP), Toulouse, France, May 15-19.

Atta, R., & Ghanbari, M. (2006). Spatio-temporal scalability-based motion-compensated 3-D subband/DCT video coding. *IEEE Transactions on Circuits and Systems for Video Technology, 16*(1), 43-55.

Bandyopadhyay, S. K., & Kondi, L. P. (2005). Optimal bit allocation for joint contour-based shape coding and shape adaptive texture coding. *International Conference on Image Processing (ICIP), I,* Genoa, Italy, September 11-14 (pp. 589-592).

Barnsley, M. F. (1988). *Fractals everywhere.* Boston: Academic Press.

Barthel, K. U., Voye, T., & Noll, P. (1993). *Improved fractal image coding.* Picture Coding Symposium, Lausanne, Switzerland, March 17-19.

Brady, N. (1999). MPEG-4 standardized methods for the compression of arbitrarily shaped video objects. *IEEE Transactions on Circuits and Systems for Video Technology, 9*(8), 1170-1189.

Brady, N., Bossen, F., & Murphy, N. (1997). Context-based arithmetic encoding of 2D shape sequences. *International Conference on Image Processing (ICIP), I,* Washington, DC, October 26-29 (pp. 29-32).

Bull, D. R., Canagarajah, N. C., & Nix, A. (1999). *Insights into mobile multimedia communications: Signal processing and its applications.* San Diego, CA: Academic Press.

CCITT. (1994). *Facsimile coding schemes and coding functions for group 4 facsimile apparatus.* CCITT Recommendation T.6.

Chen, M. J., Chen, L. G., Chiueh, T. D., & Lee, Y. P. (1995). A new block-matching criterion for motion estimation and its implementation. *IEEE Transactions on Circuits and Systems for Video Technology, 5*(3), 231-236.

Choi, S. J., & Woods, J. W. (1999). Motion-compensated 3-D subband coding of video. *IEEE Transactions on Image Processing, 8*(2), 155-167.

Chou, P. A., Lookabaugh, T., & Gary, R. M. (1989). Entropy constrained vector quantisation. *IEEE Transactions on Acoustics, Speech, and Signal Processing, 37*(1), 31-42.

Clarke, R. J. (1995). *Digital compression of still images and video.* London: Academic Press.

Crochiere, R. E., Webber, S. A., & Flanagan, F. L. (1976). Digital coding of speech in sub-bands. *IEEE International Conference on Acoustics,*

Speech, and Signal Processing, 1, April (pp. 233-236).

Daubechies, I. (1990). The wavelet transform, time-frequency localization and signal analysis. *IEEE Transactions on Information Theory, 36*(5), 961-1005.

Distasi, R., Nappi, M., & Riccio, D. (2006). A range/domain approximation error-based approach for fractal image compression. *IEEE Transactions on Image Processing, 15*(1), 89-97.

Dufaux, F., & Moscheni, F. (1995). Motion estimation techniques for digital TV: A review and a new contribution. *Proceedings of the IEEE, 83*(6), 858-876.

Ebrahimi, T. (1997). *MPEG-4 video verification model version 8.0.* International Standards Organization, ISO/IEC JTC1/SC29/WG11 MPEG97/N1796.

Eden, M., & Kocher, M. (1985). On the performance of a contour coding algorithm in the context of image coding. Part i: Contour segment coding. *Signal Processing, 8,* 381-386.

Freeman, H. (1961). On the encoding of arbitrary geometric configurations. *IRE Trans. Electronic Computers, EC-10,* 260-268.

Ghanbari, M. (1989). Two-layer coding of video signals for VBR networks. *IEEE Journal on Selected Areas in Communications, 7*(5), 771-781.

Ghanbari, M. (1991). Subband coding algorithms for video applications: Videophone to HDTV-conferencing. *IEEE Transactions on Circuits and Systems for Video Technology, 1*(2), 174-183.

Ghanbari, M. (1999). Video coding: An introduction to standard codecs. *IEE Telecommunications Series, 42.*

Ghanbari, M., & Seferides, V. (1993). Cell-loss concealment in ATM video codecs. *IEEE*

Video Coding for Mobile Communications

Transactions on Circuits and Systems for Video Technology, 3(3), 238-247.

Girod, B., Aaron, A., Rane, S., & Rebollo-Monendero, D. (2005). Distributed video coding. *Proceedings of the IEEE, 93*(1), 71-83.

Girod, B., & Farber, N. (1999). Feedback-based error control for mobile video transmission. *Proceedings of the IEEE: Special Issue on Video for Mobile Multi-media, 97*(10), 1707-1723.

Goldberger, J., & Greenspan, H. (2006). Context-based segmentation of image sequences. *IEEE Transactions on Pattern Analysis and Machine Intelligence, 28*(3), 463-468.

Greenspan, H., Goldberger, J., & Mayer, A. (2004). Probabilistic space-time video modeling via piecewise GMM. *IEEE Transactions on Pattern Analysis and Machine Intelligence, 26*(3), 384-396.

Haskell, B. G., Mounts, F. W., & Candy, J. C. (1972). Interframe coding of videotelephone pictures. *Proceedings of the IEEE, 60*(7), 792-800.

H'otter, M. (1990). Object-oriented analysis-synthesis coding based on moving two-dimensional objects. *Signal Processing, 2,* 409-428.

ISO. (1992). *Coded representation of picture and audio information—Progressive bi-level image compression.* ISO Draft International Standard 11544.

ISO/IEC 14496-2. (2001). *Coding of audio-visual objects – Part 2: Visual.* Annex F.

ISO/IEC 14496-10 & ITU-T Rec. (2003). *H.264, Advanced video coding.*

ITU-T Recommendation H.261. (1993). *Video CODEC for audiovisual services at px64 kbit/s.*

ITU-T Recommendation H.263. (1998). *Video coding for low bit rate communication, Version 2.*

Jacobs, E. W., Fisher, Y., & Boss, R. D. (1992). Image compression: A study of the iterated transform method. *Signal Processing, 29*(3), 251-263.

Jacquin, A. E. (1992). Image coding based on a fractal theory of iterated contractive image transformations. *IEEE Transactions on Image Processing, 1*(1), 18-30.

Jain, A. K. (1989). *Fundamentals of digital image processing.* Englewood Cliffs, NJ: Prentice-Hall.

Jain, J., & Jain, A. (1981). Displacement measurement and its application in interframe image coding. *IEEE Transactions on Communication, COMM-29*(12), 1799-1808.

Johnston, J. D. (1980). A filter family designed for use in quadratic mirror filter banks. *IEEE International Conference on Acoustics, Speech, and Signal Processing (ICASSP)* (pp. 291-294).

Kamaci, N., Altunbasak, Y., & Mersereau, R. M. (2005). Frame bit allocation for the H.264/AVC video coder via Cauchy-density-based rate and distortion models. *IEEE Transactions on Circuits and Systems for Video Technology, 15*(8), 994-1006.

Kampmann, M. (2002). Automatic 3-D face model adaptation for model-based coding of videophone sequences. *IEEE Transactions on Circuits and Systems for Video Technology, 12*(3), 172-182.

Karlsson, G., & Vetterli, M. (1988). Three-dimensional subband coding of video. *IEEE International Conference on Acoustics, Speech, and Signal Processing (ICASSP),* New York, April (pp. 1100-1103).

Karmakar, G. C. (2002). *An integrated fuzzy rule-based image segmentation framework.* PhD Thesis. Gippsland School of Computing and Information Technology. Monash University: Australia.

Katsaggelos, A. K., Kondi, L. P., Meier, F. W., Ostermann, J., & Schuster, G. M. (1998). MPEG-4 and rate-distortion-based shape-coding techniques. *Proceedings of the IEEE, 86*(6), 1126-1154.

Katto, J., Ohki, J., Nogaki, S., & Ohta, M. (1994). A wavelet codec with overlapped motion compensation for very low bit-rate environment. *IEEE Transactions on Circuits and Systems for Video Technology, 4*(3), 328-338.

Kaup, A. (1998). Object-based texture coding of moving video in MPEG-4. *IEEE Transactions on Circuits and Systems for Video Technology, 9*(1), 5-15.

Kim, C., & Hwang, J.-N. (2002). Fast and automatic video object segmentation and tracking for content-based applications. *IEEE Transactions on Circuits and Systems for Video Technology, 12*(2), 122-129.

Kondi, L. P. (2005). Transactions letters. A rate-distortion optimal hybrid scalable/multiple description video codec. *IEEE Transactions on Circuits and Systems for Video Technology, 15*(7), 921-927.

Kondi, L.P., Meier, F. W., Schuster, G. M., & Katsaggelos, A. K. (1998) Joint optimal object shape estimation and encoding. *SPIE Visual Communication and Image Processing*, San Jose, California, USA, January (pp. 14-25).

Kondi, L. P., Melnikov, G., & Katsaggelos, A. K. (2001). Jointly optimal coding of texture and shape. *International Conference on Image Processing (ICIP), 3,* Thessaloniki, Greece, October 7-10 (pp. 94-97).

Kondi, L. P., Melnikov, G., & Katsaggelos, A. K. (2004). Joint optimal object shape estimation and encoding. *IEEE Transactions on Circuits and Systems for Video Technology, 14*(4), 528-533.

Kunt, M., Ikonomopoloulos, A., & Kocher, R. (1985). Second generation image coding techniques. *Proceedings of the IEEE, 73*(4), 549-574.

Lee, S., Cho, D., Cho, Y., Son, S., Jang, E., Shin, J., & Seo, Y. (1999). Binary shape coding using baseline-based method. *IEEE Transactions on Circuits and Systems for Video Technology, 9*(1), 44-58.

Li, H., & Forchheimer, R. (1994). Two-view facial movement estimation. *IEEE Transactions on Circuits and Systems for Video Technology, 4*(3), 276-287.

Li, J., & Lei, S. (1997). Rate-distortion optimized embedding. In *Proc. Picture Coding Symposium,* Berlin, Germany, September (pp. 201-206).

Li, S., & Li, W. (1995). Shape-adaptive discrete wavelet transforms for arbitrarily shaped visual object coding. *IEEE Transactions on Circuits and Systems for Video Technology, 10*(5), 725-743.

Li, W., & Salari, W. (1995). Successive elimination algorithm for motion estimation. *IEEE Transactions on Image Processing, 4*(1), 105-107.

Linde, Y., Buzo, A., & Gary, R. M. (1980). An algorithm for vector quantization. *IEEE Transactions on Communication, 28*(1), 84-95.

Lynn, L. H., Aram, J. D., Reddy, N. M., & Ostermann, J. (1997). Methodologies used for evaluation of video tools and algorithms in MPEG-4. *Signal Processing: Image Communication, 9*(4), 343-365.

Man, H., de Queiroz, R., & Smith, M. (2002). Three-dimensional subband coding techniques for wireless video communications. *IEEE Transactions on Circuits and Systems for Video Technology, 12*(3), 386-397.

Meier, F. W., Schuster, G. M., & Katsaggelos, A. K. (2000). A mathematical model for shape coding with B-splines. *Signal Processing: Image Communications, 15*(7-8), 685-701.

Nanda, S., & Pearlman, W. S. (1992). Tree coding of image subbands. *IEEE Transactions on Signal Processing, 1*(2), 133-147.

Narula, A., & Lim, J. S. (1993). Error concealment techniques for an all-digital high-definition television system. In *Proc. SPIE Conf. Visual Commun. and Image Proc.,* Chicago, IL, May (pp. 304-315).

Ngan, K. N., & Chooi, W. L. (1994). Very low bit rate video coding using 3D subband approach. *IEEE Transactions on Circuits and Systems for Video Technology, 4*(3), 309-316.

Øien, G. E. (1993). L_2-optimal attractor image coding with fast decoder convergence. PhD thesis. Trondheim, Norway.

O'Connell, K. J. (1997). Object-adaptive vertex-based shape coding method. *IEEE Transactions on Circuits and Systems for Video Technology, 7*(1), 251-255.

Ordentlich, E., Weinberger, M., & Seroussi, G. (1998). A low-complexity modeling approach for embeddded coding of wavelet coefficients. *IEEE Data Compression Conference (DCC),* Snowbird, Utah, March 30-April 1 (pp. 408-417).

Pearson, D. E. (1995). Developments in model-based video coding. *Proceedings of the IEEE, 83*(6), 892-906.

Podilchuk, C., Jayant, N., & Farvardin, N. (1995). Three dimensional subband coding of video. *IEEE Transactions on Image Processing, 4*(2), 125-39.

Puri, R., & Ramchandran, K. (2002). PRISM: A new robust video coding architecture based on distributed compression principles. *Allerton Conference on Communication, Control, and Computing,* Allerton, IL, October.

Rabbani, M., & Jones, P. W. (1991). *Digital image compression techniques.* Bellingham, Washington: SPIE Optical Engineering Press.

Redmill, D. W. (1994). *Image and video coding for noisy channels.* PhD thesis. University of Cambridge. Signal Processing and Communications Laboratory.

Redmill, D. W., & Kingsbury, N. G. (1996). The EREC: An error resilient technique for coding variable-length blocks of data. *IEEE Transactions on Image Processing, 5*(4), 565-574.

Richardson, I. E. (2003). *H.264 and MPEG-4 video compression.* Chichester: John Wiley & Sons.

Said, A., & Pearlman, W. (1996). A new, fast, and efficient image codec based on set partitioning in hierarchical trees. *IEEE Transactions on Circuits and Systems for Video Technology, 6*(3), 243-250.

Salama, P., Shroff, N. B., Coyle, E. J., & Delp, E. J. (1995). Error concealment techniques for encoded video streams. *IEEE International Conference on Image Processing (ICIP),* Washington, DC, October 23-26 (pp. 9-12).

Schuster, G. M., & Katsaggelos, A. K. (1997). *Rate-distortion based video compression: Optimal video frame compression and object boundary encoding.* Boston: Kluwer Academic Publishers.

Schuster, G. M., & Katsaggelos, A. K. (1998). An optimal boundary encoding scheme in the rate distortion sense. *IEEE Transactions on Image Processing, 7*(1), 13-26.

Schuster, G. M., & Katsaggelos, A. K. (2006). Motion compensated shape error concealment. *IEEE Transactions on Image Processing, 15*(2), 501-510.

Schuster, G. M., Katsaggelos, A. K., & Xiaohuan, L. (2004). Shape error concealment using Hermite splines. *IEEE Transactions on Image Processing, 13*(6), 808-820.

Shapiro, J. M. (1993). Embedded image coding using zerotrees of wavelet coefficients. *IEEE*

Transactions on Signal Processing, 41(12), 3445-3462.

Shirani, S., Kossentini, F., & Ward, R. (2000). A concealment method for video communications in an error-prone environment. *IEEE Journal on Selected Areas in Communications, 18*(6), 1122-1128.

Sikora, T., Bauer, S., & Makai, B. (1995). Efficiency of shape adaptive transforms for coding of arbitrarily shaped image segments. *IEEE Transactions on Circuits and Systems for Video Technology, 5*(3), 254-258.

Sikora, T., & Makai, B. (1995). Shape-adaptive DCT for generic coding of video. *IEEE Transactions on Circuits and Systems for Video Technology, 5*(3), 59-62.

Slepian, J. D., & Wolf, J. K. (1973). Noiseless coding of correlated information sources. *IEEE Transactions on Information Theory, IT-19*, 471-480.

Soares, L. D., & Pereira, F. (2004). Spatial shape error concealment for object-based image and video coding. *IEEE Transactions on Image Processing, 13*(4), 586-599.

Soares, L. D., & Pereira, F. (2006). Temporal shape error concealment by global motion compensation with local refinement. *IEEE Transactions on Image Processing, 15*(6), 1331-1348.

Sohel, F. A., Dooley, L. S., & Karmakar, G. C. (2005a). A dynamic Bezier curve model. *International Conference on Image Processing (ICIP), II*, Genoa, Italy, September (pp. 474-477).

Sohel, F. A., Dooley, L. S., & Karmakar, G. C. (2005b). *A novel half-way shifting Bezier curve model*. IEEE Region 10 Conference (Tencon), Melbourne, Australia, November.

Sohel, F. A., Dooley, L. S., & Karmakar, G. C. (2006a). Accurate distortion measurement for generic shape coding. *Pattern Recognition Letters, 27*(2), 133-142.

Sohel, F. A., Dooley, L. S., & Karmakar, G. C. (2006b). *Variable width admissible control point band for vertex based operational-rate-distortion optimal shape coding algorithms*. International Conference on Image Processing (ICIP), Atlanta, GA, October.

Sohel, F. A., Dooley, L. S., & Karmakar, G. C. (2007). New dynamic enhancements to the vertex-based rate-distortion optimal shape coding framework. *IEEE Transactions on Circuits and Systems for Video Technology, 7*(10).

Sohel, F. A., Karmakar, G. C., & Dooley, L. S. (2005). An improved shape descriptor using Bezier curves. *First International Conference on Pattern Recognition and Machine Intelligence (PReMI). Lecture Notes in Computer Science, 3776*, Kolkata, India, December (pp. 401-406).

Sohel, F. A., Karmakar, G. C., & Dooley, L. S. (2006). *Dynamic sliding window width selection strategies for rate-distortion optimal vertex-based shape coding algorithms*. International Conference on Signal Processing (ICSP), Guilin, China, November 16-20.

Sohel, F. A., Karmakar, G. C., & Dooley, L. S. (2007a). Fast distortion measurement using chord-length parameterisation within the vertex-based rate-distortion optimal shape coding framework. *IEEE Signal Processing Letters, 14*(2), 121-124.

Sohel, F. A., Karmakar, G. C., & Dooley, L. S. (2007b). Spatial shape error concealment utilising image-texture. *IEEE Transactions on Image Processing* (revision submitted).

Sohel, F. A., Karmakar, G. C., & Dooley, L. S. (2007c). *Bezier curve-based character descriptor considering shape information*. IEEE/ACIS International Conference on Computer and Information Science (ICIS), Melbourne, Australia, July.

Sohel, F. A., Karmakar, G. C., Dooley, L. S., & Arkinstall, J. (2005). Enhanced Bezier curve models incorporating local information. *IEEE International Conference on Acoustics, Speech, and Signal Processing (ICASSP), IV,* Philadelphia, PA, March 18-23 (pp. 253-256).

Sohel, F. A., Karmakar, G. C., Dooley, L. S. & Arkinstall, J. (2007). Quasi Bezier curves integrating localised information. *Pattern Recognition* (in press).

Sun, Y., Ahmad, I., Li, D., & Zhang, Y.-Q. (2006). Region-based rate control and bit allocation for wireless video transmission. *IEEE Transactions on Multimedia, 8*(1), 1-10.

Sun, S., Haynor, D., & Kim, Y. (2003). Semiautomatic video object segmentation using v-snakes. *IEEE Transactions on Circuits and Systems for Video Technology, 13*(1), 75-82.

Tang, C.-W., Chen, C.-H., Yu, Y.-H., & Tsai, C.-J. (2006). Visual sensitivity guided bit allocation for video coding. *IEEE Transactions on Multimedia, 8*(1), 11-18.

Tao, H., Sawhney, H. S., & Kumar, R. (2002). Object tracking with Bayesian estimation of dynamic layer representations. *IEEE Transactions on Pattern Analysis and Machine Intelligence, 24*(1), 75-89.

Taubman, D. (2000). High performance scalable image compression with EBCOT. *IEEE Transactions on Image Processing, 9*(7), 1158-1170.

Taubman, D. S., & Marcellin, M. W. (2002). *JPEG2000: Image compression fundamentals, standards and practice.* Boston: Kluwer Academic Publishers.

Taubman, D. S., & Zakhor, A. (1994). Multirate 3-D subband coding of video. *IEEE Transactions on Image Processing, 3*(4), 572-88.

Tekalp, A. M. (1995). *Digital video processing.* Prentice Hall Signal Processing Series. Prentice Hall, Englewood Cliffs: NJ.

Tham, J. Y., Ranganath, S., Ranganath, M., & Kassim, A. A. (1998). A novel unrestricted center-biased diamond search algorithm for block motion estimation. *IEEE Transactions on Circuits and Systems for Video Technology, 8*(4), 369-377.

Toivonen, T., & Heikkilä, J. (2004). Fast full search block motion estimation for H.264/avc with multilevel successive elimination algorithm. In *Proc. International Conference on Image Processing (ICIP), 3,* Singapore, October (pp. 1485-1488).

Topiwala, P. N. (1998). *Wavelet image and video compression.* Boston: Kluwer Academic Publishers.

Vaishampayan, V. A. (1993). Design of multiple description scalar quantizers. *IEEE Transaction on Information Theory, 39*(3), 821-834.

VQEG. (1998). Final report from the Video Quality Expert Group on the validation of objective models of video quality assessment.

Wade, N., & Swanston, M. (2001). *Visual perception: An introduction (2nd ed.).* London: Psychology Press.

Wallace, G. K. (1991). The JPEG still picture compression standard. *Communications of the ACM, 34*(4), 30-44.

Wang, H., Schuster, G. M., & Katsaggelos, A. K. (2005). Rate-distortion optimal bit allocation for object-based video coding. *IEEE Transactions on Circuits and Systems for Video Technology, 15*(9), 113-1123.

Wang, Y., & Zhu, Q. (1998). Error control and concealment for video communication: A review. *Proceedings of the IEEE, 86*(5), 974-997.

Welch, T. A. (1984). A technique for high performance data compression. *IEEE Computer, 17*(6), 8-19.

Wen, J., & Villasenor, J. D. (1997). A class of reversible variable length codes for robust image and video coding. In *Proc. IEEE International Conference on Image Processing (ICIP), 2,* Washington, DC, October (pp. 65-68).

Witten, I., Neal, R., & Cleary, J. (1987). Arithmetic coding for data compression. *Communication of the ACM, 30*(6), 520-540.

Woods, J., & O'Neil, S. (1986). Subband coding of images. *IEEE Trans. Acoustics, Speech, and Signal Processing, 34*(5), 1278-1288.

Wu, H. R., & Rao, K. R. (2006). *Digital video image quality and perceptual coding.* Boca Raton, FL: CRC Press: Taylor and Francis.

Wyner, A. D., & Ziv, J. (1976). The rate-distortion function for source coding with side information at the decoder. *IEEE Transactions on Information Theory, IT-22*(1), 1-10.

Yamaguchi, N., Ida, T., & Watanabe, T. (1997). A binary shape coding method using modified MMR. In *Proc. Special Session on Shape Coding (ICIP97), I,* Washington, DC, October (pp. 504-508).

Zhu, Q.-F., Wang, Y., & Shaw, L. (1993). Coding and cell-loss recovery in DCT-based packet video. *IEEE Transactions on Circuits and Systems for Video Technology, 3*(3), 238-247.

Ziv, J., & Lempel, A. (1977). A universal algorithm for sequential data compression. *IEEE Transactions on Information Theory, IT-23*(3), 337-343.

ENDNOTES

[1] IMSI's Master Photo Collection, 1895 Francisco Blvd. East, San Rafael, CA 94901-5506, USA.

[2] Cut is defined as a visual transition created in editing in which one shot is instantaneously replaced on screen by another.

Chapter VIII
OFDM Transmission Technique:
A Strong Candidate for the Next Generation Mobile Communications

Hermann Rohling
Hamburg University of Technology, Germany

ABSTRACT

The orthogonal frequency division multiplexing (OFDM) transmission technique can efficiently deal with multi-path propagation effects especially in broadband radio channels. It also has a high degree of system flexibility in multiple access schemes by combining the conventional TDMA, FDMA, and CDMA approaches with the OFDM modulation procedure, which is especially important in the uplink of a multi-user system. In OFDM-FDMA schemes carrier synchronization and the resulting sub-carrier orthogonality plays an important role to avoid any multiple access interferences (MAI) in the base station receiver. An additional technical challenge in system design is the required amplifier linearity to avoid any non-linear effects caused by a large peak-to-average ratio (PAR) of an OFDM signal. The OFDM transmission technique is used for the time being in some broadcast applications (DVB-T, DAB, DRM) and wireless local loop (WLL) standards (HIPERLAN/2, IEEE 802.11a) but OFDM has not been used so far in cellular communication networks. The general idea of the OFDM scheme is to split the total bandwidth into many narrowband sub-channels which are equidistantly distributed on the frequency axis. The sub-channel spectra overlap each other but the sub-carriers are still orthogonal in the receiver and can therefore be separated by a Fourier transformation. The system flexibility and use of sub-carrier specific adaptive modulation schemes in frequency selective radio channels are some advantages which make the OFDM transmission technique a strong and technically attractive candidate for the next generation of mobile communications. The objective of this chapter is to describe an OFDM-based system concept for the fourth generation (4G) of mobile communications and to discuss all technical details when establishing a cellular network which requires synchronization in time and frequency domain with sufficient accuracy. In this cellular environment a flexible frequency division multiple access scheme based on OFDM-FDMA is developed and a radio resource management (RRM) employing dynamic channel allocation (DCA) techniques is used. A purely decentralized and self-organized synchronization technique using specific test signals and RRM techniques based on co-channel interference (CCI) measurements has been developed and will be described in this chapter.

INTRODUCTION

In the evolution of mobile communication systems approximately a 10-year periodicity can be observed between consecutive system generations. Research work for the current 2nd generation of mobile communication systems (GSM) started in Europe in the early 1980s, and the complete system was ready for market in 1990. At that time the first research activities had already been started for the 3rd generation (3G) of mobile communication systems (UMTS, IMT-2000) and the transition from the current second generation (GSM) to the new 3G systems will be observed this year. Compared to today's GSM networks, these new UMTS systems will provide much higher data rates, typically in the range of 64 to 384 kbps, while the peak data rate for low mobility or indoor applications will be 2 Mbps.

The current pace, which can be observed in the mobile communications market, already shows that the 3G systems will not be the ultimate system solution. Consequently, general requirements for a 4G system have to be considered which will mainly be derived from the types of service a user will require in future applications. Generally, it is expected that data services instead of pure voice services will play a predominant role, in particular due to a demand for mobile IP applications. Variable and especially high data rates (20 Mbps and more) will be requested, which should also be available at high mobility in general or high vehicle speeds in particular (see Figure 1). Moreover, asymmetrical data services between up- and downlink are assumed and should be supported by 4G systems in such a scenario where the downlink carries most of the traffic and needs the higher data rate compared with the uplink.

Figure 1. General requirements for 4G mobile communication systems

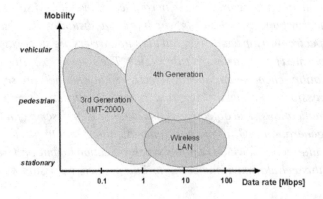

To fulfill all these detailed system requirements the OFDM transmission technique applied in a wide-band radio channel is a strong candidate for an air interface in future 4G cellular systems due to its flexibility and adaptivity in the technical system design. From these considerations, it already becomes apparent that a radio transmission system for 4G must provide a great flexibility and adaptivity at different levels, ranging from the highest layer (requirements of the application) to the lowest layer (the transmission medium, the physical layer, that is, the radio channel) in the ISO-OSI stack. Today, the OFDM transmission technique is in a completely matured stage to be applied for wide-band communication systems integrated into a cellular mobile communications environment.

OFDM TRANSMISSION TECHNIQUE

Radio Channel Behaviour

The mobile communication system design is in general always dominated by the radio channel behaviour (Bello, 1963; Pätzold, 2002). In typical radio channel situation, multi-path propagation occurs (Figure 2) due to the reflections of the transmitted signal at several objects and obstacles inside the local environment and inside the observation area. The radio channel is analytically described unambiguously by a linear (quasi) time invariant (LTI) system model and by the related channel impulse response $h(\tau)$ or alternatively by the channel transfer function $H(f)$. An example for these channel characteristics is shown in Figure 3, where $h(\tau)$ and $H(f)$ of a so-called wide-sense stationary, uncorrelated scattering-channel (WS-SUS) are given.

Due to the mobility of the mobile terminals the multi-path propagation situation will be continuously but slowly changed over time which is described analytically by a time variant channel impulse response $h(\tau, t)$ or alternatively by a frequency selective and time dependent radio channel transfer function $H(f, t)$ as it is shown in Figure 5 by an example. All signals on the various propagation paths will be received in a superimposed form and are technically characterized by different delays and individual Doppler frequencies which

Figure 2. Multi-path propagation scenario

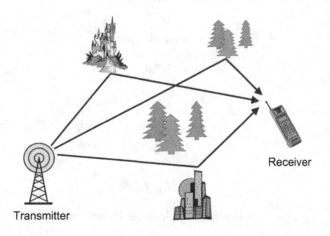

Transmitter

Receiver

Figure 3. Impulse response and channel transfer function of a WS-SUS channel

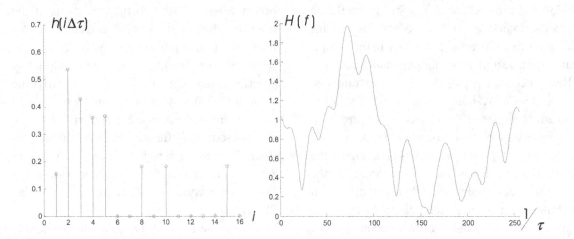

lead finally to a frequency selective behaviour of the radio channel, see Figure 4. The other two system functions, the Delay Doppler function, $v(\tau, f_D)$ and the Frequency Doppler function, $U(f, f_D)$ can be used as an alternative description of the radio channel behaviour. The Delay Doppler function $v(\tau, f_D)$ describes the variation of

the channel impulse response related to certain values of the Doppler frequency f_D. This means the channel delays change due to alteration of the relative speed between a mobile terminal and the base station. The Frequency Doppler function $U(f, f_D)$ models the same effects for the channel behaviour in the frequency domain.

Figure 4. Relationships between different system functions

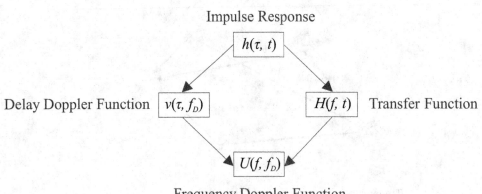

The radio channel can roughly and briefly be characterised by two important system parameters: the maximum multi-path delay τ_{max} and the maximum Doppler frequency f_{Dmax} which are transferred into the coherence time T_C and the coherence bandwidth B_C of the radio channel:

$$T_C = \frac{1}{f_{Dmax}}, \quad B_C = \frac{1}{\tau_{max}} \qquad (1)$$

Over time intervals significantly shorter than T_C, the channel transfer function can be assumed to be nearly stationary. Similarly, for frequency intervals significantly smaller than B_C, the channel transfer function can be considered as nearly constant. Therefore it is assumed in this chapter that the coherence time T_C is much larger compared to a single OFDM symbol duration T_S and the coherence bandwidth B_C is much larger than the distance Δf between two adjacent sub-carriers:

$$B_C \gg \Delta f, \quad \Delta f = \frac{1}{T_S}, \quad T_S \ll T_C \qquad (2)$$

This condition should be always fulfilled in well-dimensioned OFDM systems and in realistic time variant and frequency selective radio channels.

There are always technical alternatives possible in new system design phases. But future mobile communication systems will in any case require extremely large data rates and therefore large system bandwidth. If conventional single carrier (SC) modulation schemes with the resulting very low symbol durations are applied in this system design, it will be observed that very strong inter-symbol interference (ISI) is caused in wide-band applications due to multi-path propagation situations. This means for high data rate applications the symbol duration in a classical SC transmission system is extremely small compared to the typical values of maximum multi-path delay τ_{max} in the considered radio channel. In these strong ISI situations a very powerful equalizer is necessary in each receiver, which needs high computation complexity in a wide-band system. These constraints should be taken into consideration in the system development phase for a new

Figure 5. Frequency-selective and time-variant radio channel transfer function

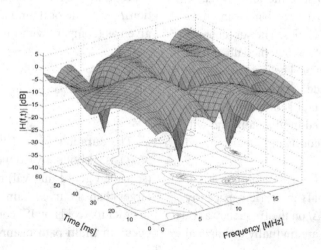

radio transmission scheme and for a new 4G air interface. The computation complexity for the necessary equalizer techniques to overcome all these strong ISI in a SC modulation scheme increases exponentially for a given radio channel with increasing system bandwidth and can be extremely large in wide-band applications. For that reason alternative transmission techniques for broadband applications are of high interest.

Alternatively, the OFDM transmission technique can efficiently deal with all these ISI effects, which occur in multi-path propagation situations and in broadband radio channels. Simultaneously the OFDM transmission technique needs much less computation complexity in the equalization process inside each receiver. The performance figures for an OFDM based new air interface for the next generation of mobile communications are very promising even in frequency selective and time variant radio channel situations.

Advantages of the OFDM Transmission Technique

If a high data rate is transmitted over a frequency selective radio channel with a large maximum multi-path propagation delay τ_{max} compared to the symbol duration, an alternative to the classical SC approach is given by the OFDM transmission technique. The general idea of the OFDM transmission technique is to split the total available bandwidth B into many narrowband sub-channels at equidistant frequencies. The sub-channel spectra overlap each other but the sub-carrier signals are still orthogonal. The single high-rate data stream is subdivided into many low-rate data streams in the several sub-channels. Each sub-channel is modulated individually and will be transmitted simultaneously in a superimposed and parallel form.

An OFDM transmit signal therefore consists of N adjacent and orthogonal sub-carriers spaced by the frequency distance Δf on the frequency axis. All sub-carrier signals are mutually orthogonal

within the symbol duration of length T_S if the sub-carrier distance and the symbol duration are chosen such that $T_S = 1 / \Delta f$. For OFDM-based systems the symbol duration T_S is much larger compared to the maximum multi-path delay τ_{max}. The k-th unmodulated sub-carrier signal is described analytically by a complex valued exponential function with carrier frequency $k\Delta f$, $\tilde{g}_k(t), \ k = 0, \ldots, N-1$.

$$\tilde{g}_k(t) = \begin{cases} e^{j2\pi k\Delta ft} & \forall t \in [0, T_S] \\ 0 & \forall t \notin [0, T_S] \end{cases} \tag{3}$$

Since the system bandwidth B is subdivided into N narrowband sub-channels, the OFDM symbol duration T_S is N times larger as in the case of an alternative SC transmission system covering the same bandwidth B. Typically, for a given system bandwidth the number of sub-carriers is chosen in a way that the symbol duration T_S is sufficiently large compared to the maximum multi-path delay τ_{max} of the radio channel. On the other hand, in a time-variant radio channel the Doppler spread imposes restrictions on the sub-carrier spacing Δf. In order to keep the resulting inter-carrier interference (ICI) at a tolerable level, the system parameter of sub-carrier spacing Δf must be large enough compared to the maximum Doppler frequency f_{Dmax}. In Aldinger (1994), the appropriate range for choosing the symbol duration T_S as a rule of thumb in practical systems is given as (compare with Equation (2)):

$$4\tau_{max} \le T_S \le 0.03 \frac{1}{f_{D,max}}. \tag{4}$$

The duration T_S as of the sub-carrier signal $\tilde{g}_k(t)$ is additionally extended by a cyclic prefix (so-called guard interval) of length T_G which is larger than the maximum multi-path delay τ_{max} in order to avoid any ISI completely which could occur in multi-path channels in the transition interval between two adjacent OFDM symbols (Peled & Ruiz, 1980).

$$g_k(t) = \begin{cases} e^{j2\pi k\Delta ft} & \forall t \in [-T_G, T_S] \\ 0 & \forall t \notin [-T_G, T_S] \end{cases} \qquad (5)$$

$$= e^{j2\pi k\Delta ft} rect\left(\frac{2t+(T_G-T_S)}{2T}\right)$$

The guard interval is directly removed in the receiver after the time synchronization procedure. From this point of view the guard interval is a pure system overhead and the total OFDM symbol duration is therefore $T = T_S + T_G$. It is an important advantage of the OFDM transmission technique that ISI can be avoided completely or can be reduced at least considerably by a proper choice of OFDM system parameters.

The orthogonality of all sub-carrier signals is completely preserved in the receiver even in frequency selective radio channels which is an important advantage of the OFDM transmission technique. The radio channel behaves linear and in a short-time interval of a few OFDM symbols even time invariant. Therefore the radio channel behaviour can be described completely by a linear and time invariant (LTI) system model characterized by the impulse response $h(t)$. The LTI system theory gives the reason for this important system behaviour that all sub-carrier signals are orthogonal in the receiver even when transmitting the signal in frequency selective radio channels. All complex valued exponential signals (e.g., all sub-carrier signals) are Eigenfunctions of each LTI system and therefore Eigenfunctions of the considered radio channel which means that only the signal amplitude and phase will be changed if a sub-carrier signal is transmitted in the linear and time invariant radio channel.

The sub-carrier frequency is not affected at all by the radio channel transmission which means that all sub-carrier signals are even orthogonal in

the receiver and at the output of a frequency selective radio channel. The radio channel interferes only amplitudes and phases individually but not the sub-carrier frequency of all received sub-channel signals. Therefore all sub-carrier signals are still mutually orthogonal in the receiver. Due to this important property the received signal which is superimposed by all sub-carrier signals can be split directly into the different sub-channel components by a Fourier transformation and each sub-carrier signal can be demodulated individually by a single tap equalizer in the receiver.

At the transmitter side each sub-carrier signal is modulated independently and individually by the complex valued modulation symbol $S_{n,k}$, where the subscript n refers to the time interval and k to the sub-carrier signal number in the considered OFDM symbol. Thus, within the symbol duration time interval T the time continuous signal of the n-th OFDM symbol is formed by a superposition of all N simultaneously modulated sub-carrier signals.

$$s_n(t) = \sum_{k=0}^{N-1} S_{n,k} g_k(t-nT) \qquad (6)$$

The total time continuous transmit signal consisting of all OFDM symbols sequentially transmitted on the time axis is described analytically by the following equation:

$$s(t) = \sum_{n=0}^{\infty} \sum_{k=0}^{N-1} S_{n,k} e^{j2\pi k\Delta f(t-nT)} rect\left(\frac{2(t-nT)+(T_G-T_S)}{2T}\right)$$

$$(7)$$

The analytical transmit signal description shows that a rectangular pulse shaping is applied for each sub-carrier signal and each OFDM symbol. But due to the rectangular pulse shaping, the spectra of all the considered sub-carrier signals are sinc-functions which are equidistantly located on the frequency axis, for example, for the k-th sub-carrier signal the spectrum is described in the following equation:

$$G_k(f) = T \cdot \mathrm{sinc}\left[\pi T\left(f - k\Delta f\right)\right] \quad \text{where}$$

$$\mathrm{sinc}(x) = \frac{\sin(x)}{x} \tag{8}$$

The typical OFDM-Spectrum shown in Figure 6 consists of N adjacent sinc-functions, which are shifted by Δf in the frequency direction.

The spectra of the considered sub-carrier signals overlap on the frequency axis, but the sub-carrier signals are still mutually orthogonal which means the transmitted modulation symbols $S_{n,k}$ can be recovered by a simple correlation technique in each receiver if the radio channel is assumed to be ideal in a first analytical step:

$$\frac{1}{T_s}\int_0^{T_s} g_k(t)\overline{g_l(t)}\,dt = \begin{cases} 1 & k = l \\ 0 & k \neq l \end{cases} = \delta_{k,l} \tag{9}$$

$$S_{n,k} = \frac{1}{T_S}\int_0^{T_s} s_n(t)\overline{g_k(t)}\,dt = \frac{1}{T_S}\int_0^{T_s} s_n(t)e^{-j2\pi k\Delta f t}\,dt \tag{10}$$

where $\overline{g_k(t)}$ is the conjugate complex version of the sub-carrier signal $g_k(t)$. The following equations show the correlation process in detail:

$$Corr = \frac{1}{T_S}\int_0^{T_s} s_n(t)\overline{g_k(t)}\,dt = \frac{1}{T_S}\int_0^{T_s}\sum_{m=0}^{N-1} S_{n,m}g_m(t)\overline{g_k(t)}\,dt$$

$$= \sum_{m=0}^{N-1} S_{n,m}\frac{1}{T_S}\int_0^{T_s} g_m(t)\overline{g_k(t)}\,dt = \sum_{m=0}^{N-1} S_{n,m}\delta_{m,k} = \underline{\underline{S_{n,k}}} \tag{11}$$

In practical applications the OFDM transmit signal $s_n(t)$ is generated in a first step and in the digital baseband signal processing part of the transmitter as a time discrete signal. Using the

sampling theorem while considering the OFDM transmit signal inside the bandwidth $B = N\Delta f$, the transmit signal must be sampled with the sampling interval $\Delta t = 1/B = 1/N\Delta f$. The individual samples of the transmit signal are denoted by $s_{n,i}$, $i = 0, 1, \ldots, N-1$ and can be calculated as follows (see Equation (7)):

$$s(t) = \sum_{k=0}^{N-1} S_{n,k}e^{j2\pi k\Delta f t}$$

$$s(i\Delta t) = \sum_{k=0}^{N-1} S_{n,k}e^{j2\pi k\Delta f(i\Delta t)}$$

$$s_{n,i} = \sum_{k=0}^{N-1} S_{n,k}e^{j2\pi ik/N} \tag{12}$$

This Equation (12) describes exactly the inverse discrete Fourier transform (IDFT) applied to the complex valued modulation symbols $S_{n,k}$ of all sub-carrier signals inside a single OFDM symbol.

The individually modulated and superimposed sub-carrier signals are transmitted in a parallel way over many narrowband sub-channels. Thus, in each sub-channel the symbol duration is quite large and can be chosen much larger as compared to the maximum multi-path delay of the radio channel. In this case each sub-channel has the property to be frequency non-selective.

Figure 7 shows the general OFDM system structure in a block diagram. The basic principles of the OFDM transmission technique have already been described in several publications like Bingham (1990) and Weinstein and Ebert (1971). In the very early and classical multi-carrier system considerations like Chang (1966) and Saltzberg (1967), narrowband signals have been generated independently, assigned to various frequency bands, transmitted, and separated by analogue filters at the receiver. The new and modern aspect of the OFDM transmission technique is that the various sub-carrier signals are generated digitally and jointly by an IFFT in the transmitter and that

Figure 6. OFDM spectrum which consists of N equidistant sinc-functions

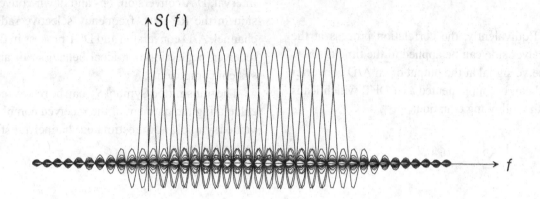

Figure 7. OFDM system structure in a block diagram

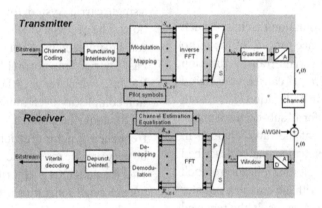

their spectra strongly overlap on the frequency axis. As a result, generating the transmit signal is simplified and the bandwidth efficiency of the system is significantly improved.

The received signal is represented by the convolution of the transmitted time signal with the channel impulse response $h(t)$ and an additive white Gaussian noise term:

$$r_n(t) = s_n(t) * h_n(t) + n_n(t) \tag{13}$$

Due to the assumption, that the coherence time T_C will be much larger than the symbol duration T_S the received time continuous signal $r_n(t)$ can be separated into the orthogonal sub-carrier signal components even in frequency selective fading situations by applying the correlation technique mentioned in Equation (10):

$$R_{n,k} = \frac{1}{T_S} \int_0^{T_S} r_n(t) e^{-j2\pi k \Delta ft} dt \qquad (14)$$

Equivalently, the correlation process at the receiver side can be applied to the time discrete receive signal at the output of an A/D converter and can be implemented as a DFT, which leads to the following equation:

$$R_{n,k} = \frac{1}{N} \sum_{i=0}^{N-1} r_{n,i} e^{-j2\pi ik/N} \qquad (15)$$

In this case $r_{n,i} = r_n(i \cdot \Delta t)$ describes the i-th sample of the received time continuous baseband signal $r_n(t)$ and $R_{n,k}$ is the received complex valued symbol at the DFT output of the k-th sub-carrier.

If the OFDM symbol duration T is chosen much smaller than the coherence time T_C of the radio channel, then the time variant transfer function of the radio channel $H(f, t)$ can be considered constant within the time duration T of each modulation symbol $S_{n,k}$ for all sub-carrier signals. In this case, the effect of the radio channel in multi-path propagation situations can be described analytically by only a single multiplication of each sub-carrier signal $g_k(t)$ with the complex transfer factor $H_{n,k} = H(k\Delta f, nT)$. As a result, the received complex valued symbol $R_{n,k}$ at the DFT output can be described analytically as follows:

$$r_n(t) = s_n(t) * h_n(t) + n_n(t)$$
$$r_{n,i} = s_{n,i} * h_{n,i} + n_{n,i}$$

$$R_{n,k} = S_{n,k} H_{n,k} + N_{n,k} \qquad (16)$$

where $N_{n,k}$ describes an additive noise component for each specific sub-carrier generated in the radio channel. This equation shows the most important advantage of applying the OFDM transmission technique in practical applications. Equation (16)

describes the complete signal transfer situation of the OFDM block diagram including IDFT, guard interval, D/A conversion, up- and down-conversion in the RF part, frequency selective radio channel, A/D conversion and DFT process in the receiver, neglecting non-ideal behaviour of any system components.

The transmitted Symbol $S_{n,k}$ can be recovered, calculating the quotient of the received complex valued symbol and the estimated channel transfer factor $\tilde{H}_{n,k}$:

$$S_{n,k} = \frac{R_{n,k} - N_{n,k}}{H_{n,k}}, \quad \widetilde{S}_{n,k} = R_{n,k} \frac{1}{\tilde{H}_{n,k}} \qquad (17)$$

It is obvious that this one tap equalization step of the received signal is much easier compared to a single carrier system for high data rate applications. The necessary IDFT and DFT calculations can be implemented very efficiently using the Fast-Fourier-Transform (FFT) algorithms such as Radix 2^2, which reduces the system and computation complexity even more.

It should be pointed out that especially the frequency synchronization at the receiver must be very precise in order to avoid any inter-carrier interferences (ICI). Algorithms for time and frequency synchronization in OFDM-based systems are described in Classen and Meyr (1994) and Mizoguchi et al. (1998), for example and will be considered in the section, *Self-Organized Cell Synchronization*.

Besides the complexity aspects, another advantage of the OFDM technique lies in its high degree of flexibility and adaptivity. Division of the available bandwidth into many frequency-non-selective sub-bands gives additional advantages for the OFDM transmission technique. It allows a sub-carrier-specific adaptation of transmit parameters, such as modulation scheme (PHY mode) and transmit power (cp. Water Filling) in accordance to the observed and measured radio channel status. In a multi-user environment the

OFDM structure offers additionally an increased flexibility for resource allocation procedures as compared to SC systems (Hanzo et al., 2003).

The important system behaviour that all sub-carrier signals are mutually orthogonal in the receiver makes the signal processing and the equalization process realized by a single-tap procedure very simple and leads to a low computation complexity.

OFDM COMBINED WITH MULTIPLE ACCESS SCHEMES

A very high degree of flexibility and adaptivity is required for new mobile communication systems and for the 4G air interface. The combination between multiple access schemes and OFDM transmission technique is an important factor in this respect. In principle, multiple access schemes for the OFDM transmission technique can be categorized according to OFDM-FDMA, OFDM-TDMA, and OFDM-CDMA (Kaiser, 1998; Rohling & Grünheid, 1997). Clearly, hybrid schemes can be applied which are based on a combination of these techniques. The principles of these basic multiple access schemes are summarized in Figure 8, where the time-frequency plane is depicted and the user specific resource allocation is distinguished by different colours.

These access schemes provide a great variety of possibilities for a flexible user specific resource allocation. In the following, one example for OFDM-FDMA is briefly sketched (cf. Galda, Rohling, Costa, Haas, & Schulz, 2002). In the case that the magnitude of the channel transfer function is known for each user the sub-carrier selection for an OFDM-FDMA scheme can be processed in the BS for each user individually which leads to a multi-user diversity (MUD) effect. By allocating a subset of all sub-carriers with the highest SNR to each user the system performance can be improved. This allocation technique based on the knowledge of the channel transfer function shows a large performance advantage and a gain in quality of service (QoS). Nearly the same flexibility in resource allocation is possible in OFDM-CDMA systems. But in

Figure 8. OFDM transmission technique and some multiple access schemes

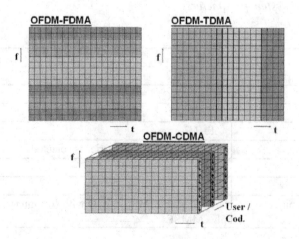

this case the code orthogonality is destroyed by the frequency selective radio channel resulting in multiple access interferences (MAI), which reduces the system performance.

TECHNICAL PROPOSAL AND EXAMPLE FOR A 4G DOWNLINK INTERFACE

Taking all these important results from the previous sections into consideration, a system design example is considered in this section. OFDM system parameters for a 4G air interface are considered and three different multiple access schemes inside a single cell are compared quantitatively. A bandwidth of 20 MHz in the 5.5 GHz domain is assumed. The assumed multi-path radio channel has a maximum delay of $\tau_{max} = 5$ µs (the coherence bandwidth is therefore $B_C = 200$ kHz). Additionally, a maximum speed of $v_{max} = 200$ km/h is assumed, which yields a maximum Doppler frequency of $f_{Dmax} = 1$ kHz and a coherence time of $T_C = 1$ ms. Table 1 shows an example for the system parameters of a 4G air interface.

For the considered OFDM based system three different multiple access concepts have been analysed and compared. The first proposal is based on a pure OFDM-TDMA structure, while the second one considers an OFDM-FDMA technique with an adaptive sub-carrier selection scheme, as described in the section, *OFDM Transmission Technique Combined with Multiple Access Schemes*. The third one is based on OFDM-CDMA where the user data are spread over a subset of adjacent sub-carriers.

In this case MAI occur and an interference cancellation technique implemented in each MT is useful. To compare the different multiple access schemes, Figure 9 shows the bit error rate (BER) for an OFDM system with the system parameters shown in Table 1.

A single cell situation has been considered in this case with a perfect time and carrier synchronization. As can be seen from this figure, the best performance can be achieved by an OFDM-FDMA system which exploits the frequency selective fading of the mobile radio channel by allocating always the best available sub-carrier to each user. Note that a channel adaptive FDMA scheme requires a good prediction of the channel transfer function which has been considered to be perfect in this comparison. If a non-adaptive FDMA technique was used (i.e., fixed or random

Table 1. Proposal of OFDM system parameters

PARAMETER	VALUE
FFT Length	$N_C = 512$
Guard interval length	$N_G = N_C/8 = 64$
Modulation technique	16-QAM
Code rate	R=1/2, m=6
FDMA	Best available sub-carrier is selected.
TDMA	
CDMA Spreading matrix	Walsh-Hadamard (L=16)
CDMA detection technique	SUD with MMSE, MUD with soft interference cancellation plus MMSE

Figure 9. BER results for a coded OFDM system employing different multiple access techniques

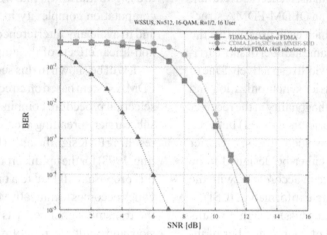

allocation of sub-carriers), the performance would be comparable to the OFDM-TDMA curve.

In the case of an OFDM-TDMA system the frequency selectivity of the radio channel can be exploited by the Viterbi decoder in conjunction with bit interleaving. A pure coded OFDM-CDMA system which utilizes an orthogonal spreading matrix with minimum mean square error (MMSE) equalization and single user detection (SUD) to exploit the diversity of the channel suffers from MAI due to loss of code orthogonality in frequency selective fading. A performance improvement can be achieved for an OFDM-CDMA scheme applying multi-user detection (MUD) techniques. By successively removing inter-code/-user interference using MUD procedure, a gain of approximately 2 dB can be achieved. But still an OFDM-FDMA system outperforms an optimized OFDM-CDMA system. Additionally, OFDM-CDMA technique has a much higher computational complexity in the MUD scheme.

TECHNICAL PROPOSAL AND EXAMPLE FOR A 4G UPLINK INTERFACE

As shown in the preceding paragraph, there are several system proposals published for an OFDM-based downlink procedure for broadcast and communication systems respectively. But by designing an OFDM uplink transmission scheme some important and additional technical questions will come up. Therefore, OFDM-based uplink systems are still under consideration and research (Rohling, Galda, & Schulz, 2004). As a contribution to this topic, an OFDM-based multi-user uplink system with M different users inside a single cell is considered in this section.

Each user shares the entire bandwidth with all other users inside the cell by allocating exclusively a deterministic subset of all available sub-carriers inside the considered OFDM system. This user specific sub-carrier selection process allows to share the total bandwidth in a very flexible way between all mobile terminals. Hence, as a relevant multiple access scheme an OFDM-FDMA structure is considered in which each user claims the same bandwidth or the same number of sub-carriers inside the total bandwidth respectively. Due to the assumed perfect carrier synchronization and resulting sub-carrier orthogonality in the receiver any multiple access interference (MAI) between different users can be avoided. The sub-carrier allocation process can either be designed to be non-adaptive or adaptive in accordance with the current radio channel state information (CSI).

Since the OFDM transmission signal results from the superposition of a large number of independent data symbols and sub-carrier signals the envelope of the complex valued baseband time signal is in general not constant but has a large peak-to-average ratio (PAR). The largest output power value of the amplifier will therefore limit the maximum amplitude in the transmit signal. Additionally, non-linear distortions due to clipping and amplification effects in the transmit signal will lead to both in-band interferences and out-of-band emissions (Brüninghaus & Rohling, 1997). Therefore, in the downlink case each base station will spend some effort and computation power to control the transmit signal amplitude and to reduce the PAR. The objective is in this case to minimize the resulting non-linear effects or even to avoid any interferences.

But for the uplink case it is especially important to design a transmit signal with low PAR to reduce computation complexity in the mobile terminal and to avoid any interference situation caused by non-linear effects of the amplification process.

It will be shown in this section that an OFDM-FDMA system based on an equidistant sub-carrier selection procedure combined with an additional sub-carrier spreading technique will reduce the resulting PAR significantly (Brüninghaus & Rohling, 1998) for the uplink procedure. Furthermore, this proposal will lead to a modulation technique which becomes technically very simple and where the transmit signal consists of a periodic extension and multiple repetition of all modulation symbols. This is the result of the duality between multi-carrier CDMA and single carrier transmission technique as described in Brüninghaus and Rohling (1998).

Figure 10 shows the general structure of an OFDM uplink signal processing in the mobile terminal, which will be considered in this section. In this block diagram there are two main components in the OFDM-based modulation scheme which will be treated in the design process of a multi-user uplink system: The sub-carrier selection technique, and a user specific spreading scheme applied to the user's selected sub-carrier subset, respectively.

The last two blocks in the block diagram show the characteristic IDFT processing and the guard

Figure 10. Block diagram of a multi-user OFDM-FDMA uplink system

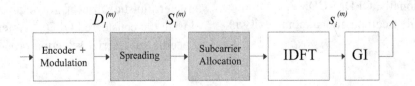

interval (GI) insertion, that are common in all OFDM-based transmitter schemes.

In the uplink system model of a multi-user, OFDM-FDMA based scheme, an arbitrary number of M different users are considered inside a single cell and each user allocates exclusively L different sub-carriers which are considered inside the entire system bandwidth for data transmission. The total number of all considered sub-carriers inside the system bandwidth of the transmission scheme is therefore $N_C = L \cdot M$.

The input data stream for each mobile user terminal m, $m = 0, ..., M - 1$, is convolutionally encoded in a first step. Afterwards, the bit sequence is mapped onto a modulation symbol vector $\vec{D}^{(m)} = \left(D_0^{(m)}, D_1^{(m)}, ..., D_{L-1}^{(m)} \right)$ of L complex valued symbols $D_l^{(m)}$ from a given modulation alphabet with 2^Q different modulation symbols inside the constellation diagram. An example for such a modulation alphabet is given in Figure 11 for a 16-QAM.

Figure 11. 16-QAM as an exemplary alphabet with modulation symbols $D_l^{(m)}$

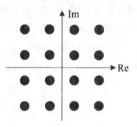

In this section, a non-differential, higher level modulation scheme is assumed for the uplink case. Each user transmits $L \cdot Q$ bits per OFDM symbol. It is assumed in this section without any loss of generality that each user transmits the same data rate or the same number of modulation symbols per OFDM signal respectively.

Sub-Carrier Allocation Process

The first important question in the OFDM-FDMA multi-user uplink system design is the user specific sub-carrier selection scheme. This process is responsible for sharing the bandwidth between M different users, see Figure 12.

There is a large degree of freedom in this system design step to allocate exclusively a subset of L specific sub-carriers to each user. This can either be done by a random or a deterministic allocation scheme. Alternatively there are proposals made for adaptive sub-carrier selection schemes to increase the resulting system capacity (Gross, Karl, Fitzek, & Wolisz, 2003; Shen, Li, & Liu, 2004; Toufik & Knopp, 2004).

In this paragraph a very specific non-adaptive sub-carrier selection procedure is proposed. In this case the allocated sub-carrier subset is equidistantly located on the frequency axis over the entire system bandwidth. This approach is shown in Figure 13 and will be pursued in the following.

In this multi-user uplink system each user m allocates exclusively in total L sub-carriers which

Figure 12. Block diagram for an OFDM-FDMA based system with sub-carrier selection process

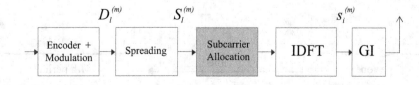

Figure 13. Equidistantly allocated subset of L sub-carriers for a single user in a multi-user environment

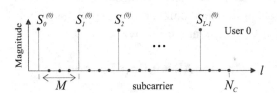

period of the resulting OFDM transmit time signal $s_i^{(0)}$ for user 0 analytically.

$$s_i^{(0)} = \frac{1}{\sqrt{L}} \sum_{l=0}^{L-1} S_l^{(0)} e^{j2\pi il/L} \quad \text{for } i = 0,1,\ldots,L-1$$

(18)

This relation in equation is simply an IDFT applied to the transmit symbols $S_l^{(0)}$, as shown in Equation (19).

$$\begin{pmatrix} s_0^{(0)} \\ s_1^{(0)} \\ \cdots \\ s_{L-1}^{(0)} \end{pmatrix} = \text{IDFT} \begin{pmatrix} S_0^{(0)} \\ S_1^{(0)} \\ \cdots \\ S_{L-1}^{(0)} \end{pmatrix}$$

(19)

are in each case placed in an equidistant way on the frequency axis. The selected L sub-carriers are modulated with L complex valued transmit symbols $S_l^{(m)}$, described and denoted by the transmit symbol vector $\vec{S}^{(m)}$. The proposed non-adaptive sub-carrier selection and modulation process does not need any radio channel state information (CSI) at the transmitter side.

Due to this specific sub-carrier selection process based on equidistantly located sub-carriers on the frequency axis the resulting OFDM uplink transmit time signal $s_i^{(m)}$ of any user has a periodic structure with period length L and consists in any case of an M-times repetition time signal, see Figure 14.

Equation (18) describes the relation between the sub-carrier transmit symbols $S_l^{(0)}$ and a single

Because the sub-carrier subset of a single user is assumed to be allocated equidistantly over all N_C sub-carriers inside the entire bandwidth (Figure 13), it can be shown that an N_C-IDFT processing of the sub-carrier transmit symbols $S_l^{(0)}$ inside the OFDM transmitter leads to the same M-times repetition of the user time signal $s_i^{(0)}$ as shown in Figure 14. The periodicity of the transmit signal is directly related to the selection process of equidistantly located sub-carrier on the frequency axis.

Figure 14. OFDM-FDMA based periodic transmit time signal with period length L and M-times repetition

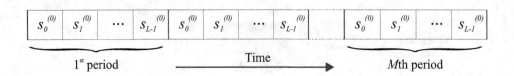

Sub-Carrier Spreading Technique

This paragraph addresses the second design element of an OFDM-FDMA based system: a spreading technique applied to the user's selected sub-carriers, see Figure 15. There are several well-known spreading techniques, which can be integrated into an OFDM-based transmission technique (Kaiser, 2002; Linnartz, 2000). Analogous to other MC-CDMA systems, described in Kaiser (2002) and Linnartz (2000), the vector $\vec{D}^{(m)}$ of L modulation symbols (see Figure 11) is spread in this case over L sub-carriers which are exclusively allocated to user m applying an unitary spreading matrix $[C]$.

This results in a transmit sub-carrier symbol vector $\vec{S}^{(m)} = \left(S_0^{(m)}, S_1^{(m)}, \ldots, S_{L-1}^{(m)} \right)$ consisting of L complex valued transmit symbols $S_l^{(m)}$, $l = 0, \ldots,$ $L-1$. The spreading operation can be denoted mathematically by the following matrix multiplication where each complex valued transmit symbol $S_l^{(m)}$ is calculated by the sum of L user specific modulation symbols $D_l^{(m)}$ weighted by L orthogonal code vectors $\vec{C}_l = \left(C_{l,0}, C_{l,1}, \ldots, C_{l,L-1} \right)$ with $l = 0, \ldots, L-1$:

The spreading Matrix $[C]$ consists of L orthogonal spreading codes. It can be designed, for example, by a Walsh-Hadamard matrix like in Kaiser (2002) and Linnartz (2000) or by a DFT matrix as described in Brüninghaus and Rohling (1997, 1998). Both matrix types fulfill the requirements for unity and orthogonality. Examples for these matrices are shown in Figure 16. In the considered multi-user uplink system, only a DFT matrix based spreading technique will be used, because of the resulting benefits in combination with an equidistant sub-carrier allocation scheme.

After the spreading process, the sub-carrier specific transmit symbols $S_l^{(m)}$ are mapped onto L sub-carrier signals which are exclusively allocated to user m. In principle, the user specific sub-carrier subset can be composed of any L out of N_C sub-carriers that have not been assigned to another user.

Figure 16. Examples for Walsh-Hadamard (left) and DFT spreading matrix

$$
\begin{pmatrix} S_0^{(m)} \\ S_1^{(m)} \\ \vdots \\ S_{L-1}^{(m)} \end{pmatrix} = \begin{bmatrix} C_{0,0} & C_{0,1} & \cdots & C_{0,L-1} \\ C_{1,0} & \ddots & & C_{1,L-1} \\ \vdots & & & \vdots \\ C_{L-1,0} & \cdots & \cdots & C_{L-1,L-1} \end{bmatrix} \cdot \begin{pmatrix} D_0^{(m)} \\ D_1^{(m)} \\ \vdots \\ D_{L-1}^{(m)} \end{pmatrix}
$$
(20)

$$
\begin{bmatrix} 1 & 1 & 1 & 1 \\ 1 & -1 & 1 & -1 \\ 1 & 1 & -1 & -1 \\ 1 & -1 & -1 & 1 \end{bmatrix} \quad \begin{bmatrix} 1 & 1 & 1 & 1 \\ 1 & e^{-j\pi/2} & e^{-j\pi} & e^{-j\frac{3}{2}\pi} \\ 1 & e^{-j\pi} & e^{-j2\pi} & e^{-j3\pi} \\ 1 & e^{-j\frac{3}{2}\pi} & e^{-j3\pi} & e^{-j\frac{9}{2}\pi} \end{bmatrix}
$$

Figure 15. Block diagram of a multi-user OFDM-FDMA uplink system with additional spreading technique

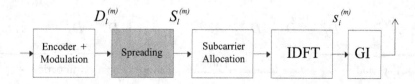

Combination of Spreading and Sub-Carrier Allocation

As explained in the previous paragraph, the spreading technique applied to the modulation symbols $D_l^{(m)}$ is considered in a way, that a DFT-Matrix can be used as spreading matrix $[C]$. Therefore, the relation between modulation symbols $D_l^{(0)}$ and sub-carrier transmit symbols $S_l^{(0)}$ are described analytically by Equation (21):

$$
\begin{pmatrix} S_0^{(0)} \\ S_1^{(0)} \\ \cdots \\ S_{L-1}^{(0)} \end{pmatrix} = \begin{bmatrix} & & \\ & [C] & \\ & & \end{bmatrix} \cdot \begin{pmatrix} D_0^{(0)} \\ D_1^{(0)} \\ \cdots \\ D_{L-1}^{(0)} \end{pmatrix} = \mathrm{DFT} \begin{pmatrix} D_0^{(0)} \\ D_1^{(0)} \\ \cdots \\ D_{L-1}^{(0)} \end{pmatrix}
$$

$$(21)$$

If this DFT-based spreading technique is combined with the earlier explained, equidistant sub-carrier selection process the transmit time signal $s_i^{(0)}$ can be calculated directly by the M-times repetition of modulation symbol vector $\vec{D}^{(0)}$ which consists of L complex valued modulation symbols, see Figure 17. Therefore, it is needless to process the DFT spreading matrix and the IFFT in the OFDM system structure explicitly, which

reduces the computation complexity in the mobile terminal, see Equation (22). Hence, a single period of the resulting time signal $s_i^{(0)}$ is directly given by the calculated modulation symbols $D_l^{(0)}$.

$$
\begin{pmatrix} s_0^{(0)} \\ s_1^{(0)} \\ \vdots \\ s_{L-1}^{(0)} \end{pmatrix} = \mathrm{IDFT} \begin{pmatrix} S_0^{(0)} \\ S_1^{(0)} \\ \vdots \\ S_{L-1}^{(0)} \end{pmatrix} = \mathrm{IDFT} \left(\mathrm{DFT} \begin{pmatrix} D_0^{(0)} \\ D_1^{(0)} \\ \vdots \\ D_{L-1}^{(0)} \end{pmatrix} \right) = \begin{pmatrix} D_0^{(0)} \\ D_1^{(0)} \\ \vdots \\ D_{L-1}^{(0)} \end{pmatrix}
$$

$$(22)$$

In almost all OFDM systems, a cyclic prefix of length N_G will be added to the transmit time signal $s_i^{(0)}$ to avoid any ISI. Therefore, the so-called guard interval is also an integral part of the multi-user uplink system described in this paragraph. Thus, the structure of the OFDM-FDMA multi-user uplink system depicted in Figure 18 can be simplified. Figure 18 shows the functionality of the overall system in detail. It becomes clear that because of the cancellation of DFT spreading and IDFT calculation these components can be completely removed in the technical realization. They are replaced by a simple repetition process of the considered user specific modulation symbols $D_l^{(0)}$.

Figure 17. Periodic transmit signal for the multi-user uplink system: Symbols $s_i^{(0)}$ and modulation symbols $D_l^{(0)}$

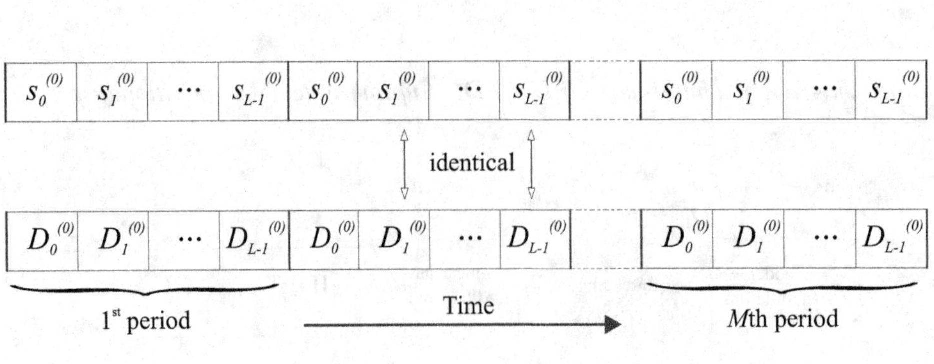

Figure 18. OFDM-FDMA based uplink system including a DFT spreading matrix applied to a set of equidistant sub-carriers

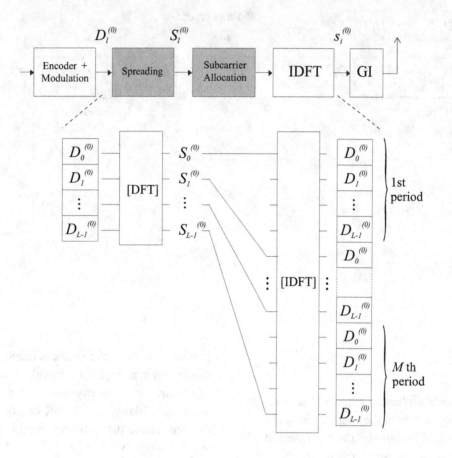

Multi-User Case

The extension from a single to an arbitrary user m is straight forward and will be described in the following. Another user m also allocates an equidistantly spaced subset of all sub-carriers which is shifted in the frequency space by m sub-carriers, see Figure 19.

Any frequency shift results in a multiplication of the transmit time signal $s_i^{(m)}$ with a complex valued signal $e^{j2\pi im/N_C}$, see Equation (23).

Figure 19. Shifting the total sub-carrier subset in a multi-user environment

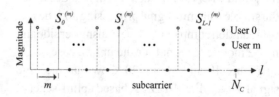

Figure 20. OFDM-FDMA uplink transmit signal for an arbitrary user m

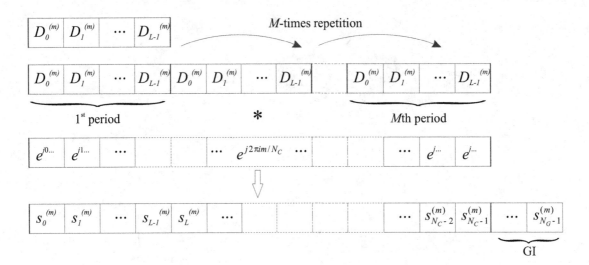

$$S_l^{(m)} \cdot \delta(l-m) \quad \bullet\!-\!\circ \quad s_i^{(m)} \cdot e^{j2\pi im/N_C} \qquad (23)$$

This yields a phase rotation of the transmit time symbols $s_i^{(m)}$ with the constant frequency $f_0 = m/N_C$. But this has no significant impact on the complexity of the transmitter structure. Also, the signal envelope of a single OFDM symbol is still constant. The simplified synthesis of the transmit time signal for the multi-user case is depicted in Figure 20. First, the vector of L modulation symbols $\bar{D}^{(m)}$ is calculated and repeated M-times on the time axis. Then, the time sequence is elementwise multiplied by the user-specific, phase rotating sequence $e^{j2\pi im/N_C}$. In the last step, the guard interval is added. Figure 20 describes the simple transmit signal processing and the low computation complexity at the transmitter side and in the mobile terminal for the uplink case.

Figure 19 and Figure 20 show that the time signal of the OFDM-FDMA based uplink scheme with DFT spreading can be considered as a blockwise single carrier periodic transmission system where a cyclic prefix is integrated into a single block as a guard interval. Therefore, the signal envelope is nearly constant and additional techniques like $\pi/4 - QPSK$ can be employed to even reduce the resulting small PAR for this single carrier system. An additional advantage of this OFDM-FDMA based uplink system is the flexible use of sub-carrier allocation process and data rate adaptation for a certain user.

OFDM-BASED AND SYNCHRONIZED CELLULAR NETWORK

In the preceding sections, several uplink- and downlink-schemes for the connection between mobile terminals and base stations were discussed. In this section, the focus will be broadened from individual links to the overall cellular network. In this context, resource allocation and synchronization of the network play an important role.

As before, the OFDM receiver in a cell has to deal with ISI effects, which occur in multi-path propagation situations in broadband radio channels. In a sufficiently designed OFDM system, these effects can be completely avoided.

Consequently, the OFDM receiver can also deal with superimposed signals which have been transmitted by several distinct and adjacent base stations (BS) in a cellular environment, if the cellular network is synchronized in time and carrier frequency. All adjacent BS operate simultaneously in the same frequency band which leads to a reuse factor of 1.

In current cellular radio networks each base station assigns resources independently and exclusively to its users. To be able to use a TDMA or FDMA multiple access scheme in a cellular environment, an off-line radio resource planning is required to avoid co-channel interference situations between adjacent cells. As a consequence only a small fraction of the available resource determined by the spatial reuse is assigned to each cell which can dynamically be accessed by its users (Zander & Kim, 2001). However, due to this fixed resource distribution among adjacent cells, a dynamic and flexible shift of resources between cells is technically difficult. Such a conventional cellular network with a fixed frequency planning is shown in Figure 21 for a time division duplex (TDD) system as an example.

By introducing the OFDM transmission technique in such a cellular environment, the limitations of fixed resource allocation can be overcome. Since the OFDM transmission technique is robust in multi-path propagation situations, a synchronized network can be established. All BS and MT are synchronized in this case and the signals from adjacent BS will be received with a mutual relative delay no longer than the guard interval. Under these synchronized network conditions each BS can use all available resources simultaneously. With this technique at hand it is possible to add an additional "macro" diversity to a cellular environment by transmitting the same signal from synchronized BS.

Synchronized networks have been intensively studied, for example, for DVB-T broadcast systems as a single frequency network (SFN). In this case the same information signal is transmitted on the same resources from different BS. In the communication case and in a synchronized network different information signals are transmitted by the adjacent BS but all received signals can be considered as co-sub-carrier interferer which

Figure 21. Conventional cellular network with fixed resource allocation

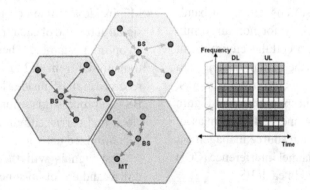

Figure 22. Flexible radio resource management by making all resources available to all BS in a synchronized OFDM-based cellular environment

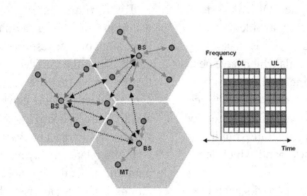

allows in general the allocation of all available resources for each BS.

A synchronized network will also be considered to implement a dynamic resource allocation scheme by assigning different sub-carriers of an OFDM-FDMA multiple access scheme to users in adjacent cells. Since the OFDM sub-carriers remain orthogonal in a synchronized network, no MAI between different user in adjacent cells will occur as long as the sub-carriers have been allocated in an exclusive way.

All resources can be accessed in this case by the MT inside a cellular network. Synchronized networks can be used to provide the needed flexibility inside a cellular environment to allocate system resources in those cells where this bandwidth is needed. Especially for non-uniformly distributed users inside a cellular environment or for hot spot situations the system capacity can be largely increased in synchronized networks. The synchronization concept is shown in Figure 22 for an OFDM-FDMA and TDD system as an example. In this case the resource management could be based on co-channel interference (CCI) measurements processed in each BS.

Self-Organized Cell Synchronization

The dynamic sharing of all available resources between adjacent cells requires a tight time and frequency synchronization of all BS and all MT inside the cellular environment. All MTs are synchronized to a single BS using a specific test signal which is transmitted in a downlink preamble. It is assumed in this paragraph that the required network synchronization is achieved without any assistance of a central controller but in a totally decentralized and self-organized way. Furthermore a TDD system is assumed. Synchronization between adjacent BS can be achieved not in a direct way but indirectly if all MT inside a single cell transmit a specific test signal at the end of each frame (or super frame) in an uplink postamble. These different test signals transmitted from all MT in adjacent cells will be received in all BS inside a local environment. Each BS can process this information to synchronize clock and carrier simultaneously to establish a synchronized network.

Test signals will therefore be used in the down- and uplink to synchronize all BS in a local environment and all MT inside a single cell, see Figure 23. The test signal itself is designed to allow an almost interference-free time and frequency offset estimation.

To generate the test signal structure, each BS selects a single pair of two adjacent sub-carriers for each frame inside the preamble as it is shown in Figure 24. The sub-carriers inside the test signal are chosen randomly and independently by each BS from a set of allowed sub-carrier pairs placed equidistantly in the frequency band and separated by a guard band of unused sub-carriers to reduce

interference in a non-synchronized situation. During the downlink preamble each BS transmits the specific test signal on the individually selected pair of sub-carriers.

In the uplink all MT inside a single cell transmit a test signal which is identical to that one they have received in the preamble at the beginning of the data frame. Each BS receives these test signals in a superimposed form from all MT inside the cell on the same sub-carrier pair. Test signals from MT in adjacent cells will be observed by the BS on distinct pairs of sub-carriers and can therefore

Figure 23. Test signal structure which is used for the synchronization of MT to a single BS during downlink and between all BS during the uplink phase

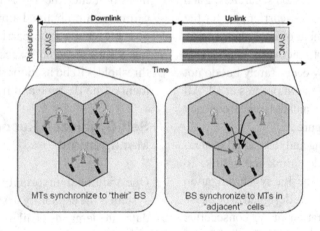

Figure 24. Sub-carrier allocation of test signals

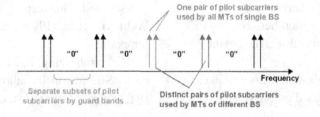

Figure 25. Time interval free of ISI and ICI between different test signals is used for the estimation of fine time and frequency offsets

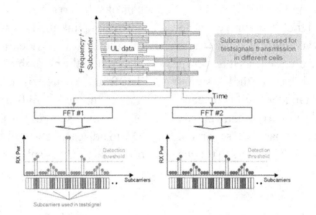

be distinguished and processed separately. Each BS selects randomly the sub-carrier pair for the test signal in the preamble of each data frame. Therefore data collisions between test signals of adjacent BS will only occur rarely but do not influence the synchronization process at all. All received test signals are evaluated in the frequency domain as shown in Figure 25.

The signal processing and test signal evaluation is identical in the downlink and uplink. To avoid ISI and ICI during the fine synchronization procedure the test signals are designed to be phase continuous for the duration of N_p consecutive OFDM symbols as it is shown in Figure 25.

In the downlink case and in the MT synchronization phase a single FFT output signal already contains the time offset information between BS and MT in a certain phase rotation between the two considered adjacent sub-carriers. The carrier frequency offset between BS and MT is given by the phase rotation between the FFT output signals of the same sub-carrier but the two consecutive FFT.

Using this synchronization technique in each BS and MT which is based on phase difference measurements the time and frequency offset estimates are obtained simultaneously for each possible received sub-carrier pair. But only those measurements which exceed a certain amplitude threshold will be used for the subsequent adjustment of the BS time and frequency offsets.

Self-Organized Resource Management

One additional important design aspect for a 4G system is the capability to serve the time-varying data rate demands of all MT efficiently, incorporating high traffic peaks at isolated BS. Therefore dynamic channel allocation (DCA) is considered as an important feature for future networks. Centralized resource management schemes in which a central unit has the complete knowledge about the resource allocation in all cells have been investigated with respect to OFDM systems in Wahlqvist et al. (1997) and Wang et al. (2003), for example.

In the following, however, it is assumed that each BS decides in the radio resource management (RRM) procedure and in the sub-carrier allocation process in a self-organizing (SO) way without any cooperation and communication between adjacent cells and without a central management unit. Therefore, the proposed system concept is termed SO-DCA.

The assumed OFDM-FDMA and TDD scheme shown in Figure 22 is only one possible way of arranging the resource management. The SO-DCA concept can be applied for any orthogonal multiple access scheme with a TDMA and/or an FDMA component. A MAC frame consists of one down link (DL) and one uplink (UL) period and has a total duration of $T_F = T_{UL} + T_{DL}$.

This paragraph introduces a suitable DCA algorithm which strongly benefits from the tight synchronization between all BS and MT inside the cellular environment. The RRM process is mainly based on continuous CCI measurements in each frame. Based on the available CCI measurements (see Figure 26), each BS decides independently

Figure 26. Each BS determines the resource allocation process by measuring the signal power on all sub-carriers inside the available bandwidth

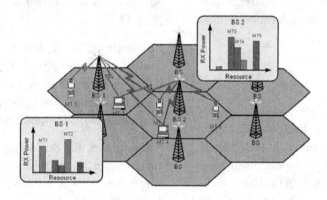

Figure 27. Resource allocation and PHY Mode selection based on the interference measurements at the MT and BS

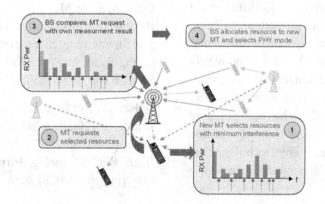

of other BS which resources will be covered for a new MT.

In order to increase the system throughput, a link adaptation (LA) procedure is further introduced. Each BS makes decisions about the modulation scheme and channel code rate (PHY mode) which can be used currently on the individual link. The choice of the applied PHY mode is derived from the radio channel measurement. The main task for the DCA algorithm is to assign a sufficient number of sub-carrier resources to a specific user (MT) to satisfy the current quality of service (QoS) demand. The sub-carrier selection process in the DCA procedure is important to allocate those sub-carriers which are less attenuated by the radio channel. This selection process is mainly based on CCI measurements in the MT and BS. The resource allocation process is summarized in Figure 27.

CONCLUSION

Some aspects for future mobile communication networks have been considered in this chapter. The OFDM transmission technique itself has a large potential due to the robust behaviour in wideband frequency selective and time variant radio channels. The combination with multiple access schemes showed good performance under realistic channel assumptions. A system proposal for an air interface structure for downlink and uplink has been discussed. Future cellular networks require high flexibility for data sources with different and time variant data rate in multi-path propagation environments. Therefore a synchronized cellular network has been proposed and the completely decentralized and self-organized time and carrier synchronization aspects have been discussed. Finally, a self-organized RRM has been proposed to establish a totally decentralized organization inside each BS. All these different techniques and technical concepts can be combined in a way to

establish a future powerful and flexible mobile communications network for the 4G.

REFERENCES

Aldinger, M. (1994). Multicarrier COFDM scheme in high bitrate radio local area networks. *Proc. of Wireless Computer Networks 94*, Den Haag, Netherlands (pp. 969-973). New York: IEEE.

Bello, P. A. (1963). Characterization of randomly time-variant linear channels. *IEEE Transactions on Communications, 11,* 360-393.

Bingham, J. (1990, May). Multicarrier modulation for data transmission: An idea whose time has come. *IEEE Communications Magazine, 28,* 5-14.

Brüninghaus, K., & Rohling, H. (1997). On the duality of multi-carrier spread spectrum and single-carrier transmission. *Zweites OFDM-Fachgespräch*, Braunschweig, Germany (pp. 210-215). Braunschweig: TU Braunschweig.

Brüninghaus, K., & Rohling, H. (1998). Multicarrier spread spectrum and its relationship to single carrier transmission. *Proc. of the IEEE VTC'98*, Ottawa, Canada (pp. 2329-2332). New York: IEEE.

Chang, R. W. (1966). Synthesis of band-limited orthogonal signals for multichannel data transmission. *Bell Syst. Tech. J., 45,* 1775-1796.

Classen, F., & Meyr, H. (1994). Frequency synchronization algorithms for OFDM systems suitable for communication over frequency selective fading channels. *Proc. IEEE VTC 94*, Stockholm, Sweden (pp. 1655-1659). New York: IEEE.

Galda, D., Rohling, H., Costa, E., Haas, H., & Schulz, E. (2002). A low complexity transmitter structure for the OFDM-FDMA uplink. *Proc. IEEE VTC'02 Spring*, Birmingham, Alabama, May (pp. 1024-1028). New York: IEEE.

Gross, J., Karl, H., Fitzek, F., & Wolisz, A. (2003). Comparison of heuristic and optimal subcarrier assignment algorithms. *Proc. of Intl. Conf. on Wireless Networks (ICWN)*, Las Vegas, Nevada (pp. 249-255). Las Vegas: CSREA Press.

Hanzo, L. et al. (2003). *OFDM and MC-CDMA for broadband multi-user communications, WLANs and broadcasting*. New York: Wiley.

Kaiser, S. (1998). *Multi-carrier CDMA mobile radio systems: Analysis and optimization of detection, decoding and channel estimation*. Fortschritt-Berichte VDI, Reihe 10, Nr. 531, VDI-Verlag, Düsseldorf, Germany.

Kaiser, S. (2002). OFDM code-division multiplexing in fading channels. *IEEE Trans. on Communications, 50*, 1266-1273.

Linnartz, J. P. (2000). Synchronous MC-CDMA in dispersive, mobile rayleigh channels. *Proc. of 2nd IEEE Benelux Signal Processing Symposium (SPS-2000)*, Hilvarenbeek, The Netherlands (pp. 1-4). New York: IEEE.

Mizoguchi, M. et al. (1998). A fast burst synchronization scheme for OFDM. *Proc ICUPC 98*, Florence, Italy (pp. 125-129). New York: IEEE.

Pätzold, M. (2002). *Mobile fading channels*. New York: Wiley.

Peled, A., & Ruiz, A. (1980). Frequency domain data transmission using reduced computational complexity algorithms. *Proc. IEEE ICASSP*, Denver, Colorado (pp. 964-967). New York: IEEE.

Rohling, H., Galda, D., & Schulz, E. (2004). An OFDM based cellular single frequency communication network. *Proc. of the Wireless World Research Forum '04*, Beijing, China (pp. 254-258). Zurich: WWRF.

Rohling, H., & Grünheid, R. (1997). Performance comparison of different multiple access schemes for the downlink of an OFDM communication system. *Proc. IEEE VTC'97*, Phoenix, Arizona (pp. 1365-1369). New York: IEEE.

Saltzberg, B. R. (1967). Performance of an efficient parallel data transmission system. *IEEE Trans. on Communications, 15*, 805-811.

Shen, M., Li, G., & Liu, H. (2004). Design tradeoffs in OFDMA traffic channels. *Proc. of IEEE ICASSP '04*, Montreal, Canada (pp. 757-760). New York: IEEE.

Toufik, I., & Knopp, R. (2004). Channel allocation algorithms for multi-carrier systems. *Proc. of the IEEE VTC '04*, Los Angeles, CA, September (pp. 1129-1133). New York: IEEE.

Wahlqvist, M. et al. (1997). Capacity comparison of an OFDM based multiple access system using different dynamic resource allocation. *Proc. IEEE VTC'97*, Phoenix, Arizona (pp. 1664-1668). New York: IEEE.

Wang, W. et al. (2003). Impact of multiuser diversity and channel variability on adaptive OFDM. *Proc. IEEE VTC 2003 Fall*, Orlando, Florida, October (pp. 547-551). New York: IEEE.

Weinstein, S. B., & Ebert, P. M. (1971). Data transmission by frequency-division multiplexing using the discrete fourier transform. *IEEE Transactions on Communication Technology, 19*, 628-634.

Zander, J., & Kim, S. L. (2001). *Radio resource management for wireless networks*. London: Artech House Publishers, Mobile Communications Series.

Chapter IX
Routing Protocols for Ad–Hoc Networks

Muhammad Mahmudul Islam
Monash University, Clayton, Australia

Ronald Pose
Monash University, Clayton, Australia

Carlo Kopp
Monash University, Clayton, Australia

ABSTRACT

Ad-hoc networks have been the focus of research interest in wireless networks since 1990. Nodes in an ad-hoc network can connect to each other dynamically in an arbitrary manner. The dynamic features of ad-hoc networks demand a new set of routing protocols that are different from the routing schemes used in traditional wired networks. A wide range of routing protocols has been proposed to overcome the limitations of wired routing protocols. This chapter outlines the working mechanisms of state-of-the-art ad-hoc routing protocols. These protocols are evaluated by comparing their functionalities and characteristics. Related research challenges are also discussed.

INTRODUCTION

An ad-hoc network consists of a set of nodes that communicate using a wireless medium over single or multiple hops and do not need any pre-existing infrastructure such as access points or base stations. Ad-hoc networks can comprise of mobile, static, or both types of nodes. Ad-hoc networks containing mobile nodes are known as MANETs (mobile ad-hoc networks). An example of ad-hoc networks with static nodes is SAHN (suburban ad-hoc network) (Kopp & Pose, 1998).

Since ad-hoc networks can be rapidly deployed, they are attractive for digital communication in battlefields, rescue operations after a disaster, and so forth. Ad-hoc networks are also useful in civil- ian forums for running demanding multimedia applications such as video conferencing.

The topology of an ad-hoc network can change dynamically due to dynamic link failure

Figure 1. Classification of ad-hoc routing protocols based on routing strategy and network structure

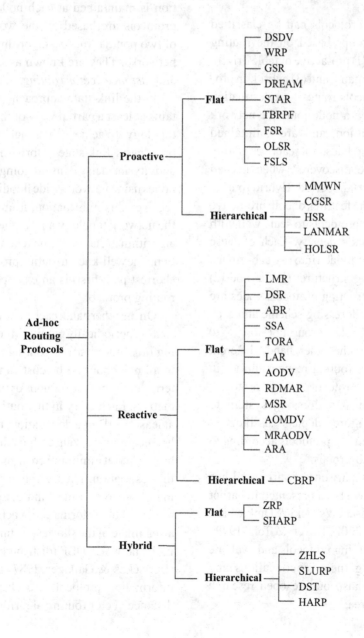

and node mobility. Its size and node density can vary unpredictably since nodes can join or leave the network, or move arbitrarily from one location to another. Due to the lack of a clear physical boundary, a wireless communication channel is usually shared by more nodes than with a cabled network. Nodes in ad-hoc networks can be constrained by computation, battery, and transmission power. Thus routing in ad-hoc networks is more challenging than in wired networks.

Ad-hoc routing protocols can be classified into three major groups based on the routing strategy. These are: (1) pro-active or table driven, (2) reactive or on-demand, and (3) hybrid. In pro-active routing protocols routes to a destination are determined when a node joins the network or changes its location, and are maintained by periodic route updates. In reactive routing protocols routes are discovered when needed and expire after a certain period. Hybrid routing protocols combine the features of both pro-active and reactive routing protocols to scale well with network size and node density. Each of these groups can be further divided into two sub-groups based on the routing structure: (1) flat and (2) hierarchical. In flat routing protocols nodes are addressed by a flat addressing scheme and each node plays an equal role in routing (Hong, Xu, & Gerla, 2002). On the other hand, different nodes have different routing responsibilities in hierarchical routing protocols. These protocols require a hierarchical addressing system to address the nodes. Figure 1 depicts classification of various ad-hoc routing protocols according to these groups and sub-groups.

Reviews and comparisons of various ad-hoc routing protocols have been presented in earlier publications (Abolhasan, Wysocki, & Dutkiewicz, 2004; Hong et al., 2002; Royer & Toh, 1999). We include more routing protocols and evaluate them by comparing their functionalities and characteristics. We also outline open research challenges in this area.

PRO-ACTIVE ROUTING PROTOCOLS

Pro-active routing protocols require each node to maintain up-to-date routing information to every other node (or nodes located within a specific region) in the network. The various routing protocols in this group differ in how topology changes are detected, how routing information is updated, and what sort of routing information is maintained at each node. These routing protocols are based on the working principles of two popular routing algorithms used in wired networks. They are known as *link-state routing* and *distance vector routing.*

In the link-state approach, each node maintains at least a partial view of the whole network topology. To achieve this, each node periodically broadcasts link-state information such as link activity and delay of its outgoing links to all other nodes using network-wide flooding. When a node receives this information, it updates its view of the network topology and applies a shortest-path algorithm to choose the next hop for each destination. The well-known routing protocol OSPF (open shortest path first) is an example of a link-state routing protocol.

On the other hand, each node in distance vector routing periodically monitors the cost of its outgoing links and sends its routing table information to all neighbours. The cost can be measured in terms of the number of hops or time delay or other metrics. Each entry in the routing table contains at least the ID of a destination, the ID of the next hop neighbour through which the destination can be reached at minimum cost, and the cost to reach the destination. Thus, through periodic monitoring of outgoing links, and dissemination of the routing table information, each node maintains an estimate of the shortest distance to every node in the network. DBF (distributed Bellman Ford) (Bertsekas & Gallager, 1987) and RIP (routing information protocol) are classic examples of distance vector routing algorithms.

Due to the limitations in communication resources such as battery power, the potentially very large number of nodes, network dynamics, and node mobility, these protocols are not well suited for ad-hoc networks. The following protocols have been proposed to alleviate the problems of traditional link-state and distance vector routing strategies.

Destination-Sequenced Distance-Vector (DSDV) Routing

DSDV (Perkins & Bhagwat, 1994) is a distance vector routing protocol that ensures loop-free routing by tagging each route table entry with a sequence number.

DSDV requires each node to maintain a routing table. This routing table lists all available destinations from that node. Each entry, corresponding to a particular destination, contains the number of hops to reach the destination and the address of the neighbour that acts as a next-hop towards the destination. Each entry is also tagged with a sequence number that is assigned by the respective destination. To maintain the consistency of the routing tables in a dynamically varying topology, each node periodically broadcasts updates to its neighbours. Updates are also broadcast to neighbours immediately when significant new information, such as link breakage, is available. In order to reduce potentially large amounts of traffic generated by these updates, two modes of updates can be employed. The first type is known as "full dump" where multiple network protocol data units may be needed to carry all available routing information to the neighbours. The other mode of update is referred to as "incremental" where only routing information changed since the last "full dump" is sent in a single network protocol data unit to the neighbours. If topological change is not rapid, "full dump" can be employed less frequently than "incremental" mode to reduce network traffic.

Updated route information, broadcast by a node X to its neighbours, contains the address of the destination Y, $HC+1$ where HC (hop count) is the number of hops to reach Y from X, the sequence number assigned to the initial updated route information broadcast by Y and the new sequence number assigned by X unique to this broadcast. Any route with the older sequence number is replaced with that of the newer sequence number. If two route updates have the same sequence number, the route with the smaller metric is chosen in order to obtain a shorter route.

Since the broadcasts of route information are asynchronous events, it is possible that a node can conceivably always receive two routes to the same destination, with a newer sequence number, one after another from different neighbours but always gets the route with higher metric first. This may lead to continuing broadcast of new route information upon receiving every new sequence number from that destination. In order to reduce the network traffic for such careless broadcasts, it has been suggested to keep track of the weighted average of the time until the route to Y with best metric is received, and delaying the broadcast of updated route information of Y by the length of the settling time.

Due to network-wide periodic and triggered update requirements, DSDV introduces excessive communication overhead. After a node or link failure DSDV may engage in prolonged exchanges of distance information before converging to shortest paths. These problems can become unacceptable if network size or node mobility increases.

Wireless Routing Protocol (WRP)

WRP (Murthy & Garcia-Luna-Aceves, 1995) is a distance vector routing protocol that aims to reduce the possibility of forming temporary routing loops in mobile ad-hoc networks. It belongs to a subclass of the distance vector protocol known as the path-finding algorithm that eliminates the counting-to-infinity problem of DBF (distributed Bellman Ford). Each node, in a path-finding algorithm, obtains the shortest-path spanning tree to all destinations of the network from each one-hop neighbour. A node uses this information along with the cost of adjacent links to construct its own shortest-path spanning tree for all destinations.

Each node in WRP maintains a distance table, a routing table, a link-cost table, and a message retransmission list. The distance table of node X is a matrix that contains the distance to each destination D via each neighbour N and predecessor P. The second-to-last hop of a destination is referred to as a predecessor. An entry in the routing table of X for destination D contains the distance between X and D, the predecessor and successor on this route, and a tag to identify if the entry is a simple path, a loop, or invalid. The neighbour of a node is referred to as the successor for a particular destination if the neighbour offers the smallest cost and loop-free path to the destination. Predecessor and successor information are needed to detect routing loops and to prevent the counting-to-infinity problem. An entry in the link-cost table of X contains the cost of the link and the number of timeouts since X has received any error-free messages from the neighbour connected to that link. The message retransmission list (MRL) contains one or more retransmission entries where each entry enables X to know which update message has to be retransmitted since a neighbour has not acknowledged it in the previous transmission.

WRP requires each node to exchange routing tables with its neighbours using update messages periodically as well as after the status of one of its links changes. When a node X transmits an update message for the first time, it lists all its neighbours so that they can send acknowledgments. If the update message is retransmitted, X obtains the list of neighbours from its MRL that have not acknowledged the update message and includes them in the retransmitted message. In this way WRP can reduce network traffic by asking the neighbours, who have sent acknowledgments for the same update message previously, not to send any more acknowledgments for the retransmitted update message.

If a node does not make any change in its routing table since the last update, it has to send an idle HELLO message to ensure connectivity. On receiving an update message, a node modifies its distance table and looks for better routes using updated information. Any new route thus found is relayed back to the node from which the update message was received. On receiving an acknowledgment for an update message, a node updates its message retransmission list.

Each time a node detects any change in a link, it checks the consistency of the predecessor information reported by all neighbours. This eliminates routing loops and ensures fast convergence after a link failure or recovery that would otherwise be impossible if the consistency check was performed only for the predecessor information reported by the neighbour connected to that link.

Fewer nodes are informed in WRP than in DSDV during a link failure. Hence WRP can find shortest path routes faster than DSDV. On the other hand, WRP requires the use of HELLO packets similar to DSDV even when there is no packet to send. Thus WRP does not allow nodes to enter into a sleep mode to conserve energy (Royer & Toh, 1999).

Multimedia Support in Mobile Wireless Networks (MMWN)

The MMWN (Kasera & Ramanathan, 1997) routing protocol maintains an ad-hoc network using a clustering hierarchy in order to reduce routing control overheads where node mobility is high or nodes do not communicate frequently.

In general each cluster contains three types of nodes: switches, (nodes V, R in Figure 2), endpoints (nodes *m, s* in Figure 2), and a location manager (nodes *I(C,A), I(D,U)* in Figure 2). A location manager of a cluster is elected from among all switches in the cluster and is responsible for performing location management, that is, location updating and location finding. Only switches and location managers can route packets. Endpoints can only be sources and destinations.

At the lowest level of the hierarchy, level-0, endpoints affiliate with switches to form a cell. Multiple cells form a cluster of level-1 and so on. For example, in Figure 2, clusters *C, D, E, F* are at level-1, clusters *A, B* are at level-2 and top level cluster *U* is at level-3. Each switch and endpoint are assumed to have a globally unique identifier,

referred to as the switch-id and endpoint-id respectively, which do not change over time. Every cluster, except the cluster at level-0, is identified by a cluster-id unique among its siblings. The cluster-id of a cluster at level-0 is denoted by the corresponding switch-id. The hierarchical address of a cluster C_k is $C_1.C_2...C_k$ where C_{i-1} is the parent cluster of C_i, where $1 \leq i \leq k$-1. For example, the hierarchical address of cluster *D* in Figure 2 is *U.A.D*. The hierarchical address of a switch is the hierarchical address of the cluster to which the switch belongs, suffixed by the switch-id. The hierarchical address of an endpoint is the hierarchical address of the switch with which the endpoint is affiliated. Unlike node identifiers, the

Figure 2. The clustering hierarchy used in MMWM

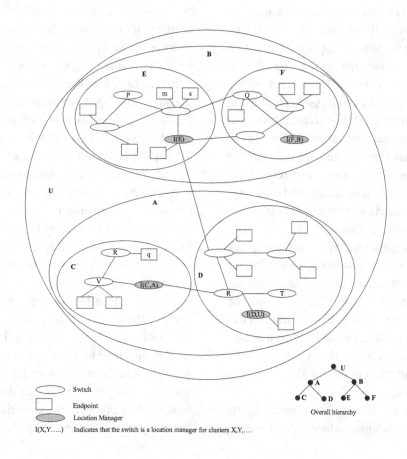

183

hierarchical addresses are autonomously acquired and may change with time.

Each endpoint is associated with two parameters referred to as the "roaming cluster" and "roaming level" for the purpose of its location updating process. The roaming cluster of an endpoint is the lowest level cluster containing the endpoint such that an update is triggered if and only if the endpoints exit this roaming cluster. The roaming level of an endpoint is the hierarchical level of its roaming cluster. For example, if the roaming level of endpoint q in Figure 2 is 2, then its roaming cluster is *U.A.* The roaming level of an endpoint may be changed dynamically based on its call frequency and speed. In general, the more mobile an endpoint is, the higher should be its roaming level.

The location update message, generated by an endpoint, contains four fields: its endpoint-id, old hierarchical address, new hierarchical address and roaming level. The update message is sent to the switch it has just affiliated with. The switch then forwards the message to the appropriate location managers. When a location manager receives an update message it trims the last n terms of the old and new hierarchical addresses contained in the message, where n is its hierarchical level, and compares the resultant hierarchical addresses. If they are not equal, then the message is forwarded to the parent location manager that repeats the check and this process continues until the message reaches a location manager such that the trimmed hierarchical addresses match.

Each location manager receiving an update message creates an association entry for the endpoint or updates the existing association entry for the endpoint. The last location manager, where the comparison resulted in equality, sends a cancel message to the previous location manager that was associated with the endpoint in order to delete the invalid entries for the endpoint's previous location.

When a switch changes its cluster, it also obtains a new hierarchical address. It then sends an aggregated update message, which contains its new hierarchical address and the list of endpoints affiliated with it, to the new location manager. The handling of this message is similar to that of those generated by endpoints.

When a cluster splits into two, the location manager of the original cluster remains with one of the new clusters and the new cluster gets a new location manager. The new location manager initially does not contain any association list. It fills up its list from the information obtained from the old location manager.

When a cluster merges with another cluster, one of the location managers resigns and sends its association list to the surviving location manager so that the new location manager can have a full list of the endpoints contained in the merged cluster.

A node wishing to obtain a hierarchical address of a remote endpoint sends a query message to the switch it is associated with. The switch searches its association list to see if the target endpoint is in its own cell. If the target resides within the same cell, the location finding procedure terminates. Otherwise the switch forwards the query message to its parent location manager that also searches its association list to find the target endpoint. If an entry is found, the query message is forwarded to the respective location manager contained in a child cluster. If no entry is found, the query message is forwarded to the parent location manager. This is how the query message makes its way up the hierarchy until it finds an entry for the target endpoint and then down the hierarchy until it reaches the location manager at level-0.

The final location manager may or may not contain the endpoint depending on the roaming level of the target endpoint. If the roaming level of the target endpoint is 0, the final location manager, that is, the final switch, is assumed to contain the target endpoint. In this case, the final switch sends a reply message to the originator of the query message containing the switch-id and the hierarchical address of the target endpoint. If the roaming level of the target endpoint is greater than 1, the final switch floods a page message containing the same information as the query message throughout the cluster of level *n*, where *n* is the roaming level of the target endpoint. When a switch receives a page message it checks if it contains the target endpoint. If the endpoint is found in the association list, a reply message is sent to the originator of the query message containing the switch-id and the hierarchical address of the target endpoint.

Since the location management is closely related to hierarchical structure of the network, messages have to travel through the hierarchical tree of the location managers. For the same reason, any change in the hierarchical cluster membership of location managers will cause reconstruction of the hierarchical location management tree and introduce complex consistency management. Thus MMWN introduces implementation problems that are potentially complex to solve (Pei, Gerla, Hong, & Chiang, 1999).

Clusterhead Gateway Switch Routing (CGSR)

CGSR (Chiang, Wu, Liu, & Gerla, 1997) is a hierarchical routing protocol that uses DSDV (Per-kins & Bhagwat, 1994) as its underlying routing algorithm but reduces the size of routing update packets in large networks by partitioning the whole network into multiple clusters. The addressing scheme used here is simpler than that of MMWM (Kasera & Ramanathan, 1997) since CGSR uses only one level of clustering hierarchy.

Each cluster in CGSR contains a clusterhead (nodes *A, B, C* and *D* in Figure 3) that manages all nodes within its radio transmission range. A node that belongs to more than one cluster works as a gateway (nodes *E, F*, and *G* in Figure 3) to connect the overlapping clusters.

CGSR requires each node to maintain two tables: a cluster member table and a routing table. The cluster member table records the clusterhead address for each node in the network and is broadcast periodically. The routing table maintains only one entry for each clusterhead, no matter how many members each clusterhead has. Thus CGSR reduces the size of the routing table as well as the size of the routing update messages. Each entry in the routing table contains the address of a clusterhead and the address of the next hop to reach the clusterhead.

A packet from a node is first sent to its clusterhead. The clusterhead then forwards the packet to its neighbouring clusterhead through the corresponding. This process continues until the packet

Figure 3. Illustration of single-level clustering hierarchy used in CGSR

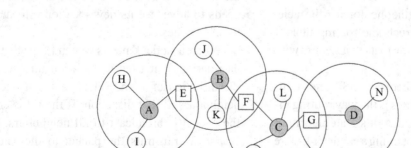

● Clusterhead   ☐ Gateway   ○ Regular node

reaches the clusterhead of the destination node. At this stage, the destination clusterhead simply forwards the packet to the destination.

Since each node only maintains routes to its clusterhead, routing overhead is lower in CGSR compared to DSDV or WRP. However, time to recover from a link failure is higher than DSDV or WRP since additional time is required to perform clusterhead reselection (Royer & Toh, 1999).

Global State Routing (GSR)

GSR (Chen & Gerla, 1998) improves the link-state algorithm by adopting the routing information dissemination method used in DBF. Instead of flooding GSR transmits link-state updates to neighbouring nodes only.

In GSR each node maintains a neighbour list, a topology table, a next-hop table, and a distance table. The neighbour list of a node X contains its neighbours that are within its radio transmission range. The topology table contains the link-state information of each destination Y as reported by Y and a timestamp indicating the time Y has generated this information. For each destination Y, the next hop table contains the next hop Z, which is a one-hop neighbour of X, to which packets must be forwarded from X destined for Y. The distance table contains the shortest distance to each destination from X in terms of the number of hops.

Whenever a node receives a routing message containing link-state updates from one of its neighbours, it updates its topology table if the timestamp is newer than the one stored in the table. After the node reconstructs the routing table it broadcasts the information to its neighbours with other link-state updates.

The key difference between GSR and traditional link-state algorithms is the way routing information is disseminated. A link-state algorithm floods a small packet containing a single link-state update whenever the link status changes. On the other hand, a node in GSR transmits longer packets containing multiple link-state updates to its

neighbours. Therefore GSR requires fewer update messages than a traditional link-state algorithm in an ad-hoc network with frequent topology changes. Thus GSR can optimise MAC (medium access control) layer throughput since frequent smaller packets incur higher MAC layer overhead than infrequent longer packets. However, as the network size and node density increase, the size of each update message becomes larger.

Distance Routing Effect Algorithm for Mobility (DREAM)

DREAM (Basagni, Chlamtac, Syrotiuk, & Woodward, 1998) uses location information using GPS (global positioning system) to provide loop-free multi-path routing for mobile ad-hoc networks.

Each node in DREAM maintains a location table that records location information of all nodes. DREAM minimises routing overhead, that is, location update overhead, by employing two principles referred to as the "distance effect" and the "mobility rate". The "distance effect" states that the greater the distance between two nodes the slower they appear to move with respect to each other. Thus nodes that are far apart need to update their location information less frequently than the nodes closer together. This is realised in DREAM by associating an age with each location update message that corresponds to how far from the sender the message can travel. The "mobility rate" states another interesting observation that the faster a node moves, the more frequently it needs to advertise its new location information to other nodes.

When a node X needs to send a packet to a destination Y, it uses its location table to find the direction of Y and selects a set of one-hop neighbours in that direction. If the set is empty the packet is broadcast to all neighbours. Otherwise X transmits the packet to the selected set of neighbours. Each neighbour repeats this process until the packet reaches Y. Y responds to each packet with an acknowledgment that is sent to X. If X does not receive an acknowledgment within a timeout period, it retransmits the packet by flooding in order to increase the possibility of reaching Y.

Source Tree Adaptive Routing (STAR)

STAR (Garcia-Luna-Aceves & Spohn, 1999) is based on a link-state algorithm that minimises the number of routing update packets disseminated into the network to save bandwidth (i.e., reduce network traffic) at the expense of not maintaining optimum routes to destinations.

STAR requires each node to maintain a source tree, which is a set of links constituting complete paths to destinations. A node knows the status of its adjacent links and the source trees reported by its neighbours. With this information the node generates a topology table and computes its own source tree. It also derives a routing table by running Dijkstra's shortest-path algorithm on its source tree. Each entry in the routing table consists of a destination address, the cost (e.g., the number of hops) of the route to destination and the next hop address towards the destination.

A node sends updates on its source tree to its neighbours only when it loses all routes to one or more destinations, when it detects new destinations, when it determines local changes to its source tree can create long-term routing loops, or when the cost of the routes exceeds a certain threshold. Instead of periodic updates for each link, the conditional dissemination of updates enables STAR to reduce the bandwidth required for link-state updates. This prevents nodes from maintaining optimum routes to destinations. The partial topology graphs of a network maintained in the nodes can change frequently as the neighbours keep sending different source trees in large and highly mobile ad-hoc networks (Abolhasan et al., 2004). In this case STAR may introduce significant memory and processing overheads.

Hierarchical Star Routing (HSR)

Pei et al. (1999) have proposed a hierarchical link-state routing protocol, referred to as HSR, designed to scale well with network size. They argue that the location management (i.e., the location updating and location finding) in MMWM is quite complicated since it couples location management with physical clustering. HSR aims to make the location management task simpler by separating it from physical clustering.

HSR maintains a hierarchical topology by clustering group of nodes based on their geographical relationship. The clusterheads at a lower level become members of the next higher level. The new members then form new clusters, and this process continues for several levels of clusters. The clustering is beneficial for the efficient utilisation of radio channels and the reduction of network layer overhead (i.e., routing table storage, processing, and transmission). In addition to the multi-level clustering HSR provides multi-level logical partitioning based on the functional affinity between nodes (e.g., tanks in a battlefield or the colleagues of the same organisation). Logical partitioning is responsible for mobility management.

An example of a three-level hierarchal clustering structure is illustrated in Figure 4. The node IDs, shown in the lowest level, are physical such as MAC (medium access control) addresses. In general each cluster contains three types of nodes: a clusterhead (nodes *1, 2, 3,* and *4* for the lowest hierarchical level), gateway node (nodes *6, 7, 8,* and *11* for the lowest hierarchical level), and internal node (nodes *5, 9, 10,* and *12* for the lowest hierarchical level). At the lowest level of the hierarchy, each node monitors the state of each link and broadcasts the observed link-state information within the cluster. The clusterhead summarises the received link-state information and sends it to the neighbouring clusterheads through gateways. The clusterheads of a level Cx become the members of the cluster of level $Cx+1$

Figure 4. An example of hierarchical clustering in HSR

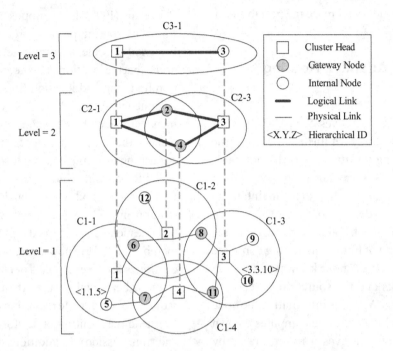

and they exchange their logical link information as well as their summarised lower level link-state information among each other. This process continues up to the highest level. A node at each level disseminates all the gathered link-state information up to this level to the nodes in the level below. In this way each node in the lowest level gets hierarchical topology information of all nodes. A hierarchical address, referred to as HID (hierarchical ID), of a node is defined as the sequence of MAC addresses of the nodes on the path from the top of the hierarchy to the node itself. For example, the HIDs of nodes *5* and *10* are *<1.1.5>* and *<3.3.10>* respectively. A gateway can have more than one hierarchical address. If node *5* wants to send a data packet to node *10* it sends the packet to its top hierarchy node *1*. Since

node *1* has a logical link, that is, a tunnel, to node *3* through the path *1→6→5→2→8→3*, it sends the packet to node *3* through this path. Finally node *3* delivers by packet to node *10* along the downward hierarchical path that is in this case its immediate neighbour. Thus a HID is enough to ensure delivery of packets from anywhere in the network to a remote destination.

In HSR nodes are also partitioned into logical partitions, that is, subnets, in order to resolve implementation problems of MMWM. In addition to the MAC addresses, nodes are assigned logical addresses of type <subnet, host>. Each subnet contains a location management server (LMS). Each member of a logical subnet knows the HID of its LMS. All nodes in a subnet have to register their logical addresses with its LMS.

Registration is both periodic and event driven. All LMSs advertise their HIDs to the top hierarchy. Optionally the LMS HIDs can be propagated downwards to all nodes. When a node wants to send a packet to a destination, it sends the packet to its network layer with the logical address of the destination. The network layer finds the HID of the destination's LMS from its LMS and sends the packet to the destination's LMS. The destination's LMS then forwards the packet to the destination. If the source and the destination know each other's HIDs, they can communicate directly bypassing their LMSs.

Though HSR requires less memory and communication overhead than any flat pro-active routing protocol, it introduces additional overhead (like any other cluster based protocol) for forming and maintaining clusters.

Topology Broadcast Based on Reverse Path Forwarding (TBRPF)

TBRPF (Bellur & Ogier, 1999) is a link-state based routing protocol that uses the concept of reverse-path forwarding to broadcast link-state updates in the reverse direction along the spanning tree formed by minimum-hop paths from all nodes to the source of the update. Unlike a pure link-state routing algorithm, which requires all nodes to forward update packets, TBRPF requires only the non-leaf nodes in the broadcast tree to forward update packets. Thus TBRPF generates less update traffic than pure link-state routing algorithms. The use of minimum-hop tree instead of a shortest-path tree makes the broadcast tree more stable and thus results in less communication cost to maintain the tree.

Each node in TBRPF maintains a list of its one-hop neighbours and a topology table. Each entry in the topology table for a link contains the most recent cost and sequence number associated with that link. With this information each node can compute a source tree that provides shortest paths to all reachable remote nodes. Moreover, for each node $src \neq i$, node i keeps record of: (1) a parent $p_i(src)$ which is the neighbour of node i and the next hop on the minimum-hop path from node i to node src, (2) a list of children $children_i(src)$ which are the neighbours of i, and (3) the sequence number $sn_i(src)$ of the most recent link-state update originating from node src. The parents $p_i(src)$, for all $i \neq src$, form a minimum-hop spanning tree directed towards src.

Node src sends an update message to other nodes by broadcasting the update message in the reverse direction along its spanning tree. A node i accepts the update message, modifies its topology table and forwards the update message to every node in $children_i(src)$ if the update message is received from $p_i(src)$ and the update message has a larger sequence number than the corresponding entry in its topology table.

If a node i detects that the parent for node src has changed, it sends a CANCEL PARENT message, which contains the identity of src, to the current parent if it is reachable. It also sends a NEW PARENT message, containing the identity of src and $sn_i(src)$, to the newly computed parent. If the new parent receives the message, it finds out all the link-state information from its topology table that originating from src and sends it to i.

When a node i detects any change in its neighbourhood, for example, appearance of a new node or loss of connectivity with an existing neighbour, it updates the link cost and the sequence number field for the corresponding link in its topology table. It sends the corresponding link-state message to all its neighbours in $children_i(i)$. Unless the change has caused a neighbour to become inaccessible, the node recomputes its list of parents. If it detects any change in its parent list, it performs the task as outlined previously.

Ogier, Templin, and Lewis (2004) have modified TBRPF where src sends only the updates of those links to i that can result in changes to i's source tree. This modification can result in less update traffic at the expense of having partial topology information at each node.

Fisheye State Routing (FSR)

FSR (Pei, Gerla, & Chen, 2000) is an improvement of GSR. GSR requires the entire topology table to be exchanged among neighbours. This can consume a considerable amount of bandwidth when the network size becomes large. FSR is an implicit hierarchical routing protocol that uses the "fisheye" technique (Kleinrock & Stevens, 1971) to reduce the size of large update messages generated in GSR for large networks. The scope of the fisheye of a node is defined as the set of nodes that can be reached within a given number of hops.

FSR, like GSR, requires each node to maintain a neighbour list, a topology table, a next hop table, and a distance table. Unlike GSR, entries in the topology table corresponding to nodes within the smaller scope are propagated to the neighbours with higher frequency. Thus the fisheye approach enables FSR to reduce the size of update messages.

In FSR each node can maintain fairly accurate information about its neighbours. As the distance (i.e., the scope of fisheye) from the node increases, the detail and accuracy of information also decreases. As a result a node may not have precise knowledge of the best route to a distant destination. However this imprecise knowledge is claimed to be compensated by the fact that the route becomes progressively more accurate as the packet gets closer to the destination.

Landmark Ad-Hoc Routing (LANMAR)

LANMAR (Gerla, Hong, & Guangyu, 2000; Guangyu, Geria, & Hong, 2000) is a combined link-state (i.e., FSR) and distance vector routing (e.g., DSDV) protocol that aims to be scalable. It borrows the notion of landmark (Tsuchiya, 1988) to keep track of logical subnets. Such subnets can be formed in an ad-hoc network with the nodes that are likely to move as a group such as brigades in the battlefield or colleagues in the same organisation.

When a network is formed for the first time, LANMAR only uses the FSR functionality. Gradually one of the nodes learns from the FSR tables that there it contains a certain number of nodes within its fisheye scope. It then proclaims itself as a landmark for that group (i.e., the subnet). When more than one node declares itself as a landmark for the same group, the node with the largest number of group members wins the election. In case of a tie, the node with the lowest ID breaks the tie.

A distance vector routing mechanism propagates the routing information about all the landmarks in the entire network. Within each subnet, a mechanism, similar to FSR, is used to update topology information. As a result, each node contains detailed topology information about all the nodes within its fisheye scope and the distance and routing vector information to all landmarks. Consequently LANMAR reduces both routing table size and control overhead for large MANETs.

When a source needs to send a packet to a destination within its fisheye scope, it uses the FSR routing table. If the destination is located outside the fisheye scope, the packet is routed towards the landmark of the destination. When the packet arrives within the scope of the destination, it is routed using FSR directly to the destination, possibly without going through the landmark.

LANMAR guarantees the shortest path from a source to a destination if the destination is located within the scope of the source. For a remote destination, though packets will reach the destination's landmark through a shortest path, the packets may travel through additional hops before the destination is reached (Hong et al.,

2002). LANMAR improves routing scalability for large MANETs with the assumption that nodes under a landmark move in groups.

Optimised Link-State Routing (OLSR)

OLSR (Jacquet, Muhlethaler, Clausen, Laouiti, Qayyum, & Viennot, 2001) optimises the link-state algorithm by compacting the size of the control packets that contain link-state information and reducing the number of transmissions needed to flood these control packets to the whole network.

In OLSR a node X selects a set of immediate (i.e., one-hop) neighbours called the multi-point relays (MPRs) of that node (see Figure 5). MPRs of X must cover (in terms of radio range) all the nodes that are two hops away from X. Every node within a two-hop neighbourhood of X must have bi-directional links with the MPRs of X. OLSR reduces the size of the control packets since in each control packet a node puts only the link-state information of the neighbouring MPRs instead of all neighbours. It minimises flooding of control traffic since only the MPRs, instead

of all neighbours, of a node are responsible for relaying network-wide broadcast traffic.

To select the MPRs, each node X periodically broadcasts HELLO messages to its one-hop neighbours. Each HELLO message contains a list of neighbours that are connected to X via bi-directional links and also the list of neighbours that are heard by X but are not connected via bi-directional links. This HELLO message can be received by all one-hop neighbours of X, but is not relayed to further nodes. Each node, receiving a HELLO message, can learn the link-state information of all neighbours up to two hops. This information is stored in a neighbour table and used to select MPRs.

Each node broadcasts specific control messages called the topology control (TC) messages. Each TC message, originating from a node X, contains the list of MPRs of X with a sequence number and is forwarded only by the MPRs of the network. Each node maintains a topology table that represents the topology of the network built from the information obtained from the TC messages.

Each node also maintains a routing table where each entry in the routing table corresponds to an optimal route, in terms of the number of hops, to a particular destination. Each entry consists of a destination address, next-hop address, and the number of hops to the destination. The routing table is built based on the information available in the neighbour table and the topology table.

Fuzzy Sighted Link-State (FSLS) Routing

FSLS (Santivez, Ramanathan, & Stavrakakis, 2001) is a link-state routing protocol that restricts the dissemination scope of routing updates in space and time similar to FSR (Pei et al., 2000) in order to scale well with network size.

Each node in FSLS sends a link-state update every $2^{i-1} \times T$ $(i = 1, 2, 3...)$ to all the nodes contained within a scope of s_i where T is the minimum

Figure 5. Multi-point relays in OLSR

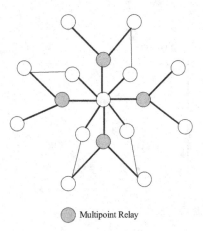

● Multipoint Relay

link-state update transmission interval and s_i is the hop distance from the node.

If s_i is set to infinity, each update message can reach the entire network and FSLS becomes similar to any standard link-state routing algorithm with the exception that a link status change is not propagated in this variant of FSLS until the current T interval finishes. If $s_i = i$, FSLS induces the same control overheads as FSR. Authors have shown that if $s_i = 2^i$, FSLS can induce the least amount of control overhead compared to other variants.

Hierarchical Optimised Link-State Routing (HOLSR)

HOLSR (Gonzalez, Ge, & Lamont, 2005) is a routing mechanism derived from the OLSR protocol.

The main improvement realised by HOLSR over OLSR is a reduction in routing control overhead, for example, topology control information, in large heterogeneous mobile ad-hoc networks. A heterogeneous mobile ad-hoc network is defined as a network of mobile nodes where different mobile nodes have different communication capabilities, for example, multiple radio interfaces with varying transmission powers.

To reduce routing control overhead, HOLSR organises mobile nodes into multiple topology levels based on their varying communication capabilities. Figure 6 illustrates the network structure formed with multiple topology levels.

Nodes having only one wireless interface with low transmission power form topology level 1. These are denoted by circles in Figure 6. Nodes that have up to two wireless interfaces can form

Figure 6. Hierarchical network structure in HOLSR (Gonzalez, Ge, & Lamont, 2005)

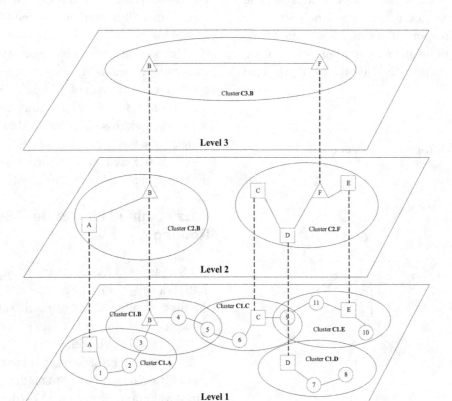

topology level 2. These nodes are designated by squares in the figure. One of the wireless interfaces of these nodes is used to communicate with the nodes of level 1. The other interface is used to relay messages at level 2 using a frequency band or a medium access control protocol different from the one used for communication at level 1. Nodes denoted by triangles in the figure represent high capacity nodes equipped with up to three wireless interfaces. The notation, for example, *C2.B*, used to name clusters in the figure means a cluster of level 2 where node *B* is the clusterhead.

Each topology level comprises of one or more clusters. Each cluster consists of a clusterhead and other mobile nodes. A node configured as a clusterhead during the HOLSR startup process invites other nodes to join its cluster by periodically sending out CIA (cluster ID announcement) messages to neighbouring nodes. To reduce the number of packet transmissions, CIA and HELLO messages are sent together. From HELLO messages, a node gets information about its immediate and two-hop neighbours. A CIA message contains two fields: *clusterhead* and *distance*. The *clusterhead* field indicates the interface address of the clusterhead and the *distance* denotes the distance in number of hops to the clusterhead. When a clusterhead generates a CIA, it sets the value of *distance* to 0. A node receiving the CIA message joins the cluster to which the clusterhead belongs, increases the value of the *distance* by 1, and then sends the CIA to its neighbours to invite them to join the cluster. Any node can receive more than one CIA from different clusterheads. In this case the node joins the cluster that is closer in terms of hop count. If the hop count values of multiple CIA messages are the same, the node joins the cluster from which it receives the first CIA. This process is repeated at each topology level.

Due to mobility, a node might find a clusterhead closer than the one it is currently connected to. In this case the node will join the closest cluster by changing its clusterhead.

Each CIA message has a timeout value. If a node does not receive any CIA message from its existing clusterhead within the timeout period of the previously received CIA message, it can consider joining another cluster provided that it receives CIA messages from other clusters.

If no CIA messages are received, that is, the network is no longer heterogeneous, the HOLSR treats the entire network as one cluster and operates as the original OLSR.

In HOLSR, a clusterhead acts as a gateway through which messages are relayed to other clusters. This requires each clusterhead to be aware of the membership information of other clusters of the same topology level. The higher the position a node possesses in the topology level, the more information it gets about the network. In this way, the nodes at the highest topology level possess full knowledge of all the nodes of the network. Since all nodes do not contain information of all other nodes of the network, the size of the routing tables of lower-level nodes in HOLSR is less than that of OLSR.

The TC (topology control) messages used in the OLSR are usually restricted within a cluster in HOLSR. If a node is located in the overlapping regions of several clusters, it passes a received TC message of one cluster to the neighbouring nodes of other clusters. This enables nearby nodes of different clusters to communicate without directly following the clustering hierarchy which in turn decreases communication delay and reduces the load on the clusterheads.

For sending data to outside clusters, the topology hierarchy is followed. Here is an example where *Node 1* in Figure 6 wants to send data to *Node 10*. *Node 1* is a member of cluster *C1.A* and *Node 2* is a member of cluster *C1.E*. Through TC and HELLO messages *Node 1* knows that *Node 10* is not located within its cluster. So it sends the data to its clusterhead *A*. *A* does not recognise *Node 10* to be located within its cluster and therefore forwards the data to its clusterhead *B*.

Table 1. Comparison of various pro-active routing protocols

	WCC	*WTC*	*RS*	*Frequency of updates*	*Critical Nodes*	*HM*	*Advantages*	*Disadvantages*
DSDV	$O(N)$	$O(D)$	F	Periodic and on-demand	No	Yes	Loop free, simple; Computationally efficient	Excessive communication overhead; Slow convergence; Tendency to create routing loops in large networks
WRP	$O(N)$	$O(h)$	F	Periodic and on-demand	No	Yes	Loop free; Lower WTC than DSDV	Does not allow nodes to enter sleep mode
MMWN	$O(m+s)$	$O(2D)$	H	On-demand	Location Manager	No	Low WCC and WTC	Complicated mobility management and cluster maintenance
CGSR	$O(N)$	$O(D)$	H	Periodic	Clusterhead	No	Lower routing overhead than DSDV & WRP; Simpler addressing scheme compared to MMWN	Higher time complexity than DSDV and WRP for a link failure involving clusterheads
GSR	$O(N)$	$O(D)$	F	Periodic	No	No	Requires less number of update messages than a normal link-state algorithm	Update messages get larger if node density and network size increase
DREAM	$O(N)$	$O(D)$	F	On-demand	No	No	Low routing overhead	Requires GPS
STAR	$O(N)$	$O(D)$	F	On-demand	No	No	Minimises the number of routing update packets disseminated in the network	May not provide optimum routes to destinations; Significant memory and processing overheads for large and highly mobile MANETs
HSR	$O(n*l)$	$O(D)$	H	Periodic	Clusterhead	No	Requires less memory and communication overhead than any flat pro-active routing protocol	Introduces additional overhead for forming and maintaining clusters like any cluster based protocol
TBRPF	$O(N)$	$O(D)$	F	Periodic and on-demand	Parent node	Yes	Lower WCC compared to pure link-state routing	Overheads increase with node mobility and network size
FSR	$O(N)$	$O(D)$	F	Periodic	No	No	Reduces the size of update messages generated in GSR in large networks	Nodes may not have the best route to a distant destination
LANMAR	$O(N)$	$O(D)$	H	Periodic	Landmark	No	Improves routing scalability for large MANETs	Assumption of group mobility, Nodes may not have the best route to a distant destination
OLSR	$O(N)$	$O(D)$	F	Periodic	No	Yes	Reduces size of update messages and number of transmissions than a pure link-state routing protocol	Information of both 1-hop and 2-hop neighbours is required
FSLS	$O(N)$	$O(D)$	F	Periodic	No	No	Reduces control overhead required in FSR or GSR.	Nodes may not have the best route to a distant destination
HOLSR	$O(N)$	$O(D)$	H	Periodic	Clusterhead	Yes	Suitable for large heterogeneous MANETs	Information of both 1-hop and 2-hop neighbours is required; Introduces additional overhead for forming and maintaining clusters

WCC: Worst Case Communication Complexity, i.e., number of messages needed to perform an update operation in worst case; *WTC*: Worst Case Time complexity, i.e. number of steps involved to perform an update operation in worst case; *RS*: Routing Structure; *F*: Flat; *H*: Hierarchical; *HM*: HELLO Messages; *N*: Number of nodes in the network; *D*: Diameter of the network; *h*: Height of the routing tree; *n*: Average number of nodes in a cluster; *l*: number of hierarchical levels; *m*: Number of location managers in MMWN; *s*: Number of switches in MMWN.

Since *B* is located at the highest topology level, it contains information of all nodes in the network. From this information it knows that *Node 10* can be reached via *F*. So it relays the data to *F*. From *F* the data is sent to *Node 10* via *E*.

Comparisons of Pro-Active Routing Protocols

Pro-active routing protocols with flat routing structures usually incur large routing overheads in terms of communication costs and storage requirements to maintain up-to-date routing information about the whole network. Hence they may not scale well as the network size or node mobility increases. However FSR and FSLS have reduced the communication overhead by decreasing the frequency of updates for far away nodes. DREAM reduces the transmission overhead by exchanging location information rather than full or partial link-state information. OLSR reduces rebroadcasting by using multipoint relays (Abolhasan et al., 2004). Hence these flat routed protocols have better scalability potential.

The hierarchical pro-active routing protocols reduce communication and storage overhead as the network size increases since in most cases only the clusterheads are required to update their views of the entire network. However in MANETs, where group mobility is usually impossible, these protocols can introduce additional complexity and overhead for cluster formation and maintenance. Consequently these protocols may not perform better than flat pro-active routing protocols. Table 1 summarises and compares the characteristics of various pro-active routing protocols.

REACTIVE ROUTING PROTOCOLS

Unlike pro-active routing protocols, reactive routing protocols find and maintain routes when needed so that routing overheads can be reduced where the rate of topology change is very high.

Route discovery usually involves flooding route request packets through the network. When a node that is a destination or has a route to the destination is reached, a route reply is sent back to the source of the request. If the links connecting the nodes are bi-directional, the reply is sent back through the path on which the route request travelled. Otherwise the reply is flooded. Thus, in the worst case the route discovery overhead grows by $O(N+M)$ when bi-directional links are available and by $O(2N)$ when only uni-directional links are possible (Abolhasan et al., 2004). Here N and M denote the total number of nodes in the network and the number of nodes in the reply path (if bi-directional links are available) respectively.

Reactive routing protocols can be classified into two groups based on the way routing information is stored at each node and carried in routing packets. These are source routing and hop-by-hop routing.

In source routing, each data packet contains a list of node addresses known as the source route that constitutes the complete path from the source to the destination. When a node wants to send data to a destination, it transmits the data packets to the first hop identified in the source route. When an intermediate node receives the packet, it simply transmits the packet to the next hop by finding it from the source route. Thus the packet propagates through the network until it reaches the destination. Source routing provides a very easy way to avoid forming loops in the network. However, the size of each packet gets bigger as the number of intermediate nodes increases for a particular source and destination pair.

On the other hand, with hop-by-hop routing, each data packet carries only the destination address and the next hop address, and each intermediate node in the routing path uses its routing table to forward the data packet to the next hop towards the destination. In this sense hop-by-hop routing is similar to pro-active routing. In this approach, each node updates its routing table when it receives updated topology information and forwards the data packets over fresher and better routes. Hence routes can be adapted to the dynamically changing topologies of mobile ad-hoc networks. The disadvantage of the hop-by-hop routing over source routing is that each intermediate node has to store and maintain routing information for each active route and may require sending periodic beaconing messages to its neighbours to be aware of its neighbourhood.

A variety of reactive protocols have been proposed based on these strategies. The rest of this section describes and compares a number of such protocols.

Light-Weight Mobile Routing (LMR)

LMR (Corson & Ephremides, 1995) maintains multiple routes to reach each destination. This feature increases the reliability of LMR since whenever a route to a particular destination fails the next available route to the destination can be used without initiating a new route construction procedure. It uses sequence numbers and inter-nodal coordination to avoid long-term loops.

Each node maintains a list of its available neighbours. When a source node needs to find routes to a destination, it initiates a route construction phase by broadcasting a query (QRY) packet to its neighbours. The QRY packet contains the address of the source, the address of the destination, a monotonically increasing sequence number maintained for each destination by the source, and the address of transmitter which is updated at each intermediate node as the QRY packet propagates through the network. The triplet <address of the source, address of the

Figure 7. Route construction using QRY and RPY packets in LMR

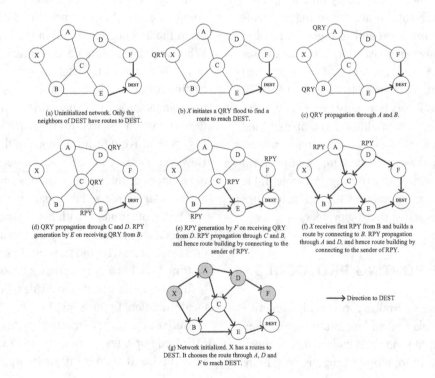

(a) Uninitialized network. Only the neighbors of DEST have routes to DEST.

(b) *X* initiates a QRY flood to find a route to reach DEST.

(c) QRY propagation through *A* and *B*.

(d) QRY propagation through *C* and *D*. RPY generation by *E* on receiving QRY from *B*.

(e) RPY generation by *F* on receiving QRY from *D*. RPY propagation through *C* and *B*, and hence route building by connecting to the sender of RPY.

(f) *X* receives first RPY from B and builds a route by connecting to B. RPY propagation through *A* and *D*, and hence route building by connecting to the sender of RPY.

(g) Network initialized. X has a routes to DEST. It chooses the route through *A, D* and *F* to reach DEST.

⟶ Direction to DEST

destination, sequence number> uniquely identifies a QRY from other queries and allows a node to remember if it has previously received the QRY. When a node receives a QRY, it rebroadcasts it to its neighbours provided that it has not received this QRY before. Thus the QRY propagates through the network and eventually reaches a node that has a route to the destination (e.g., a neighbour of the destination). This process has been illustrated in Figure 7(a)-(d).

A reply (RPY) packet is broadcast by a node which has a route to the destination, in response to the QRY packet. The RPY contains the addresses of the destination and the transmitter. The RPY is flooded back to the source in the same manner as the QRY packet with the exception that the propagation of RPY forms a directed acyclic graph that is rooted at the destination and pointed towards the origin of the RPY. Figures 7(d)-(g) illustrate this process.

When a node loses its last route to a destination due to an adjacent link failure, it enters into the route maintenance phase. If routes from other source nodes for the destination do not pass through this node, the node may enter the route construction phase if it needs to find new routes to the destination. Otherwise the node broadcasts a failure query (FQ) packet to the nodes between itself and the source node in order to inform them of the link failure and at the same time ask them if they have alternate routes to the destination. When a node receives a FQ over a link, it erases the routes containing the link. If it has any alternate route, it broadcasts an RPY. It rebroadcasts the FQ if and only if it does not have any alternate route to the destination.

LMR requires reliable delivery of its control packets. This may be an unreasonable requirement for highly dynamic networks. If reliability is not guaranteed, the protocol can suffer from temporary routing loops or may provide invalid routes temporarily in the partitioned portion of a network (Marina & Das, 2003; Park & Corson, 1997).

Dynamic Source Routing (DSR)

DSR (Johnson & Maltz, 1996) is based on the concept of source routing. Each node in DSR is required to maintain a route cache that contains the source routes to the destinations the node has learned recently. An entry in the route cache is deleted when it reaches its timeout.

When a source node needs to send a data packet to a destination node, it searches its route cache to determine if it already has a route to the destination. If there is a route to the destination, it uses the route to send the data packet. Otherwise it initiates a route discovery process by broadcasting a route request (RREQ) packet to its neighbours. The RREQ contains the address of the source, the address of the destination, a request id, and a route record. The request id is a sequence number maintained locally by the source node. The route record is the addresses of the intermediate nodes through which the RREQ will pass to reach the destination. At the source the route record does not contain anything.

When a node receives a copy of the RREQ, it checks the <source address, request id> pair in its list of recently seen route requests. If there is a match or the route record contains the address of the node, the RREQ is dropped. Otherwise the node checks whether it is the destination or contains a route to the destination. If it is not the destination or does not have a route to the destination it appends its address to the route record and rebroadcasts the RREQ to its neighbours. A copy of the RREQ thus propagates through the network until it reaches the destination or a node that has a route to the destination. Figure 8(a) illustrates the formation of a route record as it propagates through the network towards the destination.

A route reply (RREP) is generated when the RREQ reaches either the destination or an intermediate node that contains a route to the destination. A node does not generate more than one RREP for a particular source and destina-

Figure 8. Formation of route record through propagation of RREQ and RREP in DSR

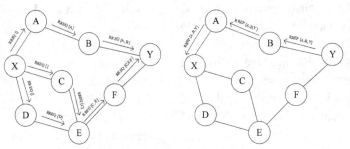

X : Source, Y: Destination, A-F: Other Nodes

(a) Formation of route record in
DSR as RREQ propagates through
the network.

(b) Propagation of RREP
through the network using
route record.

tion pair. If the node generating the RREP is the destination itself, it copies the route record from the RREQ to the RREP. If the responding node is an intermediate node, it appends its cached route to the incomplete route record and puts the complete route record in the RREP.

To send the RREP to the source, the responding node must have a route to the source in its route cache. If the node has a route entry in its route cache for the source node, it may use this route to unicast the RREP in the same way as source routing. Otherwise the responding node may reverse the route in the route record from the RREQ and use this route to send the RREP to the source. Figure 8(b) shows the propagation of a RREP using this latter scheme. This scheme, however, will work if the neighbouring nodes, listed in the route record, can communicate equally well in both directions. As an alternative approach, the responding node can piggyback the RREP on a RREQ generated to find a route to the source.

Nodes can operate in promiscuous mode to extract route records used in the overheard packets transmitted by neighbouring nodes and thus up-

date entries in their route caches without actually participating in any route discovery process.

If a node receives a packet in promiscuous mode and finds out its address in the unprocessed part of the source route multiple hops away from the current sender, it sends a gratuitous reply message to the packet's sender informing it that the packet can be forwarded to it directly bypassing the additional hops.

Each node monitors the operation of each route it is currently using through a route maintenance module. If it cannot send a packet to a neighbour, it declares the corresponding link to be broken and sends a route error (RERR) packet to the source of the associated route. The RERR contains the address of the node that detected the error and the address of the neighbour (i.e., the hop in error) to which the node failed to send packets. When the RERR is received, the hop in error is removed from the route cache and all routes that contain this hop are truncated at that point.

Caching route entries can be beneficial for networks with low mobility. In highly mobile or large networks aggressive use of route caching and lack of an efficient mechanism to purge stale routes can lead to problems like stale caches and relay storm. As a result network performance can be degraded (Marina & Das, 2001a). Moreover, the use of a source route in each packet consumes extra channel bandwidth. The size of each packet gets larger as the size of the network increases. These problems, however, have been addressed in Hu and Johnson (2000, 2001) and Marina and Das (2001a).

Associativity-Based Routing (ABR)

ABR (Toh, 1996, 1997) uses the concept of source routing similar to DSR, but selects routes based on association stability, that is, connection stability, of nodes. Routes selected in this manner are likely to be long lived, resulting in requiring fewer route reconstructions and less route control traffic. However, routes selected in this way may not be the shortest in terms of the number of intermediate nodes.

Each node generates periodic beacons to notify others of its existence. When a node receives a beacon, it increments its associativity tick with respect to the neighbour from which it received the beacon. If a node observes low associativity ticks with its neighbours, it is said to exhibit a high state of mobility, that is, low association stability. On the other hand, if a node has high associativity ticks with its neighbours, it can be considered to be in a high stability state and selected for routing.

When a source needs to find a route to a destination, it broadcasts a BQ (broadcast query) packet to its neighbours. When a node, other than the destination, receives a BQ, it checks if it has previously seen the BQ. If so, the node drops the BQ. Otherwise it appends its address and associativity ticks to the BQ, and then rebroadcasts the updated BQ to its neighbours. The next succeeding node erases the associativity tick entries from the BQ that were appended by the upstream neighbour and retains only the entry concerned with itself and

its upstream neighbour. Then it rebroadcasts the BQ to its neighbours after appending its address and associativity ticks to it. In this manner the BQ propagates through the network and eventually reaches the destination.

The destination, after receiving multiple BQs, selects the best route by examining the associativity ticks along each of the routes. If multiple routes have the same overall degree of association stability, the route with minimum number of intermediate nodes is selected. Once a route has been selected, the destination sends a REPLY packet back to the source along the selected route. As the REPLY passes through each intermediate node, it marks the embedded route in the REPLY packet as valid and regards all other possible routes to the destination as invalid in order to avoid duplicated packets arriving at the destination.

The route maintenance phase of ABR consists of new route discovery, partial route discovery, invalid route erasure, and valid route updates depending on node mobility along the route.

If a source node moves away from its downstream neighbour (i.e., the next hop neighbour towards the destination), it initiates a new route discovery.

When a destination moves, its immediate upstream neighbour (i.e., the next hop neighbour towards the source) erases its route to the destination. Then the upstream neighbour broadcasts a localised query (LQ [H]) packet, where H refers to the hop count from the upstream node to the destination, to find out if the destination is still reachable. If the destination receives the LQ packet, it selects the best partial route and responds with a REPLY packet. If the node, which initially generated the LQ [H], times out, it notifies the immediate upstream neighbour to erase the invalid route and invoke a LQ [H] process. If this process backtracks more than halfway towards the source, the source is notified to initiate a new route discovery phase. If an intermediate node moves, a similar process is invoked at other intermediate nodes between the point of failure and the source. Additionally, the immediate downstream neighbour propagates a route delete message towards the destination in order to delete corresponding route entries from the route tables of all the subsequent downstream nodes.

When a route for a particular destination is no longer needed, the source broadcasts a route delete message to its neighbours. A node, receiving the route delete message, deletes the corresponding route entries from its routing table and rebroadcasts the route delete message to its neighbours. Thus a route delete message is propagated through the network until it is received by a node that does not have any entry in its routing table for the destination corresponding to the route delete message.

ABR is suitable for small MANETs. The beaconing interval should be short enough to be able to adapt quickly to spatial, temporal, and connectivity states of the neighbouring nodes (Royer & Toh, 1999). This requirement may result in extra bandwidth and power consumption.

Signal Stability-Based Adaptive (SSA) Routing

SSA (Dube, Rais, Kuang-Yeh, & Tripathi, 1997) selects routes based on signal stability, that is, the combination of signal strength and location stability, rather than using association stability as used in ABR. Like ABR, routes selected in SSA may not be shortest in terms of the number of intermediate nodes.

Each node sends out a link layer beacon to its neighbours periodically and maintains a signal stability table where each row corresponds to the signal strength and location stability of each neighbour. When a node receives a beacon, it measures the signal strength at which the beacon was received and updates the corresponding entry in its signal stability table. If the node receives a certain number of strong beacons from a neighbour for a predefined period, it classifies the neighbour

as strongly connected. Otherwise the neighbour is regarded as weakly connected.

Each node maintains a routing table where each entry contains the next hop address for each reachable destination. When a source needs to send a packet to a destination, it looks up the destination in its routing table. If there is an entry, the data packet is forwarded using the hop-by-hop strategy. Otherwise the node initiates a route discovery process using a source routing strategy. Route search packets are forwarded to the next hop only if they are received from strongly connected neighbours and have not been previously processed. The first route search packet that arrives at the destination is considered to be the one arriving over the shortest or least congested path. The destination responds to the route search packet by sending a route reply packet to the source. When an intermediate node detects one of its neighbours is not available any more, for example, has moved out of its transmission range or shut down, it sends an error message to the source indicating which link has failed. The source then sends an erase message to erase the invalid route and initiates a new route discovery process to find a new route to the destination.

In SSA, intermediate nodes cannot reply to a route search packet. This incurs longer delays than DSR before a route can be found. Unlike ABR, SSA does not have any route repair mechanism at the point where link failure occurs. The source has to be notified to perform the route reconstruction. Therefore SSA may incur additional delays before a broken route is re-established (Abolhasan et al., 2004).

Temporally Ordered Routing Algorithm (TORA)

TORA (Park & Corson, 1997) is an improved variant of LMR. Like LMR it uses a directed acyclic graph, rooted at a destination, to represent multiple routes for a source and destination pair.

However, unlike LMR, it restricts the propagation of control messages to a very small set of nodes near the occurrence of a topological change by using the concept of link reversal proposed by Gafni and Bertsekas (1981). When a link in a directed acyclic graph breaks, the link reversal method can transform the distorted graph in finite time so that the destination becomes the only node with no outgoing links. TORA uses time stamps and internodal coordination to avoid long-term loops.

The process of route creation in TORA is similar to LMR with few exceptions. TORA uses query (QRY) and update (UPD) packets for creating new routes. The UPD packet is known as the reply packet in LMR. Unlike LMR, TORA assumes that nodes have synchronised clocks and use a height metric to establish a directed acyclic graph for each destination. The height of a node is defined by two parameters: a reference level and a delta with respect to the reference level. The height of the destination is always zero, that is, the values of the reference level and delta are both zero. The heights of other intermediate nodes increase by 1 towards the source node. This is accomplished by increasing the value of delta. Unlike LMR, a node in TORA may process multiple UPD packets for the same source and destination pair if the most recent UPD packet gives the node a lesser height. For example, in Figure 9(b), the source X may have received an UPD from node A or node C before the UPD from node B, but since the UPD from node B gives it lesser height it retains this height.

When a node loses its last downstream link (i.e., the link directed from this node to one of its neighbours) for a particular destination as a result of link failure, the node selects a new height so that the new height becomes a global maximum. This can be accomplished by defining a new reference level and a new delta, such as increasing the value of the current reference level and assigning zero to delta. This action results in link reversals, which may cause other nodes to lose their last downstream links for the destination. Such nodes also select a new height and perform link reversal with respect to their neighbours. Thus the new height is propagated outward from the point of the original failure and gets updated. This propagation continues only through the nodes, which have lost all the routes to the destination. As a result, the propagation of control messages becomes restricted to a very small set of nodes near the occurrence of a topological change.

If the node, which detected the link failure, receives the propagated new height, it determines that no route to the destination exists. The node then begins the process of erasing invalid routes to the destination by flooding a clear (CLR) packet throughout the network.

Figure 9. Route creation in TORA using QRY and UPD propagation

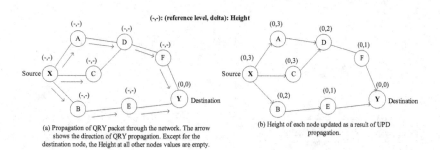

(a) Propagation of QRY packet through the network. The arrow shows the direction of QRY propagation. Except for the destination node, the Height at all other nodes values are empty.

(b) Height of each node updated as a result of UPD propagation.

TORA can falsely detect partitions because it only considers links known from previous route discovery; links that can come up later are ignored though they can be used to join the partitions (Marina & Das, 2003). It requires reliable and in-order delivery of route control packets. These requirements can degrade the network performance to such an extent that the advantage of having multiple routes can be undermined (Broch, Maltz, Hu, & Jecheva, 1998; Das, Castaneda, & Yan, 2000). Moreover it can create short-term routing loops due to the nature of its link reversal technique.

Location-Aided Routing (LAR)

LAR (Ko & Vaidya, 1998) is a flood based routing algorithm, like DSR, that uses location information in order to reduce route search space and thereby minimises route control traffic. It assumes that each node obtains its location information using a GPS (global positioning system).

In LAR a node forwards route request packets only to the nodes that reside inside the route search space (also referred to as the request zone). Any node outside the request zone ignores such packets. If route is not discovered within a suitable timeout period, the request zone is expanded. Two schemes have been considered in LAR to determine a request zone.

Consider a node source S, in Figure 10(a) and Figure 10(b), wants to find a route to destination node D.

In *Scheme 1* (Figure 10(a)) at time t_1, S determines the expected zone of D from D's location information recorded at time t_0. If where v is the average speed of D, the expected zone of D from S's viewpoint is the circular region with radius $R = v(t_1 - t_0)$, that is, S can assume that D will be in any location within that circle during the interval $(t_1 - t_0)$. Now LAR defines the request zone by the smallest rectangle that includes current location of S and the expected zone of D. S includes the four coordinates, that is, (X_S, Y_S) and (X_D, Y_D), with the route request packet. A node, such as F and G, ignores a route request packet if it does not belong to the rectangle defined by the four corners. If the location cannot be precisely detected then the radius of an expected zone becomes $e + v(t_1 - t_0)$, where e denotes the maximum error in the coordinate estimated by the source node.

In *Scheme 2* (Figure 10(b)) node S includes $DIST_s$ and (X_D, Y_D) in each route request packet. Here (X_D, Y_D) is the location of node D and $DIST_s$ denotes the distance between S's current location and (X_D, Y_D) recorded by S at time t_0. When a node F receives the request from S, it calculates its distance $DIST_F$ from (X_D, Y_D). If $DIST_s + \delta \geq DIST_F$, where δ is an error margin, F forwards the route request packet after updating $DIST_s$

Figure 10. Routing schemes of LAR

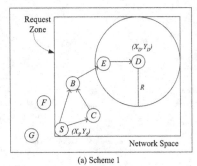

(a) Scheme 1

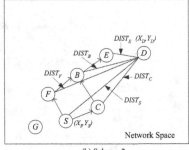

(b) Scheme 2

with $DIST_F$. If $DIST_S + \delta < DIST_F$, F discards the route request.

Though LAR aims to reduce the number of control packets in a network, it may generate control packets similar to a flooding algorithm in a highly mobile network.

Ad-Hoc On-Demand Distance Vector (AODV) Routing

AODV (Perkins & Royer, 1999) routing protocol minimises the number of required broadcasts of DSDV by creating routes on a demand basis. It uses sequence numbers to avoid long-term loops.

AODV requires each node to maintain a list of its active neighbours by sending periodic HELLO packets or by listening to data transmissions of neighbouring nodes in promiscuous mode.

When a source node needs to send a data packet to a destination node and does not contain any route to the destination, it initiates a path discovery process to find a route to the destination. For this purpose every node maintains two monotonically increasing counters: a sequence number and a broadcast id. The source broadcasts a route request (RREQ) packet to its neighbours

containing the address of the source (i.e., the address of itself), its sequence number, its broadcast id, the address of the destination, its last known sequence number of the destination, and a hop count with a value of zero. The pair <address of the source, broadcast id> uniquely identifies a RREQ. The destination sequence number is used to determine the relative freshness of two pieces of routing information generated by two nodes for the same destination, that is, the packet with the highest destination sequence number is more recent. The broadcast id is incremented by the source every time it broadcasts a RREQ.

When a node receives a RREQ, it checks if it has received a RREQ with the same <address of the source, broadcast id> pair before. If there is a match, the node drops the RREQ and thereby limits the number of broadcast packets. Otherwise it accepts the RREQ for further processing.

During further processing the node checks if it has a route to the destination with a destination sequence number greater than the destination sequence number of the received RREQ. If such a record is found, the node can respond to the RREQ by sending a route reply (RREP) packet to the source. Otherwise it rebroadcasts the RREQ

Figure 11. Creation of reverse and forward paths in AODV

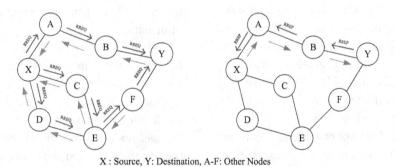

X : Source, Y: Destination, A-F: Other Nodes

(a) Propagation of RREQ and reverse path setup (shown in red coloured arrows) in AODV.

(b) Propagation of RREP and forward path setup (shown in red coloured arrows) in AODV.

to its neighbours after increasing the value of the hop count. This process repeats and eventually a RREQ is assumed to arrive at a node that is either the destination itself or has a current route to the destination.

As a RREQ travels from the source to the destination, it sets up a reverse path from the destination to the source through which the corresponding RREP will travel. To set up a reverse path, a node records in its routing table the address of the neighbour from which it received the first copy of the RREQ.

As the first copy of an RREP is sent to the source along a reverse path, each node along the path sets up a forward route entry in its routing table by recording the address of the neighbour from which it has received the RREP. This forward route entry will be used to forward the data packet from the source to the destination. If a node receives a further RREP, it updates its routing information and forwards the RREP only if the RREP contains a greater destination sequence number than the previous RREP or the same destination sequence number with a smaller hop count value. This rule ensures that the number of RREPs propagated towards the source is minimised, avoids forming loops, and restricts the source from learning multiple routes to the destination. If a route entry is not used for a certain period, it is deleted. The route discovery mechanism has been depicted in Figure 11.

When a node detects a link failure (e.g., due to node movement) it sends a link failure message to the source of the associated route along the corresponding reverse path. All the nodes receiving the link failure message erase the associated entries in their routing tables. The source may also choose to reinitiate a route discovery for the destination if required.

Since AODV does not provide any localised route repair mechanism, it introduces extra delays and consumes more bandwidth as the size of the network increases (Abolhasan et al., 2004).

Relative Distance Micro-Discovery Ad-Hoc Routing (RDMAR)

RDMAR (Aggelou & Tafazolli, 1999) minimises routing overheads by localizing query flooding into a limited area. It uses the concept of sequence numbering, similar to AODV, to prevent forming long-term loops.

If a source node does not contain a feasible route to a destination, it initiates a route discovery process. During this process the source node refers to its routing table in order to find information on its previous relative distance with the destination and the time elapsed since it last received routing information for the destination. With this information and assuming a moderate velocity and a moderate transmission range for the destination, the source node estimates the new relative distance to the destination in terms of the number of hops. The source then updates its routing table with this new relative distance. It inserts the new relative distance in the time-to-live (TTL) field of the route request (RREQ) packet so that nodes outside the range of the TTL do not process the RREQ.

The handling of RREQ and RREP (route reply) by intermediate nodes is similar to AODV. However, RDMAR allows only the destination to send an RREP packet in response to the RREQ that arrives first. It is claimed that preventing any intermediate node, which may have a route to the destination, from sending an RREP reduces the possibility of nodes receiving stale routing information.

When an intermediate node detects a link failure for a destination, it initiates a route discovery procedure by itself if it is located near to the destination. Otherwise it sends a failure notification to the source.

Each node, upon receiving the failure notification, removes the next hop information associated with the destination from its routing table. The node can then initiate a new route discovery procedure if it has kept a copy of the data packet for which the link failure was generated. Otherwise it forwards the failure notification to the neighbour towards the source. When the source receives the failure notification, it initiates a new route discovery procedure if it still needs a route to the destination.

If a source and destination pair does not have any prior communication, RDMAR behaves like a pure flooding algorithm.

Cluster-Based Routing Protocol (CBRP)

CBRP (Jiang, Li, & Tay, 1999) is a hierarchical on-demand routing algorithm that uses source routing, similar to DSR, to avoid forming loops and route packets. Like other hierarchical routing algorithms, CBRP aims to scale well with network size. It can best perform in MANETs where nodes in each cluster move together (Abolhasan et al., 2004).

CBRP groups the nodes in a network into several clusters. Each cluster has a clusterhead that coordinates data transmission within the cluster and with other clusters. When a node is switched on, it sets its state to undecided, starts a timer, and broadcasts a HELLO message. If a clusterhead gets this HELLO message it responds immediately with another HELLO message. If the undecided node gets this message within the timeout period, it sets its state to member. If the undecided node times out, but detects some bi-directional links with some neighbours, it declares itself as the clusterhead. Otherwise, it remains in the undecided state and repeats this process to become either a clusterhead or a member.

Each node maintains a neighbour table. Each entry in the neighbour table contains information about each neighbour, that is, the status of the associated link (uni-directional or bi-directional) and the state of the neighbour (clusterhead or member). A clusterhead keeps a list of its members. It also maintains a cluster adjacency table where each entry contains information about each neighbouring cluster, that is, the gateway through which the cluster can be reached and the clusterhead of the cluster.

When a source wants to send data to a destination, it broadcasts route request packets to its neighbourhood. When a clusterhead receives the request, it checks if the destination is located within its cluster. If the destination is available within its cluster, it forwards the request to the destination. Otherwise it rebroadcasts the request to all its neighbouring clusterheads. This process continues until the destination receives the request packet and responds with a reply. The propagation of route request and route reply is similar to that of DSR. If the source does not receive a reply within a timeout period, it backs off exponentially before sending a route request again.

CBRP uses route shortening to reduce the length of a route. To do so a node receiving a source route packet tries to find the farthest node in the route that is its neighbour. This situation can arise due to a topology change. If such a neighbour is found, the node sends the packet to that neighbour directly.

While forwarding a data packet, if a node detects a link failure, it sends an error message to the source and also tries to forward the packet through a local repair mechanism. In the local repair mechanism, the node checks if the next hop or the hop after the next hop can be reached through any of its neighbours. If the node succeeds, the data packet can be delivered to the destination over the repaired path.

Like other hierarchical routing protocols, cluster formation and maintenance involve additional communication and processing overhead. CBRP may provide invalid routes temporarily as a node moves from one cluster to another (Abolhasan et al., 2004).

Multi-Path Source Routing (MSR)

MSR (Wang, Zhang, Shu, Dong, & Yang, 2001) is an extension of DSR. It tries to improve end-to-end delay, average queue size, network congestion, and path fault tolerance by employing the multi-path finding capability of DSR.

MSR allows the source to receive multiple route reply packets in a single route discovery phase. Each route discovered is stored in the route cache with a unique route index so that multiple routes for a particular destination can be distinguished properly. Each route with index i for a particular destination j is assigned a weight W_i^j based on the round trip delay of that route. The weight is measured in terms of the number of packets to be sent consecutively on the same route and is used for distributing load among multiple routes for a particular destination. If d_{max}^j is the maximum delay of all the routes to destination j, d_i^j is the delay of the route with index i for destination j, U is a bound to ensure that W_i^j does not become too large and R is a factor to control switching frequency between routes, then W_i^j is calculated as follows:

$$W_i^j = Min_j\left(\left\lceil \frac{d_{max}^j}{d_i^j} \right\rceil, U\right) \times R$$

MSR sends periodic probe packets along the active routes to measure their round trip delays as well as to test their validity.

Ad-Hoc On-Demand Multi-Path Distance Vector (AOMDV) Routing

AOMDV (Marina & Das, 2001b, 2003) extends AODV to support multi-path routing in mobile ad-hoc networks. It adds some extra fields in routing tables and control packets, and requires few new rules to be followed during a route discovery phase in order to compute loop-free and link-disjoint multiple routes. Link-disjoint routes do not contain any common link among the multiple routes between a source and destination pair.

Every node maintains a variable called the advertised hop count for each destination in order to achieve loop-freedom. This variable is added in each RREQ (route request) or RREP (route reply) and in the routing table with the usual fields that are used for AODV. When a node initiates a RREQ or RREP with a particular destination sequence number, its advertised hop count field is set to the length of the longest available path to the destination expressed in terms of the number of hops. The advertised hop count remains unchanged until the associated destination sequence number is changed. The rules for loop-freedom state that if a node receives a RREQ (RREP) for a particular destination with a destination sequence number: (a) higher than the one stored in its routing table, it should update its routing information with the information obtained from the received RREQ (RREP); (b) equal to the one stored in its routing table, it can re-send the received RREQ (RREP) if the advertised hop count in the RREQ (RREP) is greater than the corresponding value in its routing table; and (c) equal to the one stored in its routing table, it can update its routing table with the information contained in the received RREQ (RREP) if the advertised hop count in the RREQ (RREP) is less than the corresponding value in its routing table.

For link-disjointness, each node maintains a route list in its routing table for each destination. A route list for a particular destination contains entries with next hop, last hop, and hop count information for the destination. The next hop refers to a downstream neighbour via which the destination can be reached. The last hop is the node immediately preceding the destination. The hop count measures the distance from the node to the destination via the associated next and last hops. If a node can ensure that all paths to a destination from itself differ in their next and last hops, then link-disjointness among those paths can be achieved. AOMDV uses this observation to ensure link-disjointness among multiple routes for the same source and destination pair. For this purpose AOMDV adds a last hop field in each RREQ and RREP.

During route discovery AOMDV allows all copies of an RREQ to be examined for potential alternate reverse paths. When an intermediate node receives an RREQ, it creates a reverse path if the RREQ satisfies the rules for loop-freedom and link-disjointness. It then checks if it has one or more valid next hop entries for the destination. If such an entry is found, the node generates an RREP and sends it back to the source along the reverse path. If the intermediate node does not find such an entry and has not previously broadcast any copy for this RREQ, it rebroadcasts the RREQ.

When the destination receives RREQ copies, it follows the same rules for creating reverse paths. However, unlike intermediate nodes, it generates an RREP for every copy of RREQ that arrives via a loop-free path. This feature increases the possibility of finding more disjoint routes.

Multiple-route ad-hoc on-demand distance vector (MRAODV) routing (Higaki & Umeshima, 2004) is another extension of AODV that has similar aims to AOMDV. During the propagation of route request packets in AOMDV the links through which route request packets arrive are stored as backward links in order to establish potential reverse paths for the route replies. Unlike AOMDV, MRAODV switches the direction of some of these backward links so that the number of multiple routes can be increased. However this method of link reversal may not produce link-disjoint routes. Moreover, it is unclear if this method can preserve loop-freedom.

Ant-Colony Based Routing Algorithm (ARA)

ARA (Gunes, Sorges, & Bouazizi, 2002) adopts the food searching behavior of ants to find routes. When ants search for food, they start from their nest and walk towards the food. While walking they leave behind a transient trail by depositing pheromone, a substance that ants can smell. The concentration of pheromone on a certain route indicates its usage and allows other ants to follow the most commonly used route. In the course of time the concentration of pheromone is reduced due to diffusion. Like AODV, ARA uses sequence numbers to avoid forming loops. However, unlike AODV, ARA can find multiple routes between a source and destination pair.

During route discovery, a FANT (forward ant) packet is propagated through the network similar to RREQ in AODV. When a node receives a FANT for the first time, it calculates a pheromone value depending on the number of hops the FANT has traveled to reach the node. It creates an entry in its routing table with the calculated pheromone value, the address of the neighbour from which the FANT was received, and the address of the source from which the FANT originated. This entry in the routing table creates a pheromone track towards the source. Then the node forwards the FANT to its neighbours. Sequence numbers, similar to AODV, are used to avoid duplicate FANTs and prevent forming loops.

Once a FANT reaches the destination, the destination creates a BANT (backward ant) from the extracted information of the FANT and returns the BANT to the source. The BANT performs a similar task to the FANT, that is, establishing a pheromone track towards the destination. Unlike, the FANTs, the propagation of the BANTs enables ARA to establish multiple paths between a source and destination pair. When the source receives a BANT, a path is established between the source

Table 2. Comparison of various reactive routing protocols

	WCC [RD]	WCC [RM]	WTC [RD]	WTC [RM]	RS	MR	PB	*Advantages*	*Disadvantages*
LMR	$O(2N)$	$O(2a)$	$O(2D)$	$O(2D)$	F	Yes	No	Multiple routes	Requires reliable delivery of routing control packets; Can suffer from temporary routing loops
DSR	$O(2N)$	$O(2N)$	$O(2D)$	$O(2D)$	F	Yes	No	Intermediate nodes do not store route information; Can provide multiple paths	Stale caches and relay storm problems may arise in large and highly mobile MANETs; Additional communication overhead due to source routing
ABR	$O(N+r)$	$O(a+r)$	$O(D+c)$	$O(b+c)$	F	No	Yes	Stable routes; Localised route repair mechanism	Suitable for small MANETs; Frequent beacons may result in extra bandwidth and power consumptions;
SSA	$O(N+r)$	$O(a+r)$	$O(D+c)$	$O(b+c)$	F	No	Yes	Stable routes	Introduces more delays than DSR to find routes; Does not have any localised route repair mechanism
TORA	$O(2N)$	$O(2a)$	$O(2D)$	$O(2D)$	F	Yes	No	Localised route maintenance	Can falsely detect partitions; Requires reliable and in-order delivery of route control packets; Temporary routing loops
LAR	$O(2e)$	$O(2e)$	$O(2d)$	$O(2d)$	F	Yes	No	Limits the propagation of routing control packets	Flooding is used if no location information is available; Behaves like a flooding algorithm in highly mobile MANETs
AODV	$O(2N)$	$O(2N)$	$O(2D)$	$O(2D)$	F	No	Yes*	Adaptable to highly dynamic topologies; Multicast routing capability	Requires HELLO messages; Does not support multiple routes; Intermediate nodes need to store routing information; May not scale well with network size
RDMAR	$O(2e)$	$O(2e)$	$O(2d)$	$O(2d)$	F	No	No	Limits the propagation of routing control packets	Flooding is used if nodes do not have any prior communication; Suited for MANETs having low to moderate topological changes
CBRP	$O(2m)$	$O(2a)$	$O(2D)$	$O(2b)$	H	No	No	Reduces communication; Localised route maintenance	Introduces additional overhead for forming and maintaining clusters; Temporary routing loops
MSR	$O(2N)$	$O(2N)$	$O(2D)$	$O(2D)$	F	Yes	Yes#	Multi-path routing and load balancing	Requires periodic probe packets to gather information
AOMDV	$O(2N)$	$O(2N)$	$O(2D)$	$O(2D)$	F	Yes	Yes*	Link-disjoint multi-path routing	Requires periodic HELLO messages
MRAODV	$O(2N)$	$O(2N)$	$O(2D)$	$O(2D)$	F	Yes	Yes*	May provide more multiple paths than AOMDV	Requires periodic HELLO messages; May not produce link-disjoint routes; May not preserve loop-freedom
ARA	$O(N+r)$	$O(a+r)$	$O(D+c)$	$O(D+c)$	F	Yes	No	Multiple routes; Localised route maintenance	Route discovery is based on flooding

WCC: Worst Case Communication Complexity, i.e. number of messages needed to perform a route discover or an update operation in worst case; *WTC*: Worst Case Time complexity, i.e. number of steps involved to perform a route discovery or an update operation in worst case; *RD*: Route Discovery; *RM*: Route Maintenance; *RS*: Routing Structure; *F*: Flat; *H*: Hierarchical; *MR*: Multiple Routes; *PB*: Periodic Beacons; *N*: Number of nodes in the network; *D*: Diameter of the network; *a*: Number of affected nodes; *D*: Diameter of the network; *c*: Diameter of the directed path of RREP, BANT; *d*: Diameter of the localised region; e: Number of nodes in the localised region; *r*: Number of nodes in the route reply path; *m*: Number of clusters in CBRP; *: Beacons in terms of HELLO Messages; #: Sends periodic probe packets along active routes;

and the destination through which data packets can be sent.

Each time a node relays a data packet to the next hop toward the destination, it increases the pheromone value of the corresponding entry in its routing table. If a link of a route is not used, its pheromone value decreases over time.

While forwarding a data packet, if a node detects a link failure, it first checks its routing table to find an alternate route to the destination of that data packet. If no route is found, it requests its neighbours to find a route to the destination. If the neighbours fail to find an entry in their routing tables for the destination, the request backtracks until it reaches the source node. The source then can initiate a new route discovery phase if needed.

Since the route discovery process is based on flooding, ARA may not scale well as the numbers of nodes and flows increase (Abolhasan et al., 2004).

Comparisons of Reactive Routing Protocols

Most of the reactive routing protocols use a flat routing structure. Nodes using flat reactive routing protocols usually flood route discovery packets through the entire network to find a feasible route to the destination. LAR and RDMAR can reduce the number of route discovery packets by limiting the search space within a calculated region. However if an estimated location of a remote node is not known a priori, these protocols behave like a pure flooding based algorithm. In ABR and SSA routing overheads are minimised by selecting stable routes. Routes selected in this manner are likely to be long lived and consequently would require fewer route reconstructions and less route control traffic. However, routes selected in this way may not be the shortest in terms of the numbers of hops. ABR, TORA, and ARA provide localised route repair mechanisms to reduce delays, and

limit route control packets that could otherwise be increased if alternate routes were required to be found by the source nodes.

CBRP reduces control overhead by applying a hierarchical structure to the network, since during route discovery only the clusterheads exchange routing information (Abolhasan et al., 2004). CBRP further minimises delay and the number of control packets by providing a localised route repair mechanism. However, like most other hierarchical routing protocols, CBRP may incur excessive processing and communication overheads for cluster formation and maintenance in MANETs. Therefore CBRP is most suitable for medium-sized networks with slow to moderate mobility (Abolhasan et al., 2004). Table 2 compares the main characteristics of various reactive routing protocols.

HYBRID ROUTING PROTOCOLS

These protocols combine the features of both pro-active and reactive routing strategies to scale well with the increase in network size and node density. This is usually achieved by maintaining routes to nearby nodes using a pro-active routing strategy and determining routes to far-away nodes using a reactive routing strategy. Description and comparison of a number of such protocols are provided in the rest of this section.

Zone Routing Protocol (ZRP)

ZRP (Haas, 1997; Haas & Pearlman, 1998) utilises both pro-active and reactive routing strategies in order to gain benefits from the advantages of both types.

Each node in ZRP has a routing zone centred at itself. The radius of the zone is expressed in terms of the number of hops. Nodes within the same zone can use any pro-active routing protocol to maintain routing information. If a source needs to send packets to a destination, it looks in its routing table to find out if the destination is within its routing zone. If so, the packet can be routed using any pro-active routing protocol.

If the source does not find an entry for the destination in its routing table, it uses a route request/route reply cycle of any reactive routing protocol to determine a route to the required destination. Each zone contains some nodes on its border, referred to as border nodes. A route request packet is propagated from one zone to another through these border nodes until is reaches the zone of the required destination.

Zone-Based Hierarchical Link-State (ZHLS) Routing

Unlike ZRP, ZHLS (Joa-Ng & Lu, 1999) divides the network into non-overlapping zones and employs a hierarchical structure to maintain routes. Unlike other hierarchical protocols, ZHLS does not require any clusterheads so avoids traffic bottlenecks, single points of failure, and complicated mobility management. It is pro-active if the destination resides within the same zone of the source. Otherwise it is reactive, since location search is employed to find the zone ID of the destination. Thus it reduces communication overheads compared to any pure reactive routing protocol such as DSR and AODV.

ZHLS defines two levels of topologies: node level and zone level. The node level topology provides the information on how the nodes are connected through physical links. If there is at least one physical link connecting two zones, a virtual link is assumed to exist between those zones. The zone level topology tells how zones are connected by these virtual links.

Initially a node knows its physical position by using a GPS and determines its zone ID by mapping its physical location to the zone map. With this zone ID, the node starts the intra-zone clustering and then the inter-zone clustering

procedures to build its routing tables. To have a preprogrammed zone map may not be feasible in networks where physical boundaries of the zones are dynamic.

During intra-zone clustering, each node asynchronously broadcasts a link request to which neighbouring nodes respond with their node IDs and zone IDs. When all the link request responses are received, the node broadcasts a node LSP (link-state packet) containing the node IDs of its neighbours of the same zone and the zone IDs of the neighbours of different zones. The node LSP is propagated only within its zone. After receiving all node LSPs of the same zone, each node knows the node level topology of that zone and can use a shortest path algorithm to build its intra-zone routing table.

Nodes that receive link request responses from their neighbouring zones are called the gateway nodes. During inter-zone clustering, these gateway nodes broadcast zone LSPs throughout the network. A zone LSP contains the list of all the neighbouring zones from which it is originated. After each node receives all zone LSPs, it can build an inter-zone routing table.

When a source node wants to send data to a destination, it checks if the destination exists in its intra-zone routing table. If so, the packet can be routed to the destination like any other link-state routing protocol using the information from intra-zone routing tables of the intermediate nodes. If the destination resides in a different zone, the source sends a location request according to its inter-zone routing table. The gateway of each zone will receive this request and check if the destination exists in its intra-zone routing table. If so the gateway replies with a location response containing the zone ID of the destination. When the source receives this reply, it specifies the zone ID and the node ID of the destination in the data header and routes the data according to its inter-zone routing table.

Scalable Location Update Routing Protocol (SLURP)

Like ZLHS, SLURP (Woo & Singh, 2001) organises nodes into a number of non-overlapping regions. However it does not employ a global route discovery mechanism and thereby reduces the cost of maintaining routing information.

SLURP assigns a home region for each node in the network. It assumes that each node uses a GPS to know its current location, contains a list of IDs of other nodes in the network, and uses a one-to-many static mapping function f, that is, $f(Node\ ID) \rightarrow Region\ ID$, to determine the home region of other nodes. This function is known to all nodes in the network and is assumed to generate always the same home region for a specific node ID.

Each node always informs the nodes, currently present in its home region, the identity of the region it is located in by unicasting a location update message towards its home region. Once the location update message reaches the home region, it is broadcast to all nodes in the home region.

If a source node wants to send data to a destination node, it needs to find the current location of the destination. The home region of the destination contains the information about the current location of the destination. To get this information the source determines the home region of the destination using a static mapping function and then unicasts a location discovery packet to that home region using the most forwarding with fixed radius (MFR) (Hou & Li, 1986) geographic routing algorithm. In MFR the source sends the route discovery packet to one of its neighbours, which is closest to the destination in terms of physical distance. This process repeats until the route discovery packet reaches a node in the home region of the destination. This node then generates a location reply packet containing the ID of the destination's current region.

There can be cases when a home region may contain only one node and the node moves to another region from its home region. In this case the source node will not receive a reply. To address this issue, the source sends the location discovery packet, after a certain period, to the surrounding regions of the home region of the destination node, with the hope that the destination has registered itself in any of those regions. This process of expanding the search region continues until a threshold is reached.

Once the current location of the destination is found, the data packets are routed to the destination's current region using MFR. Once one of the nodes in the destination's current region gets the packets, it can route them to the destination using DSR if it contains a cached route to the destination. Otherwise that node floods the packets, using a method similar to route discovery in DSR, within its current region. Eventually the packets reach the destination. Thus SLURP uses MFR to get data packets to the current region of the destination and then DSR to get the packets to the destination.

Distributed Spanning Tree (DST) Based Routing Protocol

DST (Radhakrishnan, Racherla, Sekharan, Rao, & Batsell, 1999, 2003) uses spanning trees in regions where the topology is stable and a flooding-like scheme in highly dynamic regions of the network.

In DST, nodes are grouped into a number of disjoint trees. If two nodes of two trees come within the transmission range of each other but are likely to move away shortly, they form a bridge between the two trees. Otherwise those trees merge to form a larger tree and one of the roots of the trees becomes the root of the newly merged tree.

To determine a route, the authors have proposed two different routing techniques: hybrid tree-flooding and distributed spanning tree shuttling. In hybrid tree-flooding packets are sent to all possible neighbours and adjoining bridges. When a node receives a packet, it stores the packet for a period, known as the *holding-time*, after which the packet is deleted. The rationale for the holding-time is that for systems, which are becoming more stable and connected, it might be useful to buffer packets and route them as the connectivity of the network increases over time. In distributed spanning tree shuttling, the routing happens in several steps. In the first step, if the source node is not the root, the packet is sent to the root. In the second step the packet is re-broadcast along the tree edges toward the leaf nodes. When a packet reaches a leaf node, the packet is sent upwards, which is the third step, until it reaches a certain height known as the *shuttling level*. Once a node in the shuttling level is reached, the packet is sent along the adjoining bridges and the algorithm continues from the second step.

Radhakrishnan et al. (1999, 2003) claim that shuttling mechanism uses fewer messages compared with tree-flooding. However, DST is prone to a single point of failure due to the reliance on the root node to configure the tree (Abolhasan et al., 2004). Moreover, the holding time may introduce extra delays for packets that may be unsuitable for some time sensitive applications.

Hybrid Ad-Hoc Routing Protocol (HARP)

HARP (Nikaein, Bonnet, & Nikaein, 2001) is a tree-based hybrid routing protocol. The trees are connected via gateway nodes, that is, the neighbouring nodes belonging to different trees, to form a forest. Unlike DST, HARP does not require the trees to have root nodes. The trees are also referred to as zones. Similar to ZHLS, the zones in HARP do not overlap. However, unlike ZHLS, HARP does not rely on a static zone map. Moreover, it does not require a clusterhead to coordinate data and control packet transmissions.

For zone creation, HARP relies on the distributed dynamic routing algorithm (DDR) (Nikaein, Labiod, & Bonnet, 2000). DDR con-

sists of six cyclic time-ordered phases: preferred neighbour election, forest construction, intra-tree clustering, inter-tree clustering, zone naming, and zone partitioning. It uses only beacons, which are periodic messages exchanged between a node and its neighbours, to perform each of these phases. Thus it avoids global broadcasting throughout the network. In the beginning, each node starts the preferred neighbour election procedure to choose a preferred neighbour. The preferred neighbour of a node is the node that has the most neighbours. Then a forest is constructed by connecting each node to its preferred neighbour and vice versa. After that the intra-tree clustering algorithm is used to build up the structure of the zone and to build up the intra-zone routing table. Next the inter-tree clustering algorithm is executed to determine the connectivity of neighbouring zones. The gateway nodes keep this information in their inter-zone routing tables. Each constructed tree is assigned a name by executing the zone naming algorithm. Then the network is partitioned into a number of non-overlapping zones.

In HARP, routing is performed on two levels: intra-zone and inter-zone depending on whether the destination belongs to the same zone or in a different zone. The intra-zone routing mechanism uses a pro-active approach to find the destination within a zone. For inter-zone routing, HARP uses a reactive approach. During the route discovery phase, route request packets propagate from zone to zone via the gateway nodes. Inside a zone, the route request packets follow the tree structure provided by DDR. Therefore, unlike ZRP, HARP limits the flooding of route request packets to a subset of nodes. After this limited flooding, several paths may be discovered for a given destination. The destination chooses the most suitable path and unicasts a path reply packet to the source.

HARP assigns each path a refresh time, after which a new route discovery phase is triggered to avoid path failure, as the network topology may change over time. If a link failure occurs in the meantime, the node that detected the link failure sends a path error message to the source and holds the traffic for a period. The rationale for this holding period is that HARP applies the pro-active approach inside the zone and there is a chance of receiving new routing information embedded in the periodic beacons. This feature may increase the percentage of data received successfully at the destination. When the source receives the path error message it initiates a new route discovery procedure.

Sharp Hybrid Adaptive Routing Protocol (SHARP)

SHARP (Ramasubramanian, Haas, & Sirer, 2003), unlike other hybrid routing protocols, adapts between pro-active and reactive routing strategies by adjusting the radii of pro-active zones dynamically.

SHARP creates pro-active zones automatically around the destinations that receive data from many sources. SHARP has borrowed techniques from DSDV and TORA to build an efficient pro-active routing protocol known as the SPR (SHARP pro-active routing) protocol.

SPR performs routing by building and maintaining a directed acyclic graph (DAG) rooted at the destination in each zone. Therefore nodes within a pro-active zone maintain routes pro-actively only to the destination node. The topology forming mechanism of SPR does not guarantee that a node can always find the shortest path, in terms of hop count, to a destination. SPR requires the destination to periodically initiate a DAG construction process. In addition to this, SPR uses the failure recovery mechanism of TORA to restore the DAG in response to link-failures. Each node in a pro-active zone periodically broadcasts update packets to its neighbours. Each update packet contains the height (used in TORA) of the transmitted node and also acts as a HELLO beacon.

The nodes that are not located in the pro-active zone of a given destination use a reactive routing protocol, that is, AODV in this case, to establish routes to that destination. Once a data packet enters the pro-active zone of a destination using AODV, it is routed using SPR.

In SHARP, each destination monitors the network characteristics as well as the data traffic characteristics. Monitoring network characteristics enables SHARP to vary the radius of the pro-active zone in order to reduce per-packet routing overhead (that is, route discovery, update, and maintenance costs). Monitoring data traffic characteristics is used to adjust the radius of the pro-active zone in order to reduce loss rate and delay jitter in the application layer.

In order to estimate per-packet routing overhead of SPR, the nodes in a pro-active zone measure the average link lifetime (λ) of their immediate links and the number of immediate neighbours (n). These values are sent to the destination D of a pro-active zone periodically. Let f_u be the frequency at which each node within the pro-active zone generates update packets, and let f_c be the frequency at which D performs DAG constructions. If N_r^D is the number of nodes around D with radius r then the total fixed overhead of SPR pro-active routing component is N_r^D $(f_u + f_c)$ packets/second. The average frequency of event-triggered updates at a node can be approximated by

$$\frac{1}{2(\beta_n - 1)\lambda} \text{ where } \beta_n = \sum_{i=1}^{n} \frac{1}{i}$$

Therefore, the total routing overheads of SPR at node D is

$$N_r^D \left(f_u + f_c + \frac{1}{2(\beta_n - 1)\lambda} \right) \qquad (1)$$

On the other hand the per-packet routing overhead of AODV for each source S outside the pro-active zone at a distance h hops from D can be approximated by

$$N^S_{h-r}\left(\frac{h-r}{\lambda}\right) \tag{2}$$

Here N^S_{h-r} is the number of nodes around S with radius $h\text{-}r$.

D estimates the incremental difference in the overheads of the pro-active and reactive routing components using Equation (1) and Equation (2) respectively. If the reduction in the overhead of the reactive component is more than *up_threshold* times the increase in the overhead of the pro-active component, r is increased by 1. Conversely if the reduction in the overhead of the pro-active component is more than *down_threshold* times the increase in the overhead of the reactive component, then r is decreased by 1. Otherwise r is kept fixed.

To adjust the radius of the pro-active zone in order to reduce loss rate and delay jitter, the destination measures loss rate and delay jitter from its observed traffic pattern. If the perceived loss rate (or the delay jitter) is more than a threshold, r is increased by 1. On the other hand r is decreased by 1 if the measured loss rate (or delay jitter) is less than a threshold.

It has been shown by simulation that SHARP performs well compared to a pure reactive or pro-active routing protocol in an ad-hoc network where most of the nodes are sources with few destinations and moderate mobility. Since SHARP performs some additional steps and requires extra control packets compared to that of the constituent protocols in their original forms, there is a possibility that these extra overheads may not be compensated by the overall network performance improvement if the number of destinations or the mobility of the nodes increases for the same network.

Comparisons of Hybrid Routing Protocols

Hybrid routing protocols try to minimise routing overheads compared to any reactive and pro-active routing protocol as network size increases. However, since most of the hybrid routing protocols are hierarchical in structure, they are likely to incur excessive processing and communication overheads for forming and maintaining clusters/zones/trees in highly mobile MANETs. Large MANETs with group mobility are likely to be best served by these protocols.

DST relies on a single node to configure the associated tree. Hence it can introduce extra routing delays, and excessive processing and communication overheads when the root node of a tree becomes unreachable by its children.

In HARP, nodes have to pass their traffic through a subset of neighbours known as preferred neighbours. These preferred nodes have to transmit more routing and data packets than any other nodes and hence may get less sleep time to conserve energy than others (Abolhasan et al., 2004). Since many nodes would want to communicate with the same preferred neighbour, channel contention would increase around that preferred neighbour. In networks with high node density and traffic, channel contention around preferred neighbours would increase waiting for transmission channel. This can significantly reduce overall throughput of the network since packets have to be dropped when buffers become full (Abolhasan et al., 2004).

On the other hand, ZHLS, SLURP, and SHARP do not rely on fixed nodes to perform any critical operation such as cluster formation. Therefore these protocols may have lower processing and communication overheads for large MANETs than other hybrid routing protocols. Table 3 shows the comparison of various hybrid routing protocols based on their major characteristics.

Table 4 provides an overall comparison of the pro-active, reactive, and hybrid routing strategies.

Table 3. Comparison of various hybrid routing protocols

	WCC[I]	WCC [I',RD]	WCC [I',RM]	WTC [I]	WTC [I',RD]	WTC [I'RM]	RS	PB	*Advantages*	*Disadvantages*
ZRP	O(n)	O(N+r)	O(N+r)	O(d)	O(2D)	O(2D)	F	Yes*	Reduced communication compared to pure pro-active routing algorithms; Faster route discovery within a zone than any pure reactive routing protocol	For large values of routing zone it may behave like a pure reactive routing protocol; Overlapping zones
ZHLS	O(N/M)	O(N+r)	O(N+r)	O(d)	O(2D)	O(2D)	H	No	Non overlapping zone and hence capable of supporting frequency reuse	Requires GPS, Requires static zone maps;
SLURP	O(x+N/M)	O(2y)	O(2x)	O(2d)	O(2D)	O(2D)	H	No	Home region reduces cost of maintaining routing information by eliminating global route discovery	Requires static zone maps;
DST	O(n)	O(2N)	O(N)	O(h)	O(2D)	O(2D)	H	No	Holding time reduces retransmissions	Requires trees to have root nodes; Single point of failure
HARP	O(n)	O(N+r)	O(N+r)	O(d)	O(2D)	O(2D)	H	Yes	Applies early route maintenance to avoid extra delay caused by path failure during data transmission; Does not require zone map; Does not require the trees to have roots	Preferred neighbours may become bottlenecks and hence degrade network performance
SAHRP	O(n)	O(2N)	O(y)	O(d)	O(2D)	O(D)	F	Yes	Automatically finds the balance point between pro-active and reactive routing.	Moderate mobility; More overhead than pure AODV and TORA

WCC: Worst Case Communication Complexity, i.e. number of messages needed to perform a route discovery or an update operation in worst case; *WTC*: Worst Case Time complexity, i.e. number of steps involved to perform a route discover or an update operation in worst case; *RD*: Route Discovery; *RM*: Route Maintenance; *I*: Intra zone; *I'*: Inter zone; *RS*: Routing Structure; *F*: Flat; *H*: Hierarchical; *PB*: Periodic Beacons; *N*: Number of nodes in the network; *D*: Diameter of the network; *d*: Diameter of a zone, home region or cluster or tree; *n*: Number of nodes in a zone, home region, cluster or tree; *r*: Number of nodes in the route reply path; *M*: Number of zones, home regions or clusters; *x*: Number of nodes in the path to the home region; *h*: Height of the tree; *y*: Number of nodes from the source of route error to the source node; *: If pro-active and reactive routing protocols using beacons are used;

Table 4. Comparison of pro-active, reactive, and hybrid routing protocols (Abolhasan et al., 2004)

	Pro-active	Reactive	Hybrid
Routing Structure	Both flat and hierarchical	Usually flat	Usually Hierarchical
Availability of Routes	Always available.	Determined when needed. Sometimes overheard routes are stored for a limited time (e.g. in DSR).	Always available within if source and destination reside within the same zone/cluster/tree.
Volume of control traffic	Usually high. Exceptions such as FSLS and HOLSR.	Usually lower than pro-active routing.	In most cases lower than pro-active and reactive routing protocol.
Storage requirement	Usually high	Usually lower than pro-active routing protocols	Usually lower than pure pro-active and reactive routing protocols if the size of zones/clusters/trees can be properly determined in large networks.
Delay for route discovery	Predetermined	Higher than pro-active routing protocols	Similar to pro-active routing protocols if source and destination are located within the same zone/cluster/tree. Otherwise usually higher than pro-active but lower than reactive.
Mobility support	Low to moderate mobility is supported. Group mobility is usually required for hierarchical structured routing.	Can support higher mobility than pro-active routing protocols.	Usually supports lower level of mobility than reactive routing protocols since routing structure is mostly hierarchical in this approach.
Scalability	Usually up to 100 nodes. FSLS and HOLSR may scale higher.	Source routing protocol does not scale well, usually up to few hundred nodes. Hop by hop routing scales better than source routing.	1000 or more.

OPEN RESEARCH CHALLENGES

Ad-hoc networks have been one of the popular research fields in ubiquitous communications for more than a decade. Researchers have explored various ad-hoc routing protocols at various levels of detail. However, no routing protocol has become a winner for all scenarios. The following aspects have been considered to make routing protocols more efficient and robust:

- Quality of service (QoS)
- Flexibility and scalability
- Security
- Energy efficiency
- Multicast
- Antennas

QoS routing aims to guarantee certain performance for a flow in terms of bandwidth, delay, jitter, successful delivery probability, and so forth. However due to the dynamics of nodes (e.g., limited power, mobility, etc.), fluctuating link characteristics, limited radio bandwidth, varying level of radio channel contention along multiple hops (Islam, Pose, & Kopp, 2005a, 2005b, 2005c), and varied user demand, routing with guaranteed end-to-end QoS is not a trivial issue.

Since no single routing protocol has been proven to be a winner, a routing protocol has to be flexible enough to switch between various pro-active and reactive routing protocols based on characteristics of the network. SHARP adapts between pro-active and reactive routing strategies by adjusting dynamically the radii of pro-active zones based on link lifetime. However other network parameters can be considered to make the routing protocol more flexible to scale adaptively with the heterogeneity of nodes' capabilities, network size, user demand, and so forth.

Usually the routing protocols assume node in the network would follow the protocol description properly. However, due to the lack of clear physical boundary, wireless networks are at more risk from attackers than their cabled counterpart. Multi-hop ad-hoc networks make the security problem even more challenging since these may lack central trusted servers. Several intrusion prevention and detection schemes (Yang, Luo, Ye, Lu, & Zhang, 2004) have been proposed. These solutions come at a cost. Implementational complexity, processing, and communication costs are involved that need to be minimised for fully automated and power conserved ad-hoc networks.

Nodes in an ad-hoc network are assumed to route traffic for other nodes if needed. These nodes may have to rely on limited battery power if they are mobile or not connected to a main power source. Examples of such nodes are nodes in a sensor network. In these situations the issue of energy efficiency becomes one of the most important problems. There are routing protocols, for example, Singh, Woo, and Vaidya (1998), that try to conserve power. However, it is still an open research challenge to determine how to conserve power and perform routing at the same time without compromising overall network performance.

Multi-cast routing enables one-to-many communication. For ad-hoc networks multi-cast routing must be able to cope with network size and node mobility (Royer & Toh, 1999). Moreover issues like uni-directional links (Gerla, Lee, Park, & Yi, 2005) and power conservation (Wan, Calinescu, & Yi, 2004) have to be considered.

Ad-hoc routing protocols usually assume each node is equipped an omni-directional antenna. In depth investigation is needed to see the effect of using smart, or multiple omni- or directional antennas at the routing layer of ad-hoc networks.

CONCLUSION

Ad-hoc network topologies can change dynamically. Similarly performance of wireless links can vary unpredictably. Hence routing in ad-hoc networks is much more difficult than in conventional networks. In this chapter we have provided a description of various routing protocols suitable for ad-hoc networks. We have evaluated their major characteristics and differences. The study suggests that no single routing protocol or class of protocols is best suited for all scenarios. In this chapter we have looked at various hybrid routing protocols with the aim of combining the most appropriate features of each for ad-hoc network situations. While this leads to various practical approaches to ad-hoc routing, there are still many open research questions to be answered before generally applicable routing protocols can be developed to suit ad-hoc networks, especially when they are used to support demanding multimedia applications. We have outlined some of these research challenges.

REFERENCES

Abolhasan, M., Wysocki, T., & Dutkiewicz, E. (2004). A review of routing protocols for mobile ad hoc networks. *Ad Hoc Networks, 2*(1), 1-22.

Aggelou, G., & Tafazolli, R. (1999). RDMAR: A bandwidth-efficient routing protocol for mobile ad-hoc networks. *Proceedings of the 2nd ACM International Workshop on Wireless Mobile Multimedia (WOWMOM)*, Seattle, Washington, August 20 (pp. 26-33). ISBN 1-58113-129-1. New York: ACM Press.

Basagni, S., Chlamtac, I., Syrotiuk, V. R., & Woodward, B. A. (1998). A distance routing effect algorithm for mobility (DREAM). *Proceedings of the 4th International Conference on Mobile Computing and Networking (MobiCom)*, Dalla, Texas, October 25-30 (pp. 76-84). ISBN 1-58113-035-X. New York: ACM Press.

Bellur, B., & Ogier, R. G. (1999). A reliable, efficient topology broadcast protocol for dynamic networks. *Proceedings of the 18th Annual Joint Conference of the IEEE Computer and Communications Societies (InfoCom)*, New York, March 21-25 (pp. 178-186). ISBN 0-7803-5417-6, 1. New York: IEEE.

Bertsekas, D., & Gallager, R. (1987). *Data networks* (pp. 297-333). NJ: Prentice-Hall Inc.

Broch, J., Maltz, D., Hu Y.-C., & Jecheva, J. (1998). A performance comparison of multi-hop wireless ad hoc network routing protocols. *Proceedings of the 4th ACM/IEEE International Conference on Mobile Computing and Networking (MobiCom)*, Dallas, Texas, October 25-30 (pp. 85-97). ISBN 1-58113-035-X. New York: ACM Press.

Chen, T.-W., & Gerla, M. (1998). Global state routing: A new routing scheme for ad-hoc wireless networks. *Proceedings of the IEEE International Conference on Communications (ICC)*, Atlanta, Georgia, June 7-11 (pp. 171-175). New York: IEEE.

Chiang, C., Wu, H., Liu, W., & Gerla, M. (1997). Routing in clustered multihop, mobile wireless networks. *Proceedings of the IEEE Singapore International Conference on Networks (SICON)*, Singapore, April 14-17 (pp. 197-211). New York: IEEE.

Corson, M. S., & Ephremides. (1995). A distributed routing algorithm for mobile wireless networks. *ACM/Baltzer Wireless Networks, 1*(1), 61-81.

Das, S. R., Castaneda, R., & Yan, J. (2000). Simulation-based performance evaluation of routing protocols for mobile ad hoc networks. *ACM/Baltzer Mobile Networks and Applications (MONET), 5*(3), 179-189.

Dube, R., Rais, C. D., Kuang-Yeh, W., & Tripathi, S. K. (1997). Signal stability-based adaptive routing (SSA) for ad-hoc mobile networks. *IEEE Personal Communications, 4*(1), 36-45.

Gafni, E., & Bertsekas, D. (1981). Distributed algorithms for generating loop-free routes in networks with frequently changing topology. *IEEE Transactions on Communications, 29*(1), 11-18.

Garcia-Luna-Aceves, J. J., & Spohn, M. (1999). Source-tree routing in wireless networks. *Proceedings of the 7th Annual International Conference on Network Protocols (ICNP)*, Toronto, Canada, October 31-November 3 (pp. 273-282). ISBN 0-7695-0412-4. Los Alamitos, CA: IEEE Computer Society Press.

Gerla, M., Hong, X., & Guangyu, P. (2000). Landmark routing for large ad hoc wireless networks. *Proceedings of the IEEE Global Telecommunications Conference (GLOBECOM)*, 3, San Francisco, November 27-December 1 (pp. 1702-1706). New York: IEEE.

Gerla, M., Lee Y.-Z., Park, J.-S., & Yi, Y. (2005). On demand multicast routing with unidirectional links. *IEEE Wireless Communications and Networking Conference, 4*, 2162-2167.

Gonzalez, L. V., Ge, Y., & Lamont, L. (2005). HOLSR: A hierarchical proactive routing mechanism for mobile ad hoc networks. *IEEE Communication Magazine, 43*(7), 118-125.

Guangyu, P., Geria, M., & Hong, X. (2000). LANMAR: Landmark routing for large scale wireless ad hoc networks with group mobility. *Proceedings of the 1st Annual Workshop on Mobile and Ad Hoc Networking and Computing (MobiHoc)*, Boston, MA, August 11 (pp. 11-18). Piscataway, NJ: IEEE Press.

Gunes, M., Sorges, U., & Bouazizi, I. (2002). ARA: The ant-colony based routing algorithm for

MANETs. *Proceedings of the International Conference on Parallel Processing (ICPP) Workshops*, Vancouver, Canada, August 18-21 (pp. 79-85). Washington, DC: IEEE Computer Society.

Haas, Z. J. (1997). A new routing protocol for the reconfigurable wireless networks. *Proceedings of the 6th IEEE International Conference on Universal Personal Communications (ICUPC)*, 2, San Diego, October 12-16 (pp. 562-566). IEEE Computer Society.

Haas, Z. J., & Pearlman, M. R. (1998). The performance of a new routing protocol for the reconfigurable wireless networks. *Proceedings of the IEEE International Conference on Communications (ICC)*, 1, Atlanta, Georgia, June 7-11 (pp. 156-160). New York: IEEE.

Higaki, H., & Umeshima, S. (2004). Multiple-route ad-hoc on-demand distance vector (MRAODV) routing protocol. *Proceedings of the 18th International Parallel and Distributed Processing Symposium (IPDPS)*, Santa Fe, New Mexico, April 26-30. New York: IEEE.

Hong, X., Xu, K., & Gerla, M. (2002). Scalable routing protocols for mobile ad hoc networks. *Kluwer Wireless Networks, 16*(4), 11-21.

Hou, T.-C., & Li, V. (1986). Transmission range control in multihop packet radio networks. *IEEE Transactions on Communications, 34*(1), 38-44.

Hu, Y.-C., & Johnson, D. (2000). Caching strategies in on-demand routing protocols for wireless ad hoc networks. *Proceedings of the 6th International Conference on Mobile Computing and Networking (MobiCom)*, Boston, MA, August 6-11 (pp. 231-242). ISBN 1-58113-197-6. New York: ACM Press.

Hu, Y.-C., & Johnson, D. (2001). Implicit source routes for on-demand ad hoc network routing. *Proceedings of the 2nd ACM International Symposium on Mobile Ad Hoc Networking and Computing (MobiHoc)*, Long Beach, CA, October

4-5 (pp. 1-10). ISBN 1-58113-428-2. New York: ACM Press.

Islam, M. M., Pose, R., & Kopp, C. (2005a). Challenges and a solution to support QoS for real-time traffic in multi-hop ad-hoc networks. *Proceedings of the 2nd IEEE and IFIP International Conference on Wireless and Optical Communications Networks (WOCN)*, Dubai, UAE, March 6-8. ISBN 0-7803-9019-9. New York: IEEE.

Islam, M. M., Pose, R., & Kopp, C. (2005b). MAC layer support for real-time traffic in a SAHN. *International Conference on Information Technology: Coding and Computing (ITCC)*, Las Vegas, NV, April 4-6 (pp. 639-645). Washington, DC: IEEE Computer Society.

Islam, M. M., Pose, R., & Kopp, C. (2005c). Making SAHN-MAC independent of single frequency channel and omnidirectional antennas. *IASTED International Conference on Networks and Communication Systems (NCS 2005)*, Krabi, Thailand, April 18-20, 6pp. ISBN: 0-88986-490-X. ACTA Press.

Jacquet, P., Muhlethaler, P., Clausen, T., Laouiti, A., Qayyum, A., & Viennot, L. (2001). Optimized link state routing protocol for ad hoc networks. *Proceedings of the IEEE National Multi-Topic Conference (INMIC)*, Lahore, Pakistan, December 28-30 (pp. 62-68). New York: IEEE.

Jiang, M., Li, J., & Tay, Y. C. (1999). *Cluster based routing protocol*. Draft-ietf-manet-cbrp-spec-01.txt, Internet Draft, August.

Joa-Ng, M., & Lu, I.-T. (1999). A peer-to-peer zone-based two-level link state routing for mobile ad-hoc networks. *IEEE Journal on Selected Areas in Communications, 17*(8), 1415-1425.

Johnson, D., & Maltz, D. (1996). Dynamic source routing in ad hoc wireless networks. In T. Imielinski, & H. Korth (Eds.), *Mobile computing* (vol. 353, pp. 151-181). USA: Kluwer Academic Publishers.

Kasera, K. K., & Ramanathan, R. (1997). A location management protocol for hierarchically organized multihop mobile wireless networks. *Proceedings of the IEEE 6ʰ International Conference on Universal Personal Communications (ICUPC)*, 1, San Diego, CA, October 12-16 (pp. 158-162). IEEE Communications Society.

Kleinrock, L., & Stevens, K. (1971). *Fisheye: A lens like computer display transformation.* Technical Report, UCLA Computer Science Department.

Ko, Y. B., & Vaidya, N. H. (1998). Location-aided routing (LAR) in mobile ad hoc networks. *Proceedings of the 4ʰ Annual ACM/IEEE International Conference on Mobile Computing and Networking (MobiCom)*, Dallas, Texas, October 25-30 (pp. 66-75). ISBN 1-58113-035-X. New York: ACM Press.

Kopp, C., & Pose, R. (1998). Bypassing the home computing bottleneck: The suburban area network. *3ʳᵈ Australasian Computer Architecture Conference (ACAC), 20*(4), Perth, Australia (pp. 87-100). ISBN: 981-3083-93-X. Springer-Verlag.

Marina, M. K., & Das, S. R. (2001a). Performance of route caching strategies in dynamic source routing. *Proceedings of the International Workshop on Wireless Networks and Mobile Computing (WNMC) in Conjunction with International Conference on Distributed Computing Systems (ICDCS)*, Phoenix, Arizona, April 16-19 (pp. 425-432). Los Alamitos, CA: IEEE Computer Society.

Marina, M. K., & Das, S. R. (2001b). On-demand multipath distance vector routing in ad hoc networks. *Proceedings of the International Conference for Network Protocols (ICNP)*, Riverside, CA, November 11-14 (pp. 14-23). Washington, DC: IEEE Computer Society.

Marina, M. K., & Das, S. R. (2003). *Ad-hoc on-demand multipath distance vector routing.* Technical Report, Computer Science Department, Stony Brook University.

Murthy, M., & Garcia-Luna-Aceves, J. J. (1995). A routing protocol for packet radio networks. *Proceedings of the 1ˢᵗ Annual International Conference on Mobile Computing and Networking (MobiCom)*, Berkeley, CA, November 13-15 (pp. 86-95). ISBN 0-89791-814-2. New York: ACM Press.

Nikaein, N., Bonnet, C., & Nikaein, N. (2001). HARP: Hybrid ad hoc routing protocol. *Proceedings of the International Symposium on Telecommunications (IST)*, Tehran, Iran, September 1-3.

Nikaein, N., Labiod, H., & Bonnet, C. (2000). DDR: Distributed dynamic routing algorithm for mobile ad hoc networks. *Proceedings of the 1ˢᵗ Annual Workshop on Mobile and Ad Hoc (MobiHoc)*, Boston, MA, August 11 (pp. 19-27). Piscataway, NJ: IEEE Press.

Ogier, R., Templin, F., & Lewis, M. (2004, February). *RFC 3684 on topology dissemination based on reverse-path forwarding (TBRPF).* Draft-ietf-manet-tbrpf-11.txt, Internet Draft, Network Working Group.

Park, V. D., & Corson, M. S. (1997). A highly adaptive distributed routing algorithm for mobile wireless networks. *Proceedings of the 16ʰ Annual Joint Conference of the IEEE Computer and Communications Societies (InfoCom)*, 3, Kobe, Japan, April 7-11 (pp. 1405-1413). ISBN 0-8186-7780-5. New York: IEEE.

Pei, G., Gerla, M., & Chen, T. (2000). Fisheye state routing in mobile ad hoc networks. *Proceedings of the ICDCS Workshop on Wireless Networks and Mobile Computing*, Taipei, Taiwan, April 10 (pp. 71-78). IEEE Computer Society.

Pei, G., Gerla, M., Hong, X., & Chiang, C. C. (1999). A wireless hierarchical routing protocol with group mobility. *Proceedings of the IEEE Wireless Communications and Networking Conference (WCNC)*, 3, New Orleans, September 21-24 (pp. 1538-1542). Piscataway, NJ: IEEE Press.

Perkins, C., & Bhagwat, P. (1994). Highly dynamic destination-sequenced distance-vector routing (DSDV) for mobile computers. *Proceedings of the ACM Conference on Communications Architectures, Protocols and Applications (SIGCOMM)*, London, UK, August 31-September 2 (pp. 234-244). New York: ACM Press.

Perkins, C. E., & Royer, E. M. (1999). Ad-hoc on-demand distance vector routing. *Proceedings of the 2nd IEEE Workshop on Mobile Computing Systems and Applications (WMCSA)*, New Orleans, February 25-26 (pp. 90-100). IEEE Computer Society.

Radhakrishnan, S., Racherla, G., Sekharan, C. N., Rao, N. S. V., & Batsell, S. G. (1999). DST-A routing protocol for ad hoc networks using distributed spanning trees. *Proceedings of the IEEE Wireless Communications and Networking Conference, (WCNC)*, 3, New Orleans, September 21-24 (pp. 1543-1547). Piscataway, NJ: IEEE Operations Center.

Radhakrishnan, S., Racherla, G., Sekharan, C. N., Rao, N. S. V., & Batsell, S. G. (2003). Protocol for dynamic ad-hoc networks using distributed spanning trees. *Kluwer Wireless Networks, 9*(6), 673-686.

Ramasubramanian, V., Haas, Z. J., & Sirer, E. G. (2003). SHARP: A hybrid adaptive routing protocol for mobile ad hoc networks. *Proceedings of the 4th ACM International Symposium on Mobile Ad Hoc Networking and Computing (MobiHoc)*, Annapolis, MD, June 1-3 (pp. 303-314). ISBN 1-58113-684-6. New York: ACM Press.

Royer, E. M., & Toh, C.-K. (1999). A review of current routing protocols for ad hoc mobile wireless networks. *IEEE Personal Communications, 6*(2), 46-55.

Santivez, C., Ramanathan, R., & Stavrakakis, I. (2001). Making link-state routing scale for ad hoc networks. *Proceedings of the 2nd ACM International Symposium on Mobile Ad Hoc Networking*

Computing (MobiHoc), Long Beach, CA, October 4-5 (pp. 22-32). ISBN 1-58113-428-2. New York: ACM Press.

Singh, S., Woo, M., & Vaidya, N. H. (1998). Power-aware routing in mobile ad hoc networks. *Proceedings of the 4th ACM/IEEE International Conference on Mobile Computing and Networking (MobiCom)*, Dallas, Texas, October 25-30 (pp. 181-190). ISBN 1-58113-035-X. New York: ACM Press.

Toh, C. K. (1996). A novel distributed routing protocol to support ad-hoc mobile computing. *Proceedings of the 15th IEEE International Performance, Computing, and Communications Conference (IPCCC)*, Phoenix, Arizona, March 27-29 (pp. 480-486). New York: IEEE.

Toh, C. K. (1997). Associativity-based routing for ad hoc mobile networks. *Journal of Wireless Personal Communications, 4*(2), 103-139.

Tsuchiya, P. F. (1988). The landmark hierarchy: A new hierarchy for routing in very large networks. *Computer Communication Review, 18*(4), 35-42.

Wan, P.-J., Calinescu, G., & Yi, C.-W. (2004). Minimum-power multicast routing in static ad hoc wireless networks. *IEEE/ACM Transactions on Networking, 12*(3), 507-514.

Wang, L., Zhang, L. F., Shu, Y. T., Dong, M., & Yang, O. W. W. (2001). Adaptive multi-path source routing in wireless ad-hoc networks. *Proceedings of the IEEE International Conference on Communications (ICC)*, St. Petersburg, Russia, June 11-15 (pp. 867-871). New York: IEEE.

Woo, S.-C. M., & Singh, S. (2001). Scalable routing protocol for ad hoc networks. *Journal of Wireless Networks, 7*(5), 513-529.

Yang, H., Luo, H., Ye, F., Lu, S., & Zhang, L. (2004). Security in mobile ad hoc networks: Challenges and solutions. *IEEE Wireless Communications, 11*(1), 38-47.

Chapter X
Basics of Ubiquitous Networking

Kevin Park
University of Auckland, New Zealand

Jairo A. Gutiérrez
University of Auckland, New Zealand

ABSTRACT

Networking is no longer a luxury but rather has reached a stage where it could be regarded as a commodity, if not a necessity. Improvements in networking technologies and devices have resulted in many services and facilities that were considered a dream just a decade ago. The demands of users and organisations are driving networks to provide four "A's" networking, or the "anytime, anywhere, by anything and anyone" networking, referred to as "ubiquitous networking". This chapter provides an overview of ubiquitous networking, commencing with a discussion on how ubiquitous networking environments will change our lives, followed by a section on the importance of networking infrastructure, along with the applications and services necessary to maximise the benefits of a ubiquitous networking environment. Furthermore, this chapter addresses the global evolution of ubiquitous networking including some of New Zealand's attempts in this increasingly important field.

INTRODUCTION

The exponential growth of the Internet has diminished the difficulties associated with communication between distant places, allowing people to participate in the digital economy regardless of their geographical limitations. Additionally, developments in wireless technologies are freeing people from using wires for communicating. For example, the conveniences of wireless connections have converted mobile phones in a commodity, rather than a luxury item (Weatherall & Jones,

2002). The worldwide penetration of handheld devices through 2005 is the more than 511 million, with 310 million of those in Asia, as reported in the *Statistics for Mobile Commerce* (Retrieved April 26, 2006 from: http://www.epaynews.com/statistics/mcommstats.html#7). A report by McKinsey cited in the same Web site forecasts a mobile phone penetration in Europe of 85% for the year 2005. Technological advances of wireless technologies are truly leading us to a world that is capable of delivering "anytime, anywhere, by anything and anyone" networks (Ministry of Internal Affairs and Communications, Japan, 2005a).

The idea of "anywhere, anytime, by anything and anyone" (or 4As) networking is at the core of a new emerging networking technology, referred to as a "ubiquitous networking". The origin of the term "ubiquitous" is Latin, meaning "being everywhere, especially at the same time" (Phobe.com, 2003). The concept of ubiquitous networking originated from the concept of ubiquitous computing, which was aimed to "make many computers available throughout the physical environment, while making them effectively invisible to the user" (Weiser, 1993; Wikipedia, 2005). Yuhan (2003) distinguishes the term "ubiquitous computing" from "ubiquitous networking" by stating *"ubiquitous computing requires good network connections, but not necessarily ubiquitous networking"*. Additionally, Weiser (1993) highlights four important "networking" issues when focusing on ubiquitous computing, namely: wireless media access, wide-bandwidth range, real-time capabilities for multimedia over standard networks, and packet routing.

In the Tokyo Ubiquitous Network Conference, four main objectives of ubiquitous networking were stated, indicating that ubiquitous networking should be: (1) freed from networking constraints concerning capacity, location, and different link ups; (2) freed from the constraints of terminal limitation; (3) freed from the constraints of limited service and contents; and (4) freed from the constraints of network risk (Ministry of Interal Affairs and Communications, Japan, 2005b).

The world of ubiquitous networking creates new business opportunities, from the development of services and applications that maximise these earlier-mentioned objectives of ubiquitous networking. Evolving networking technologies will change our daily lives, both in social and economical terms. For example, according to Kitamura (2002), the potential demand that can be generated in Japan with ubiquitous networking environments exceeds 10 trillion yen (Kitamura, 2002). The potential demand includes, but is not limited to services such as ubiquitous health/concierge systems, ubiquitous automobile systems and ubiquitous education/learning systems.

In the following sections of this chapter the concept of ubiquitous networking will be explained in detail. The next section addresses the importance of networking infrastructure in ubiquitous networking, followed by the section, which provides an overview of business models and proposed applications and services. The fourth and fifth sections covers some approaches to ubiquitous networking around the world, including New Zealand. The sixth section addresses the underlying issues in achieving a ubiquitous networking environment, while the last section concludes the chapter.

NETWORK INFRASTRUCTURE FOR UBIQUITOUS NETWORKING

The core of a ubiquitous networking environment is the underlying networking infrastructure that is capable of meeting the requirements of users. In general, networking technologies can be divided into two broad categories: (1) wired networks and (2) wireless networks. The key functional distinction between the two types of networks is the speed, where wired networking technologies are capable of providing much higher speed communications compared to wireless networking technologies. The technologies supporting wireless networks will always be "resource-poor" when compared with those supporting wired networks (Satyanarayanan, 1996). The following is the list of key wired and wireless networking technologies. Detailed descriptions of them are beyond the scope of this chapter.

- Fibre technologies (e.g., FTTH, FTTC)
- Wireless LANs (802.11g, 802.16)
- Wireless MANs (Fixed wireless, LMDS)
- Wireless WANs (3G/4G, GPRS, UMTS, GPS, mesh networks)
- Short-distance wireless communication technologies (Wireless PAN) (e.g., Bluetooth)
- Satellite systems (e.g. Motorola's Iridium)
- Sensor networks

Figure 1 (sourced from Beltran & Roggendorf, 2005) provides a model for understanding the issue of resource allocation in wireless networks and it gives a useful way of appreciating the entities involved in providing ubiquitous services. The columns in the model depict the specifics of the different wireless technologies up to the packet layer (IP) which is widely seen as the first unifying layer and provides full transparency to the upper layers. The lowest levels of the layer model are used to describe the specific properties of the particular wireless channel technology (Beltran & Roggendorf, 2005).

Wired vs. Wireless Technologies

Drew (2003) argues that wired networks are necessary while considering wireless networks as a supplement. However, in a ubiquitous network environment, the role of wireless networks matches the importance of wired networks. In a ubiquitous networking environment, wired networks act as the backbone providing high to very high speed

Figure 1. A layer model for resource allocation (Beltran & Roggendorf, 2005)

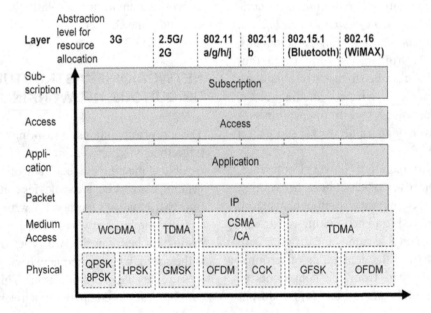

connection to end users. However, wired networks are very restricted in terms of mobility. Wireless networks address this deficiency, and provide high mobility to users although, as stated before, with lower speeds than wired networks. Thus, in a ubiquitous networking environment, wired and wireless networking technologies form a symbiotic relationship, by compensating for the deficiencies of each other. Murakami (2004) provides a clear overview of the appropriate utilisation of wired and wireless networking technologies in a given sector of a ubiquitous networking environment (see Figure 2).

In a ubiquitous networking environment, the networking technologies are not limited to providing "person to person" communications; they also need to address "person to object" and "object to object" communications (Ministry of Internal Affairs and Communications, Japan, 2005b). The convergence of wired and wireless technologies will provide an appropriate use of networking technologies for those three different types of communications, maximising the potentials of ubiquitous networking. That convergence is a key characteristic of environments which support the spontaneous appearance of

Figure 2. The ubiquitous network (Murakami, 2004)

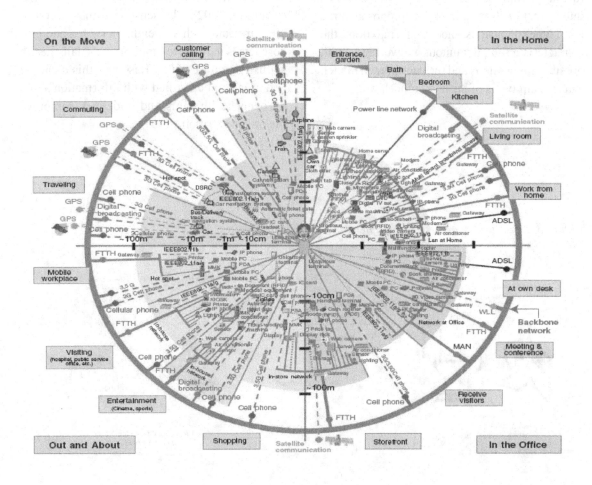

entities, in other words "persons" or "objects" that move into range and initiate participation in the ubiquitous network (Kindberg & Fox, 2002). The next section discusses a number of suggested business models along with the potential applications and services that could be offered as part of a ubiquitous networking environment.

BUSINESS MODELS AND APPLICATIONS FOR UBIQUITOUS NETWORKING

As mentioned in the introduction, ubiquitous networking is a promising environment capable of delivering services that were considered a dream only a few years ago. The development of appropriate business models is already in place turning ubiquitous networking environments into a reality. Nagumo (2002) proposed three main types of business models that maximise the characteristics of ubiquitous networking, with possible applications and services in each model. Three main types of business models are:

- Concierge type business model
- Knowledge asset management type business model
- Wide area measurements business model

Among these three types of business models, the "concierge type" business model delivers the most promising applications and services to everyone.

Healthcare Toilet

In Japan, Matsushita Electrical Industrial Co., Ltd. has developed a new service called the "Healthcare Toilet" (see Figure 3). The key idea of this service is to provide a health monitoring system, using a toilet seat equipped with sensors and networking devices. The objective is to provide a seamless service to users, without interrupting one's life (Nagumo, 2002). The sensors in the toilet seat measure data such as weight, body fat, and glucose levels in urine to monitor one's daily health status (Nagumo, 2002). Based on this data, the end user will be notified with information about his or her health status, and guidelines to improve his or her nutrition.

Figure 3. Health management using a networked toilet (Nagumo, 2002)

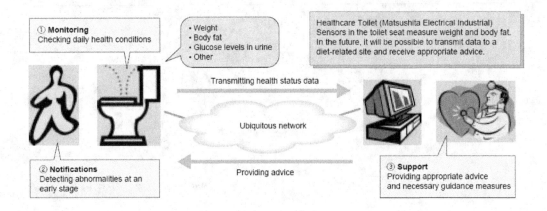

Ubiquitous Automobile Network Services

Another example is found in the automobile industry with the introduction of ubiquitous automobile network services. This type of service is already in place with the use of GPS (global-positioning system) equipped cars that allow drivers to find the shortest path to a destination, to locate the current position of the vehicle, and to find the routes

Figure 4. Example of a GPS system in an automobile (Langa, 2002)

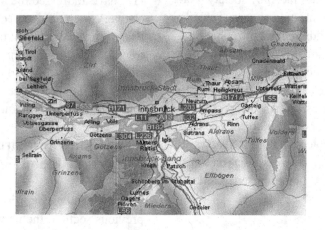

Figure 5. Services in a ubiquitous networking environment (Nagumo, 2002)

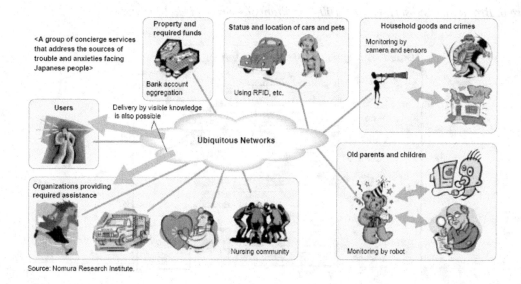

with low levels of traffic (Langa, 2002) (see Figure 4). In addition to the services that GPS provides, the ubiquitous automobile network is capable of providing services such as monitoring functions and warning messages regarding the vehicle condition, automatic toll-fee collection, and car theft warning and prevention (Kitamura, 2002).

These two examples are just a glimpse of what ubiquitous networking environments can deliver. As suggested by Nagumo (2002), a ubiquitous networking environment is capable of overcoming the troubles and anxieties faced by society (Nagumo, 2002). For example, taking care of elderly relatives or young children can be troublesome for workers who cannot stay with them all the time. In a ubiquitous networking environment, the use of surveillance cameras and sensors can reduce the potential risks, with the ability to send alerts to family members and emergency centres when anomalies are detected. Figure 5 shows an example of a group of concierge services that can be delivered in a ubiquitous networking environment.

GLOBAL EVOLUTION OF UBIQUITOUS NETWORKING

The promising benefits and attractive proposals of applications and services for ubiquitous networking environments have sparked the interest of many countries around the world and have also attracted the attention of the International Telecommunications Union (ITU), the organisation that is well known for its standardisation efforts in the telecommunications field. Recently, the ITU have addressed the idea of the "Ubiquitous Network Society" as part of their "New Initiatives Programme" which aims to identify the emerging trends in the telecommunications environment (ITU, 2005a). In that ITU programme South Korea and Japan were selected to illustrate early implementations of ubiquitous networking.

Networking Infrastructure in South Korea and Japan

Why these two countries? As discussed in the previous section, the networking infrastructure

Figure 6. Broadband subscribers per 100 inhabitants (Source: ITU, 2005a)

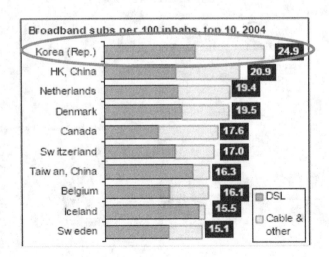

is the critical factor for ubiquitous networking. According to the ITU, South Korea is the world's broadband leader by a large margin (24.9 broadband subscribers per 100 inhabitants in 2004, see Figure 6); along with a high number of mobile subscribers that even outnumbers fixed line subscribers (ITU, 2005b). Similar statistics apply to Japan (ITU, 2005c). The high penetration rate of broadband and the widely use of wireless technologies around the country allows facilitates the implementation of "anywhere, anytime by anything, anyone" ubiquitous networking. The success factors for these two countries are explained in detail in the ITU's case studies (ITU, 2005a, 2005b).

Approaches to Ubiquitous Networking in South Korea

In South Korea, the Ministry of Information and Communication has the intention of realising their "digital home plan", in which digital home appliances with communications capabilities are installed in apartment houses as a total, integrated system (Murakami, 2004). The ministry is about to invest 2 trillion won (2 billion $US) for 10 million households for the four years ending in 2007 (Murakami, 2004). In the private sector, plans for ubiquitous networking are also emerging: the Dongtan Ubiquitous Networking city plan, supported by the Samsung group, involves 40,000 households (Murakami, 2004). At the university level, a number of institutions have successfully implemented ubiquitous networking environments, usually labelled "u-campuses". For example, at Sukmyung Woman's University, students can download "credit-card" functionality to their PDA or mobile phone and use the device as a medium for payment (Jung, 2004).

The strong focus and national level of support towards ubiquitous networking in South Korea and Japan are driving other countries to adopt similar strategies. In Europe, the project Amigo is addressing the idea of ubiquitous networking,

stating that it is an *"integrated project that will realize the full potential of home networking to improve people's lives"* (Amigo Project, 2004). Active participants in this project include companies from various countries, among them: France, Germany, Spain, and the Netherlands. However, these efforts are limited to the commercial sector without the stronger government and country-wide support found in South Korea and Japan. According to the ITU, Italy and Singapore are the two other countries that are actively participating in achieving a ubiquitous networking environment (ITU, 2005d, 2005e), with relatively well-established infrastructure throughout their territories. The case of New Zealand will be discussed in the next section.

APPROACHES TO UBIQUITOUS NETWORKING IN NEW ZEALAND

In New Zealand, the concept of ubiquitous networking is not highly recognised. The networking infrastructure in New Zealand is far behind other countries, in terms of broadband penetration rate, use of mobile communications (See Figure 7), and pricing (both wired and wireless). Despite this situation, a number of attempts have been made to provide some "ubiquitous services". Three main services will be discussed in this section.

Wireless Hotspots

The "Wireless Hotspot" service from Telecom New Zealand can be regarded as an effort to provide ubiquitous networking environments. Telecom provides "Wireless Hotspot" services around New Zealand (TNZ, 2004) in major cities, in places such as major hotels and cafés. The "Wireless Hotspot" service allows users to utilise their devices for Internet connectivity at the places where the service is available. However, this number of "Wireless Hotspots" is very small compared to the "Wireless Hotspots"

Figure 7. New Zealand's position on broadband penetration rate and mobile Internet use (Source: ITU, 2005c)

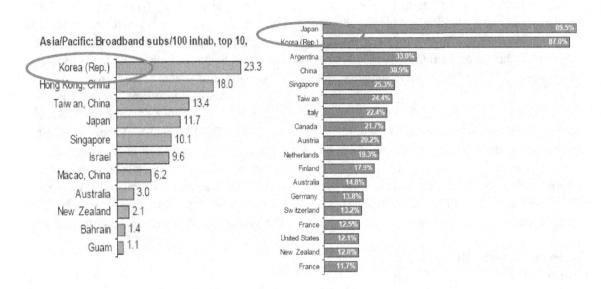

available in countries like South Korea (10,000 "Hotspots") (ITU, 2005b), and the service is only limited to customers of "Xtra", the largest Internet service provider in New Zealand. Telecom New Zealand recently allowed users to purchase the prepaid cards for the wireless Internet access in the "Wireless Hotspots".

TXT-a-Park

Another example of a ubiquitous networking service in New Zealand is the trial "txt-a-park" system, pioneered by Vodafone and the Auckland City Council. The Auckland City Council claims that the trial "aims to obtain customer feedback and operational performance data about the new parking payment technology, with a view to replacing Auckland City's existing pay and display machines under the council's parking asset replacement plan" (Auckland City Council, 2005). Auckland City Council has invested approximately $100,000 on this service, which

is funded from revenue generated by pay and display machines. Each TXT-a-Park machine has a GPRS dial-up modem, allowing the council to configure the machines remotely (Wellington City Council, 2005).

To use the TXT-a-Park service, the user first needs to press the "TXT-a-Park" button from the parking machine (See Figure 8) to get a cost and code for one- or two-hour parking (Vodafone New Zealand, 2005). To pay for the parking, the user

Figure 8. Txt-a-park service from Vodafone / Auckland and Wellington City Council (Auckland City Council, 2005; Wellington City Council, 2005)

sends a text message to a given number using the obtained code as content. If the text message successfully goes through, the parking machine will notify the user with a beep and ask for confirmation to accept or to cancel the transaction. If the user accepts the transaction the machine will print a parking ticket and confirmation will be sent to the user's mobile phone. The service is only available to Vodafone customers who have either pre-paid or monthly mobile phone plans.

RoamAD's Metro Wi-Fi Networks

RoamAD's proprietary solution on Metro Wi-Fi (802.11b/g) networks provides ubiquitous wireless

networking services to end users in a metropolitan area (RoamAD, 2002). The solution is based on a multipoint-to-multipoint wireless 802.11b/g star grid topography created using RoamAD's proprietary software, which consists of node software operating on intelligent network nodes (INNs) and intelligent network servers (INS). The proprietary software allows use of existing 802.11b/g wireless networking devices, and is also capable of creating wireless networks with a coverage area of 150 square kilometres (Watson, 2002). The first functional demonstration network in New Zealand covers 3 kilometres of downtown Auckland, and allows users within this area to use always-on wireless broadband con-

Figure 9. RoamAD's proprietary Metro Wireless Network (RoamAD, 2005)

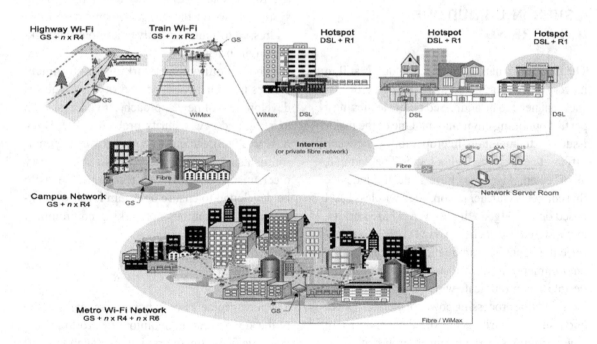

KEY

R1	1 Radio INN (Operating RoamAD INN Software)		**R2**	2 Radio INN (Operating RoamAD INN Software)
R4	4 Radio INN (Operating RoamAD INN Software)		**R6**	6 Radio INN (Operating RoamAD INN Software)
INS	Intelligent Network Server		**GS**	Gateway Server (Operating RoamAD Software)
INN	Multi-radio Intelligent Network Node assembled from commodity hardware			

nections to the Internet, their office network and the PSTN (Wireless Developer Network, 2002). The aspect which separates the RoamAD's technology from traditional wireless hotspots is that RoamAD's technology is not reliant on existing terrestrial network infrastructure (Watson, 2002). This technology is also used in campus environments, providing ubiquitous wireless networking services to students near or on campus (Brislen, 2005). An example of RoamAD's metro network is illustrated in Figure 9.

Despite these discussed attempts, New Zealand is still far from achieving a complete ubiquitous network environment. However, early attempts to provide services utilising ubiquitous networking principles look promising, and it is expected that more services will develop in parallel with the improvements in the networking infrastructure.

ISSUES IN UBIQUITOUS NETWORKING

The ubiquitous networking technology is still in its very early stages, and there are numerous issues that need to be addressed before achieving a perfect operating environment. One of the major issues is to maintain interoperability between different networking technologies. For example, an office employee may have a Bluetooth device that connects with her laptop, use a wireless LAN based on 802.11g, a wireless WAN based on 3G, and a wired connection using ADSL. To maximise the benefits from a ubiquitous networking environment, these various technologies should be able to communicate without any disruptions. Additionally, processing power of mobile devices and issue of security is one of other concerns for true ubiquitous networking environment. Currently, significant research emphasis is given to the security and middleware side of ubiquitous networking to address this, and it is highly related with improvements in the processing power of mobile devices.

Network Selection and Billing

Selection of networks in a ubiquitous networking environment is one of the main operating issues with this technology (Beltran & Roggendorf, 2004a). For example, in a ubiquitous networking environment, a cordless phone may substitute your mobile phone when you are outside the house. Choosing the best network based solely on the user requirements complicates the selection of the "ideal" network for a particular connection time and location. The user-initiated selection of a provider also generates the issue of billing. Currently customers "subscribe" to the desired services, and get billed based on the usage. However, in a ubiquitous networking environment, there is no need to "subscribe" for a desired service, but rather users have the capability to employ ad-hoc type services when needed as depicted in Figure 10. This adds complexity to existing billing systems however these requirements need to be addressed to achieve a truly ubiquitous networking environment.

The ubiquitous networking environment creates new challenges in security and requires development of new approaches to address both existing and new security problems (Van Dyke & Koc, 2003). Heterogeneous networking environments add a complexity to existing security mechanisms, and different techniques needs to be developed to ensure optimum levels of security in the ubiquitous networking environment (Privat, 2005).

Middleware

The advancements in handheld devices are one of the key drivers of ubiquitous networking, and these devices are improving its capabilities at exponential rates. However, due mainly to their size restrictions, these devices suffer from a number of limitations. These limitations include but are not limited to: inadequate processing capability, restricted battery life, limited memory

Figure 10. Issue of interoperability and network selection (Beltran & Roggendorf, 2004b)

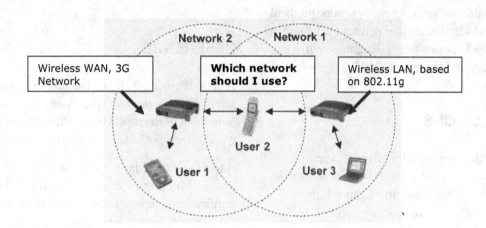

space, slow expensive connections, and confined host bandwidth (Sharmin, Ahmed, & Ahamed, 2006). To address these limitations, middleware can play an essential role. For example, rather than delegating processing responsibility to the light-weight handheld devices, core processing can be performed by the middleware applications. Currently developed middleware applications are capable of providing services such as security, data backup services, resource discovery services, and ad-hoc communication services, to list a few (Sharmin, Ahmed, & Ahamed, 2006). Given that middleware is the most viable solution to minimise limitations of handheld devices, a large number of middleware applications are under research by both academics and practitioners (Murphy, Picco, & Roman, 2001; Sharmin, Ahmed, & Ahamed, 2006; Yau, Wang, & Karim, 2002).

Security

Security has always been a critical issue within the area of networking, and this is not an exception in ubiquitous networking environments. In fact, security in this type of environment requires more emphasis than what has been perceived in traditional networks. The convenience of hand-held devices, such as PDAs, means that people are storing personal data on these devices, which means that more stringent security mechanisms to protect this data are required. The typical characteristics of handheld devices also create security concerns, such as (Raghunathan, Ravi, Hattangady, & Quisquater, 2003):

- Mobile communications uses a public transmission medium, which creates opportunity for hackers to eavesdrop communications more easily than with secured private connections.
- Mobile devices are vulnerable to theft, loss, and corruptibility.
- Processing power limitations on mobile devices can imply restrictions on security features (e.g., Algorithm selection).

To address these issues, various methods are proposed. The common approach is the use of security protocol, middleware, and hardware-driven security. Many application vendors and researchers are working together to develop a more light-weight security protocols for handheld devices, and standards such as wireless application protocol (WAP) and wired equivalent privacy (WEP) are in wide-spread use nowadays (Raghunathan, Ravi, Hattangady, & Quisquater, 2003).

CONCLUSIONS

During the Tokyo Ubiquitous Networking Conference, Mr. Taro Aso, Minister of Internal Affairs and Communications stated that *"We aim to realize this ubiquitous network society by 2010. At the end of last year, we formulated policy for realizing a ubiquitous network society which consists of three pillars: preparations for ubiquitous networks, enhancement of ICT applications, and the setting up of an environment conducive to usage"* (Aso, 2005). It is clear that "anywhere, anytime by anything and anyone" networking will not arrive overnight. It will be an incremental process that requires commitment, careful planning and preparation. Viable implementations will initially take place at a relatively small scale, in apartment buildings or university campuses before extending to a wider environment. Eventually, a world-wide level of ubiquitous networking will be reached, allowing people to use the same device everywhere.

Developments in networking have changed our everyday lives, especially with the introduction and popularity of the Internet. High demand from individual users and organisations has brought the concept of ubiquitous networking to the front as seen in the cases of South Korea and Japan. The importance of the underlying networking infrastructure is one of the key success factors for establishing ubiquitous networking environments along, of course, with successful applications and services. Ubiquitous networking is not without problems, and there are a number of critical issues that need to be resolved. However, the technology offers great promise, and it will eventually reach

us, converting our "4As" networking dreams into reality.

REFERENCES

Amigo Project. (2004, December). *Ambient intelligence for the networked home environment: Summary.* Retrieved May 2, 2006, from http://www.hitech-projects.com/euprojects/amigo/

Aso, T. (2005, May). *Message from the organizers.* Minister of Internal Affairs and Communications, Japan. Tokyo Ubiquitous Network Conference, Tokyo, Japan, May 16-17. Retrieved May 5, 2006, from http://www.wsis-japan.jp/ministeriac_e.html

Auckland City Council. (2005, March). *Transport: Parking – Princess Street pay and display trial.* Retrieved May 4, 2006, from http://www.aucklandcity.govt.nz/auckland/transport/parking/princesst.asp

Beltran, F., & Roggendorf, M. (2004a, October). *Innovative pricing and charging in next-generation wireless networks* (pp. 1-4). Auckland, New Zealand: PING Research Group.

Beltran, F., & Roggendorf, M. (2004b, December). *An incentive-compatible pricing scheme for competitive access to wireless networks* (pp. 1-6). Auckland, New Zealand: PING Research Group.

Beltran, F., & Roggendorf, M. (2005). A simulation model for the dynamic allocation of network resources in a competitive wireless scenario. *Proceedings of the Second International Workshop, MATA 2005,* Montreal, Canada, October 17-19. Lecture Notes in Computer Science 3744. Auckland, New Zealand: PING Research Group.

Brislen, P. (2005, April). *RoamAD sells metro Wi-Fi to Perth ISP: Kiwi company transitions from carrier to vendor.* Computerworld New

Zealand. Retrieved May 3, 2006, from http://computerworld.co.nz/news.nsf/UNID2B8B78E695883CAFCC256FE80026A6D0?OpenDocument&Highlight=2,roamad

Drew, Jr., W. (2003, February). Wireless networks: New meaning to ubiquitous computing. *The Journal of Academic Librarianship, 29*(2), 1-11.

ITU [International Telecommunication Union]. (2005a). *New initiatives programmes, Feb 2005.* Retrieved May 3, 2006, from http://www.itu.int/osg/spu/ni/

ITU [International Telecommunication Union]. (2005b, April). *ITU workshop on ubiquitous network societies: The case of the Republic of Korea* (pp. 10-45). Retrieved May 10, 2006, from http://www.itu.int/osg/spu/ni

ITU [International Telecommunication Union]. (2005c, April). *ITU workshop on ubiquitous network societies: The case of the Japan* (pp. 13-45). Retrieved May 10, 2006, from http://www.itu.int/osg/spu/ni

ITU [International Telecommunication Union]. (2005d, April). *ITU workshop on ubiquitous network societies: The case of the Italian Republic* (pp. 25-50). Retrieved May 10, 2006, from http://www.itu.int/osg/spu/ni

ITU [International Telecommunication Union]. (2005e, April). *ITU workshop on ubiquitous network societies: The case of the Republic of Singapore* (pp. 21-39). Retrieved May 10, 2006, from http://www.itu.int/osg/spu/ni

Jung, K. U. (2004, December). *Community, S&A research group, research group notice board: Ubiquitous campus in Korea.* Ubiquitous IT Korea Forum. Retrieved May 12, 2006, from http://www.ukoreaforum.or.kr/bbs/view.php?code=c_rg01&page=1&number=2&keyfield=&key=

Kindberg, T., & Fox, A. (2002). System software for ubiquitous computing. *IEEE Pervasive Computing*, January, 70-81.

Kitamura, M. (2002, September 1). *NRI papers: Using ubiquitous networks to create new services based on the commercial and public infrastructure* (pp. 1-14). NRI Papers No. 54. Japan: Nomura Research Institute.

Langa, F. (2002, August). Langa letter: A real-life GPS road test. *InformationWeek*. Retrieved May 11, 2006, from http://www.informationweek.com/story/showArticle.jhtml?articleID=6502601&pgno=1

Ministry of Internal Affairs and Communications, Japan. (2005a, May). *Towards realisation of ubiquitous networking society.* Tokyo Ubiquitous Networking Conference, Tokyo, Japan, May 16-17. Retrieved May 12, 2006, from http://www.wsis-japan.jp/about_e.html

Ministry of Internal Affairs and Communications, Japan. (2005b, May). *Tokyo Ubiquitous Networking Conference: Program (Session 5) Ubiquitous Network Society*, Tokyo, Japan, May 16-17. Retrieved May 12, 2006, from: http://www.wsis-japan.jp/session5_e.html

Murakami, T. (2004, August 1). *NRI papers: Ubiquitous networking: Business opportunities and strategic issues* (pp. 1-24). NRI Papers No. 79. Japan: Nomura Research Institute.

Murphy, A. L., Picco, G. P., & Roman, G. C. (2001, May). Time: A middleware for physical and logical mobility. *Proceedings of the 21st International Conference on Distributed Computing Systems.*, April (pp. 524-533).

Nagumo, T. (2002, June 1). *NRI papers: Innovative business models in the era of ubiquitous networks* (pp. 1-13). NRI Papers No. 49. Japan: Nomura Research Institute.

Phobe.com. (2003). *Yer latin lesson for today: Today's latin lesson.* Retrieved May 14, 2006, from http://www.phobe.com/octopusmotor/latin.html

Privat, G. (2005, April). *Ubiquitous network & smart devices: New telecom services & evolution of human interfaces.* ITU Ubiquitous Network Societies Workshop, April 6. Retrieved May 20, 2006, from http://www.itu.int/osg/spu/ni/ubiquitous/Presentations/8_privat_applications.pdf

Raghunathan, A., Ravi, S., Hattangady, S., & Quisquater, J. (2003). Securing mobile appliances: New challenges for the system designer. *Proceedings of the Design, Automation and Test in Europe Conference and Exhibition, 1* (pp. 10176).

RoamAD. (2005, January). *Products & services: Introduction.* Retrieved May 16, 2006, from http://www.roamad.com/prodservintro.htm

Satyanarayanan, M. (1996). Fundamental challenges in mobile computing. *Fifteenth ACM Symposium on Principles of Distributed Computing, 1*(1), Philadelphia, PA, May, 7pp.

Sharmin, M., Ahmed, S., & Ahamed, S. I. (2006). MARKS (middleware adaptability for resource discovery, knowledge usability and self-healing) for mobile devices of computing environments. *Proceedings of Third International Conference on Information Technology: New Generation (ITNG 2006),* Las Vegas, NE, April (pp. 306-313).

Telecom New Zealand. (2004, June). *Productfinder: Internet and data, Telecom's wireless hotspot service.* Retrieved May 18, 2006, from http://www.telecom.co.nz/content/0,3900,204163-1487,00.html

Van Dyke, C., & Koc, C. K. (2003, February). On ubiquitous network security and anomaly detection. *Proceedings of the Applications and the Internet Workshops,* January 27-31 (pp. 374-378). Electrical & Computer Engineering, Oregon State University.

Vodafone New Zealand. (2005, March). *Business services: Paying with your mobile – TXT-a-Park.* Retrieved May 20, 2006, from http://www.vodafone.co.nz/promos/txt-a-park/txt_a_park.jsp?item=txt_a_park

Watson, D. (2002, September). *Auckland wireless net lures Asians: A firm planning a large wireless network around downtown Auckland is already claiming interest from Asia and the Pacific islands in its distance-boosting technology.* Computerworld New Zealand. Retrieved May 20, 2006, from http://computerworld.co.nz/news.nsf/UNID/CC256CED0016AD1ECC256C3C007A6299?OpenDocument&Highlight=2,roamad

Weatherall, J., & Jones, A. (2002, February). Ubiquitous networks and its applications. *IEEE Wireless Communications, 9*(1), 18-29.

Weiser, M. (1993, July). Some computer science problems in ubiquitous computing. *Communications of the ACM, 36*(7), 75-84.

Wellington City Council. (2005, March). *Technology & software, TXT-a-Park.* Retrieved May 21, 2006, from http://www.wellington.govt.nz/innovation/details/txtapark.html

Wikipedia. (2005, May). *Article: Ubiquitous computing.* Wikipedia: The Free Encyclopaedia. Retrieved May 5, 2006, from http://mobileman.projects.supsi.ch/glossary.html

Wireless Developer Network. (2002, August). *Wireless developer network – Daily news: RoamAD announces 802.11b breakthrough with metropolitan-wide Wi-Fi network.* Wireless Developer Network. Retrieved May 9, 2006, from http://www.wirelessdevnet.com/news/2002/238/news1.html

Yau, S., Wang, Y., & Karim, F. (2002). Development of situation aware application software for ubiquitous computing environments. *26th Annual International Computer Software and Application Conference* (pp. 233-238).

Yuhan, A. H. (2003). *Ubiquitous computing and its network requirements.* Samsung Advanced Institute of Technology. Retrieved May 13, 2006, from http://www.t-engine.org/aucnc2003/yuhan.pdf

Chapter XI
Wireless Security and Privacy Issues

Joarder Kamruzzaman
Monash University, Australia

ABSTRACT

Security and privacy protection are very strong requirements for the widespread deployment of wireless technologies for commercial applications. The primary aim of this chapter is to present an overview of the security and privacy issues by highlighting the need to secure access to wireless networks and the loss that might accrue from the breach of a network. The vulnerabilities of the IEEE 802.11 and Bluetooth networks are discussed, and a paradigm for secure wireless network is presented. The legal framework guiding the privacy issues in wireless communications is also presented.

INTRODUCTION

Wireless connectivity offers large organizations as well as individual users many advantages including mobility and flexibility, increased productivity, and low installation cost. The third generation wireless communications bring multimedia communication, mobile commerce, and many innovative applications and services in diverse areas including finance, industry, entertainment, and military. Wireless LAN (WLAN) allows users to move into an organization without losing network connectivity. This rapid development of technology and applications has created a diminishing boundary between wireless and wireline networks with an ease of access to the Internet through mobile devices. As the technology progresses towards maturity, new security problems arise that are specific to wireless environment. Wireless devices are usually portable and constrained by bandwidth, memory, and processing power. These constraints demand less computational complexity of encryption algorithms and fewer numbers of messages involved in security protocol.

Wireless access is inherently less secure, mainly due to the communication medium which is open to the intruders, and mobility adds higher risk than those encountered by the fixed wired networks. The motivating idea of going wireless, being able to connect "anywhere anytime", has the potential to attract an increasing number of users and hence also intruders. Wireless devices, for example, cellular phone and personal digital assistance (PDA) with Internet access, were not initially designed with security as one of the highest consideration. Increasing use of wireless and mobile technology for data, voice, and video communication without the appropriate security mechanism in place has made it easy for attackers to intrude into a wireless network and potentially into other parts of the network placing enterprise data in jeopardy. In this chapter, we present an overview of security problems and privacy issues related to the wireless environment and the technologies available to enhance the security standard/feature in wireless network.

WIRELESS SECURITY THREATS

Risk and Challenges

Most conveniently, a wireless access point can be plugged into a wireline network and made available for use within a certain range. This easy installation of wireless connectivity often does not implement any security mechanism in place. The result is that, on the top of various types of attacks common in wireline networks, wireless communication is prone to many types of malicious attacks specifically targeted to wireless networks. In April 2002, a security flaw in wireless network forced a major telecommunication company in New Zealand to shut down its mobile e-mail services (Grifin, 2002). In many countries, wireless computer communications share the same frequency as many other applications, for example, garage door opener, cordless phones, and other short distance applications. This

causes a main problem—interception of signals. A strong antenna, properly tuned, would be able to intercept this signal from a few kilometers away. Encryption on wireless data can improve this situation, but encryption is not always used in wireless communication and in many cases, the encryption implemented in wireless devices is not strong enough to prevent an attack. The three key threats to wireless security are discussed next.

Confidentiality

Confidentiality of data transmission is one of the fundamental security requirements for organizations as well as individuals. Since wireless transmission is of broadcast nature and easily interceptable, meeting this security requirement is more difficult in wireless network than wireline network. The physical security countermeasure is far less ineffective in this case. Interception of data means compromising proprietary information, network IDs and passwords, configuration data, and encryption keys. Most often, confidentiality features of wireless LAN technology are not enabled and the hackers exploit the numerous vulnerabilities in the IEEE 802.11 technology security (Cam-Winget, Housley, Wagner, & Walker, 2003; Housley & Arbaugh, 2003; Stubblefield, Ioannidis, & Rubin, 2002). This type of attack is further made easy by the use of easily available tools on the Internet, like wireless packet analyzers, such as AirSnort (2006) and WEPcrack (2006). Another risk to the loss of confidentiality through simple eavesdropping is broadcast monitoring. When an access point is connected to a hub, instead of a switch, it leaves all network traffic vulnerable to unauthorized monitoring. Wireless access point is just one of the entry points to a wireline network. In general, if an attacker is successful in intercepting transmission, he might gain further access to the wireline network and eventually launch further attack on servers, workstation, and other connected devices.

Integrity

The loss of data integrity issues in wireless networks is similar to those in wired networks. The data can be modified in transit or in storage. Because the existing security features of the 802.11 standard do not provide strong mechanism for message integrity, other kinds of active attacks that compromise system integrity are also possible. The WEP (wired equivalent privacy) based integrity mechanism is simply a linear cyclic redundancy check (CRC). Message modification attacks are possible when cryptographic checking mechanisms such as message authentication codes and hashes are not used.

Availability

In a wireless environment, a denial of network availability involves some form of DoS attack, such as jamming. Jamming occurs when a malicious user deliberately emanates a signal from a wireless device in order to block legitimate wireless signals. Wireless 802.11b network operates between 2.4GHz to 2.5GHz. An 802.11 device uses a technique called direct sequence spread spectrum (DSSS) to spread the transmission over multiple frequencies rather than transmitting at a single frequency. There are commercially available devices capable of emitting signals causing interference to signals generated by 802.11 devices. It has also been found that a Bluetooth device or a cordless phone operating at 2.4 GHz is located close to 802.11b devices (approximately within 10 meters) can cause jamming type of DoS attack (Peikari & Fogie, 2003). In certain circumstances, signal interference from other devices can slow down the effective data rate as the sending device may have to resend data to minimize the effect of interference.

War Driving

A new type of attack on wireless networks, called war driving, is to discover the wireless access points. The technique is similar to phone hacking, known as war dialing, where the attacker dials all the phone numbers in a given sequence to search for a modem. In war driving, instead of dialing a modem, one drives around looking for wireless networks. The attacker's car is equipped with an antenna, either inside or attached to the roof, which is connected to a laptop or a handheld device. The attacker could be driving, parking, or riding in a bus or a public transport. Special software, like NetStumbler (2006) or ORiNOCO (Lucent, 2000), installed in the laptop or PDA captures the beacon frames sent at a regular interval by the access points. The beacon frames are analyzed by the software and a number of information (i.e., Basis Service Set ID, WEP enabled or not, type of device, MAC address, transmission channel number, signal strength, GPS location, etc.) are presented on the screen. Wireless access points have virtually opened back doors in corporate networks that can be exploited by the hackers for intrusion. To ensure security, network auditing tools need to be deployed in the wireless section of the network. In most cases, once firewalls are installed, network administrators tend to think that their networks are secured. No matter how strongly firewalls are installed, hackers can gain access to corporate network through the security holes of an inappropriately configured wireless section without having to go through firewalls.

Another phenomenon associated with wardriving is war chalking. The word is derived from the fact that, during the post world-war time, homeless people used to put chalk marks to indicate to other people of the places that offered free food. Today war-chalking means to indicate free Internet access. It uses a series of well-defined symbols on different structures to indicate nearby access point availability and specific wireless setting. The other users then go to those locations and use the required setting as indicated by the symbols to connect to the Internet through a wireless access.

Airborne Virus

Viruses have not been widely considered as a security threat in smaller devices, like PDAs, Palm, or Pocket PCs because of their limited memory and processing power. However, with millions of PDAs, smart phones, and other handheld devices, the threat of wireless virus attack is growing. The Phage virus was the first to attack Palm OS platform which, when executed, infected third-party application programs and other PCs. Timofonica, a Visual Basic Script-based virus, hit a wireless network in Madrid and sent an SMS message across the GSM phone network (Peikari & Fogie, 2003). A virus in Internet-enabled smart phones in Japan caused the phones to call the emergency number resulting in the denial of emergency services. PDAs that share the same operating system as PCs are particularly susceptible to this new type of viruses. A virus for handheld devices may be designed targeting desktop PCs which might be infected during data synchronization, which in turn could directly affect the local network. Because of the easy susceptibility of a wireless network, an airborne virus may spread very quickly and cause enormous damage to both corporate network and ordinary users.

Rogue Access Points

These are the access points connected to a network without the permission of network administrators. It appears to be a legitimate access point to wireless clients to send data through it and can be used to intercept the communication between an authentic access point and clients. A malicious user can also gain access to the wireless network through access points (AP) that are configured to allow access without authorization. In many cases, such access points are deployed by employees for convenience without authorization from network administrators. Since such employees are not concerned of security issues, these access points are often deployed without proper security configurations.

WEP AND ITS VULNERABILITIES

The wired equivalent privacy (WEP) is the encryption standard for 802.11-based WLANs and requires the WAP (wireless access points) and the wireless network interface cards to be configured with a shared pass-phrase. At the beginning it was thought to offer impenetrable security against attacker, however, serious flaws were discovered later. WEP uses the RC4 encryption algorithm (Rueppel, 1992) developed by Ron Rivest of RSA with a variable length key to encrypt the data. WEP uses 40-bit key, however, some vendors have implemented products with 104-bit and even 128-bit keys. With the addition of the 24-bit initialization vector (IV), the actual key used in the RC4 algorithm is 152 bits for the 128 bits WEP key. RC4 employs a streaming cipher that creates a unique key for each packet of encrypted data. Figure 1 illustrates WEP encryption and decryption process. Briefly, the WEP encryption process consists of the following steps:

- Each data packet of the plaintext is passed through a cyclic redundancy check (CRC) algorithm (Ramabadran & Gaitonde, 1988) for integrity checking and produces a checksum (CS). WEP uses CRC-32.
- The CS is appended to the end of the data packet.

Figure 1. Wired equivalent privacy: (a) encryption; (b) decryption

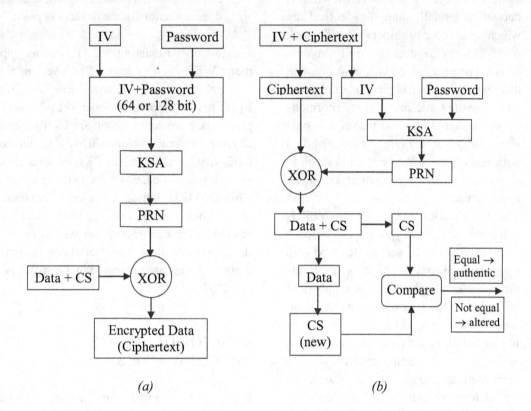

(a)   (b)

- A 24-bit initialization vector (IV) is generated and WEP key is appended to it. Using a series of IV in combination with WEP key, the key scheduling algorithm (KSA) creates the RC4 state values. These values serve as seeds to generate pseudo-random numbers (PRN) creating the key stream.
- The key stream is then XOR-ed with data packet+CS to produce ciphertext.
- The IV is appended as plaintext to the ciphertext to the receiving party.
- The IV is then used by the receiver to decrypt the data. The receiver, using IV and WEP key, generates the key stream and XOR-ed

with the ciphertext to retrieve data packet + CS.
- The data packet is then passed through the CRC algorithm to extract a new CS and compared with the received CS. If the checksum values match, the packet is considered valid, otherwise dropped.

A number of security problems exist with WEP that can compromise the confidentiality and integrity of data even with moderate computational power. Attackers with little technical knowledge on WEP can launch an attack using the tools available on the Internet. The following summarizes the WEP security issues:

1. The most obvious problem is the use of static WEP keys—many users in a wireless network potentially share the identical key which poses security vulnerability. The lack of key management in the WEP protocol does not present an easy way to change and distribute keys. For a large number of users, changing the static key requires enormous effort. A shared key also means it is possible to eavesdrop a large amount of data with less effort. A lost key would result in many devices sharing the same key being compromised.

2. The Initialization Vector IV in WEP is relatively short for cryptographic purpose (Walker, 2001) and is sent as clear text with the encrypted packet. This short IV limits the range of possible random numbers to around 16 million. However, due to the nature of randomness, the number may repeat in just 5,000 packet transmission (Peikari & Fogie, 2003) creating a collision. In a perfect stream cipher, IV value should not repeat. Moreover, the 802.11 standard does not specify how the IVs are set, and some vendors may adopt a poor scheme which can expose the key stream easily.

3. The cryptographic key length is only 40-bit which is inadequate for any system. Given other security vulnerabilities of WEP, even 128-bit key length is not adequate for most applications.

4. WEP provides no cryptographic integrity protection. The checksum on data packet is done by the 802.11 MAC protocol using CRC. The combination of noncryptographic checksums with stream ciphers often introduces vulnerabilities. In an active attack, the attacker can modify a packet after decryption and change the corresponding CRC.

5. WEP implementation of RC4 has key scheduling problem. The predictability in the first few bytes of the plain text can leads to an attack on the key scheduling algorithm.

6. Only the device is authenticated, no user authentication is done. Again, the client does not authenticate the access point.

Most of the existing 802.11 systems implement WEP in hardware. The overcome the vulnerabilities of WEP, the IEEE Task Group I (TGi) is developing stronger WLAN security protocols. Two such protocols are TKIP (temporal key integrity protocol) and CCMP (counter mode CBC MAC) protocols. The TKIP is devised as an interim solution based on the existing hardware while CCMP is envisaged as a long-term solution without being tied to existing hardware. CCMP selected advanced encryption system (AES) as the encryption algorithm. Detail descriptions of these protocols are presented by Cam-Winget et al. (2003).

BLUETOOTH AND ITS VULNERABILITIES

Bluetooth is the result of many companies (e.g., IBM, 3Com, Microsoft, Agere, Motorola, Nokia, Intel, Ericsson, Toshiba, etc.) working together for a common standard and is a relatively new paradigm of wireless communications in which no fixed infrastructure exists such as base stations or access points. The goals of Bluetooth are to build an inexpensive, easy to use, and decentralized wireless technology (Bisdikian, 2001). Bluetooth forms an ad-hoc network whereby devices form random network configurations on the fly. Bluetooth evolved as a wireless PAN (personal area technology) technology that offers fast and reliable transmission for both voice and data connecting devices like PDAs, cell phones, printers, faxes, and so forth, together in a small wireless environment such as an office or home. Users can walk into a room and be connected without wires and configuration changes, and roam around while remain connected.

Bluetooth is designed to operate in the 2.4GHz unlicensed ISM (industrial, scientific, medical applications) range. Bluetooth uses a frequency hopping spread spectrum (FHSS) by spreading over 79 channels and hopping up to 1,600 times per second. The reduced transmission power, short-range coverage, and FHSS produce a relatively interference free operation for Bluetooth. Bluetooth devices do not need to be manually configured; they can automatically locate each other which is the true strength. Bluetooth network topologies are established on a temporary and random basis. When a Bluetooth device requests a file or service, it becomes a "master" device and controls the data transmission. The master device determines what frequencies are to use and in which order. Up to eight Bluetooth devices may be networked together in a master-slave relationship, called a piconet. Devices in a Bluetooth piconet operate on the same channel and follow the same frequency hopping sequence. Several devices can be internetworked over a greater distance using several piconets in a series. This feature allows to form dynamic topology as a device moves within the network.

Bluetooth has certain security features built in to provide authentication, authorization, and confidentiality. It does that through the use of PIN. The slave and master devices compare the PINs. Bluetooth has the following three different modes of security (Karygiannis & Owens, 2002):

- Mode 1: Nonsecure mode. No security procedure is initialized.
- Mode 2: Service level security mode. Security procedures are initiated after the channel establishment. A security manager controls access to services and devices. Authorization is supported in this mode.
- Mode 3: Link level security mode. Security procedures are initiated before the channel establishment. Encryption, authorization, as well as uni-directional or mutual authentication are supported in this mode.

Bluetooth Security Vulnerability

Like IEEE 802.11, Bluetooth encryption procedure is based on a stream cipher. The key stream is XOR-ed with the data packet. This key stream is produced using a cryptographic algorithm based on linear feedback shift registers (LFSR). The key size varies from 8 bits to 128 bits and is negotiated between the master and slave devices. The encryption function takes the following items as inputs: the master identity, the random number, a slot number, and an encryption key. Bluetooth specification allows the following three encryption modes:

- Mode 1: No encryption on any traffic.
- Mode 2: Broadcast traffic is unencrypted but traffic to individual device is encrypted.
- Mode 3: All traffic is encrypted.

In addition, Bluetooth incorporates two levels of trust model. Trusted devices have permanent relationship and have full access to all services. Untrusted devices have restricted access as they maintain no permanent relationship.

The vulnerabilities of Bluetooth standard security can be listed as follows:

1. The PIN code used in Bluetooth devices varies between 1 to 16 bytes, typically 4-digit. The PIN size is relatively short which makes it easy to guess. This PIN is used for key generation. Therefore, a longer PIN is desirable, but people tend to select a short key which increases security risk.
2. The encryption key length is negotiable between the master and slave devices. A shorter key length may compromise security.
3. The master key is shared.
4. Only the device authentication is performed. There is no provision for user authentication.

5. When a Bluetooth device which is a part of a piconet is compromised, it increases the possibility of comprise of the wider network.

6. Only the individual links are encrypted, but no end-to-end security service implemented.

7. Not all the security services required for secure data communication are provided, for example, non-repudiation and data integrity are not incorporated.

SECURE WIRELESS NETWORK PARADIGM

M. Ashley (2002) suggests a four-layer approach that can provide a high degree of protection for wireless network. The approach consists of the following levels:

a. *Wireless Deployment and Policy.* At this level the following points should taken into practice:
   - conduct extensive site survey regularly and deploy a minimum number of WAPs that is needed to provide adequate coverage;
   - set WAP broadcasting power to lowest practical level;
   - set wireless NIC to infrastructure mode; and
   - provide directional antennas for wireless devices.

Physical deployment of wireless devices is one of the most important factors to consider in ensuring security. In general, when it comes to security issues, the lesser is the better. To minimize the chance of signal interception, we should avoid broadcasting to those locations where it is not needed. The best practice would be to deploy the minimum number of WAPs and set their broadcasting power to the lowest required level.

The wireless network interface card should be set to infrastructure mode to allow it to communicate with a WAP only, not just with any wireless device. Extensive site survey should be conducted regularly to determine the appropriate location of the access points, and directional antennas should be provided for wireless devices for better containment of signals.

b. *Wireless Access Control.* At this level, the following practices are suggested:
   - Configure WEP for highest level of encryption.
   - Develop and implement a service set identifiers (SSID) management scheme to change SSID regularly.
   - Do not broadcast SSID.
   - Disable DHCP and use static IP address.
   - Verify MAC address of wireless devices.
   - Disable SNMP community password on all WAPs. Rather deploy SNMPv3 on access points and other devices.

To ensure maximum security, an encryption setting should be set for the strongest encryption available and arrange the management scheme to change the pass-phrase regularly. The default values of SSID should be changed and "broadcast SSID feature" should be disabled. MAC verification provides a method to identify any unauthorized devices on the wireless network. Many vendors provide automatic MAC address verification process for restricting access to wireless network by maintaining an MAC ACL (access control list) that are stored and distributed across APs. Since DHCP automatically provides an IP address to anyone and thereby makes unauthorized penetration easier, it is a good practice to disable DHCP.

c. *Perimeter Security.* The followings practices are suggested at this stage.

- Install wireless firewall and intrusion prevention system (IPS).
- Encrypt traffic using VPN.
- Direct all traffic through VPN server.
- Maintain and enforce VPN routing and access policies.
- Use RADIUS protocol.

All 802.11b enabled devices can deploy a personal firewall that provides a certain degree of additional protection against rouge access points. An IPS can monitor the network traffic and can detect and terminate any malicious intrusion into the wireless system. Virtual private network (VPN) creates end-to-end connections between user and server as tunnels through public network where data streams are encrypted. It requires VPN client software at the client's laptop, VPN server software running on corporate network and configuring all wireless clients for VPN tunnel terminating at the server. VPN is a good security solution for a large enterprise giving employees access to corporate network from home or remote site. Additionally, user authentication to the VPN gateway can be done using RADIUS, LDAP or ACTIVE directory.

d.   *Application Security.* The following are suggested as the best practice at this level:
- Implement application level user authentication.
- Maintain and enforce access rights.

Employing security measures at the application level and enforcing access rights to system resources and applications will provide added security and should always be practiced.

MAINTAINING AND AUDITING SECURITY

Security auditing on a regular basis is an essential tool for monitoring and assessing the security measure of a wireless network and to determine the countermeasure to maintain the highest possible security level. A network administrator needs to know which access points are actually installed and their locations, whether those are properly configured and equipped with the latest patches. The wireless network needs to be audited periodically using wireless network analyzers and other tools, as access points can easily be added and modified. The wireless auditing tool needs to look at the actual wireless signals, as the needed information may not be available from the wired side. Many commercial products are available that can serve as an effective tool to conduct security auditing and troubleshooting. Some of such commercially available tools are Ethereal (2006), NetStumbler (2006), Kismet (2006), and so forth. They can determine if wireless products are transmitting correctly and on the correct channels, check for rogue access points and detect and terminate any unauthorized access.

PRIVACY ISSUES IN WIRELESS NETWORK

The laws governing the privacy issues in the United States, the European Union and other parts of the world were written before the widespread adoption and deployment of wireless technologies. Therefore, these laws do not explicitly address wireless networks. However, the existing laws can be applied to wireless technologies as well.

In the U.S., the Electronic Communication Privacy Act (ECPA, 2006) protects the privacy of voice and data communication in both wireline and wireless networks. It outlines the restrictions applied to both law enforcement authorities and private organizations regarding the use of private data, monitoring, and interception of any communication which are also to be applied to wireless communications. Except the exceptions granted by the ECPA, interception of any radio communication as well as manufacturing and distribution of devices primarily built for interception purposes are prohibited. Although any highly publicized interception of wireless network data is not known yet, interception of mobile phone conversation and pager was widely publicized in the U.S. Since similar technologies can also be used to intercept any radio traffic, privacy of wireless network data can be compromised in transit.

The exceptions granted by the ECPA for interception of radio communication mainly fall into three major categories: (1) a communication service provider is allowed to intercept communication for billing, service quality assurance, and any purpose necessary only for providing the service; (2) the government law enforcement agency or its associated private organizations may intercept data within proper legal framework; (3) interception of data is allowed with prior consent between the concerned parties. For example, the privacy commission of each country sets rules whether organizations and employers can monitor an employee's communication. In some countries, monitoring e-mail and other communications of employees is prohibited without their consent. An employee should be fully informed about an organization policy regarding network use. However, in many cases, an employee may be required to consent to monitoring as a condition for employment. Privacy commission also oversees the enforcement of privacy protection required by other agencies. If the data regarding a patient's personal medical information or an individual's financial information is compromised due to the vulnerabilities of the wireless network, the organization owning and operating the network will lawfully liable.

Restricted use of encryption technology may also contribute to less protection of personal data and compromise of privacy. For example, the U.S. imposes restrictions on the export of encryption technologies to certain extent and prohibition to some countries. Similarly, some countries impose restrictions or limit the use of encryption technologies while some countries require the users to acquire licenses from government agencies. The restriction on the use of encryption technologies will limit its use in wireless data transmission resulting in less protection and privacy of personal data.

CONCLUSIONS

The goal of wireless communications is to achieve communication "anywhere, anytime". The mobile networks, wireless local area network, and wireless personal area network are all converging to fulfill the requirement of ubiquitous services. However, security threats and privacy protection remain two major concerns in the large scale deployment of wireless technology. The current security mechanism has been proved inadequate for secure operation of wireless communications. This chapter presents an overview of these issues and interested readers are directed to more detailed resources.

REFERENCES

AirSnort. (2006). Retrieved December 10, 2006, from http://airsnort.shmoo.com

Ashley, M. (2002). A guide to wireless network security. *Latis Networks*, 1-9.

Bisdikian, C. (2001). An overview of Bluetooth wireless technology. *IEEE Communications Magazine*, 86-90.

Cam-Winget, N., Housley, R., Wagner, D., & Walker, W. (2003). Security flaws in 802.11 data link protocols. *Communication of the ACM, 46*(5).

Electronic Communication Privacy Act. (2006). Retrieved December 10, 2006, from http://www.usiia.org/legis/ecpa.html

Ethereal. (2006). Retrieved December 10, 2006, from http://www.ethereal.com

Grifin, P. (2002). Security flaws shut down Telecom's mobile email. *New Zealand Herald,* April 28.

Housley, R., & Arbaugh, W. (2003). Security problems in 802.11-based networks. *Communications of the ACM, 46*(5).

Lucent Technologies. (2000). *ORiNOCO Manager Suite Users Guide.*

Karygiannis, T., & Owens, L. (2002). *Wireless Network Security.* National Institute of Standard and technology, US Department of Commerce, Special Publication 800-48.

Kismet. (2006). Retrieved December 10, 2006, from http://kismetwireless.net

NetStumbler. (2006). Retrieved December 10, 2006, from http://www.netstumbler.com

Peikari, C., & Fogie, S. (2003). *Maximum wireless security.* SAMS Publishing.

Ramabadran, T., & Gaitonde, S. (1988). A tutorial on CRC application. *IEEE Micro, 8*(4), 62-75.

Rueppel, T. (1992). *Stream ciphers.* SIMM.

Stubblefield, A., Ioannidis, J., & Rubin, A. (2002). Using the Fluhrer, Mantin and Shamir attack to break WEP. *Proceedings of the Network and Distributed Systems Security Symposium* (pp. 17-22).

Walker, J. (2001, May). *Unsafe at any key size: An analysis of the WEP encapsulation.* IEEE 802.11 doc 00-362, grouper.ieee.org/groups/802/11.

WEPcrack. (2006). Retrieved December 10, 2006, from http://www.wirelessdefence.org/contents/WEPcrackmain.htm

Chapter XII
Security of Mobile Devices for Multimedia Applications

Göran Pulkkis
Arcada University of Applied Sciences, Finland

Kaj J. Grahn
Arcada University of Applied Sciences, Finland

Jonny Karlsson
Arcada University of Applied Sciences, Finland

Nhat Dai Tran
Arcada University of Applied Sciences, Finland

ABSTRACT

This chapter surveys security of mobile computing devices with focus on multimedia applications. Mobile computing devices are handheld devices such as PDAs (personal digital assistants) and smartphones with smaller size, processing, storage, and memory capacity compared to PCs (personal computers). The portability and various wireless network connection interfaces of the handheld devices greatly increase the risks of loss and theft of the device, exposure of confidential data, as well as the opportunities for unauthorized device, network, and network service access. The initial part of the chapter concentrates on discussing these basic security issues. Security requirements for wired and mobile multimedia network applications are outlined and network protocols (SIP, SRTP) for secure multimedia streaming services are presented. Mobile device security can be based on IPSec VPN technology and secure mobility is especially important in videoconferencing. Current research on roaming security and testbeds for mobile multimedia are also presented. In an appendix, commercially available mobile security solutions, mostly for basic mobile security requirements, are listed.

INTRODUCTION

Users of the Internet have become increasingly more mobile. At the same time, mobile users want to access Internet wireless services demanding the same quality as over a wire. Emerging new protocols and standards, and the availability of WLANs, cellular data and satellite systems are making the convergence of wired and wireless Internet possible. Lack of standards is however still the biggest obstacle to further development. Mobile devices are generally more resource constrained due to size, power, and memory. The portability making these devices attractive greatly increases the risk of exposing data or allowing network penetration.

Multimedia applications in mobile devices require support for continuous-media data types, high network and memory bandwidth, low power consumption, low weight and small size, and QoS (quality of service). Also security features like authentication and authorization for multimedia content as well as secure connectivity to multimedia sources with possibilities to verify the integrity and guarantee the confidentiality of delivered multimedia content are required (Havinga, 2000). A mobile user must also be able to roam between different networks, also between different types of networks (WLAN, cellular, etc.), and still maintain an ongoing secure multimedia application session.

A standard for a mobile multimedia system architecture has also been proposed (MITA, 2002). Two technologies, MBMS (Multimedia Broadcast/Multimedia Service [see MBMS, 2004]), and DVB-H (digital video broadcasting handheld [see ETSI EN 302 304, 2004]), are being developed for delivery of multimedia content to mobile devices.

To fulfill all security requirements for multimedia applications in a mobile environment, while still maintaining QoS, is a challenging issue. In this chapter, the security requirements and proposed solutions for fulfilling these require-ments are discussed. Attention is paid to ongoing related research. However, in order to achieve a secure mobile multimedia environment, also basic mobile device security issues must be seriously taken into account.

BACKGROUND

A mobile computer is a computing device intended to maintain its functionality while moving from one location to another. Different types of mobile computers are:

- Laptops
- Sub-notebooks
- PDAs (personal digital assistants)
- Smartphones

These devices can be divided in two groups: handheld devices and portable PCs. Handheld devices, such as PDAs and smartphones, are pocket-sized computing devices with smaller computing, memory, and display capacity compared to basic desktop computers. Portable PCs such as laptops and sub-notebooks, however, don't significantly differ from the desktops on this area.

Mobile computing and mobility are generic terms for describing the ability to use mobile devices for connecting to and using centrally located applications and/or services over a wireless network. Mobile multimedia can be defined as a set of protocols and standards for exchanging multimedia information over wireless networks. Mobile multimedia user services are usually divided in three categories (MBMS, 2004):

- streaming services like real-time video and audio streams, TV and radio programs, and on-demand video services;
- file download services; and
- carousel services, for example, news delivery with timed updating of text, image, and video objects.

For setup of sessions using multimedia streaming services in IP-based networks is used a signaling protocol, the Session Initiation Protocol (SIP), which is an Internet standard adopted by the Internet Engineering Task Force (IETF). SIP is a text-based client server protocol, in which servers respond to SIP requests sent by clients. SIP entity types are user agents (SIP clients), proxy servers, redirect servers, and registrar servers. Two or more participants can establish a session consisting of multiple media streams. SIP also provides application level mobility, which includes personal, terminal, session, and service mobility. *Personal mobility* means that a user ID provides global accessibility. *Terminal mobility* means that a mobile end-user device can maintain streaming media sessions while moving within and between IP subnets. *Session mobility* means users can change terminals during streaming media sessions and still maintain their sessions. *Service mobility* means that users maintain their streaming media session while moving, changing terminal devices, and changing network service providers (Rosenberg et al., 2002).

SIP does not define any protocol for media transport. However, streaming services typically use the Real-time Transport Protocol (RTP) over UDP. RTP defines a standardized packet format for delivering audio and video over the Internet (Schulzrinne, Casner, Frederick, & Jacobson, 2003).

Multimedia applications for handheld devices require support for continuous-media data types, high network and memory bandwidth, low power consumption, low processing power, low memory capacity, low weight and small size, and quality of service. Therefore, for multimedia applications, most of the research has currently been focused on smoothness of the multimedia data delivery and processing. Most application developers have chosen not to take the security issues into account. Security should be a natural part of the application development process. However, implementation of security in handheld device applications is complex and furthermore often tends to distract application developers from developing needed multimedia functionality and to decrease the performance (Havinga, 2000; Ong, Nahrstedt, & Yuan, 2003).

BASIC MOBILE DEVICE SECURITY

Vulnerability Analysis

Security risks associated with mobile devices are almost the inverse of the risks associated with desktop computers. Physical access to desktops can be controlled using physical access control mechanisms in form of guarded buildings and door locks. Electronic access can be controlled using personal firewalls. The physical access control and network intrusion protection mechanisms for desktop computers and wired networks are far away from complete. However, they are considered "good enough" and thus the most serious security concerns for desktop computers are related to data transmission between desktop computers. Mobile devices are small, portable, and thus easily lost or stolen. The most serious security threats with mobile devices are thus unauthorized access to data and credentials stored in the memory of the device.

Physical Access

Mobile phones and PDAs are small portable devices that are easily lost or stolen. Most platforms for mobile devices provide simple software-based login schemes for protecting device access with a password. Such systems can however easily be bypassed by reading information from the device memory without logging in to the operating system. This means that critical and confidential data stored unencrypted in the device memory is an easy target for an attacker who has physical access to the device (Symantec, 2003).

Malicious Software

Smartphones and PDAs have not actually been preferred targets for malware developers until 2004. On the time of writing, malware is still not the most serious security concern. However, the continuous increase of the number of mobile device users worldwide is changing the situation. Current malicious software is mainly focused on Symbian OS and Windows based devices. Malicious software in handheld devices may result in (Olzak, 2005):

- Loss of productivity;
- Exploitation of software vulnerabilities to gain access to recourses and data;
- Destruction of information stored on a SIM card; and/or
- Hi-jacking of air time resulting in increased costs.

Even though malicious software currently does not cause serious threats for the mobile device itself, they cause a threat for a computing network to which the mobile device is connected. Viruses are easily spread to an internal computer network from a mobile device over wireless connections such as infrared, Bluetooth, or WLAN.

Wireless Connection Vulnerabilities

Handheld devices are often connected to the Internet through wireless networks such as cellular mobile networks (GSM, GPRS, UMTS), WLANs, and Bluetooth networks. These networks are based on open air connections and are thus by their nature easy to access. Many Bluetooth networks and especially WLANs are still unprotected. This means that any device, within the network coverage, can access the wireless network. Furthermore, confidential data transmitted over an unprotected wireless network can easily be captured by an intruder.

Security Policy

Examples of security policy rules for mobile device end users as well as for administrators of mobile devices in corporate use are proposed (Taylor, 2004-2005, Part V):

- Make sure that the cell phone, PDA, or smartphone is password protected.
- Use secure remote access VPN to connect to the corporate network for the purpose of checking e-mail.
- Keep a firewall and an anti-virus client with up-to-date anti-virus signatures installed on the handheld if connecting the corporate network.
- Use the security policies on the handheld firewall that are recommended by the corporate security team.

Examples of rules proposed for corporate administrators of mobile devices (Taylor, 2004-2005, Part V):

- End users must first agree to the End-User Rules of Behavior.
- All handheld users are to be setup with a secure remote access VPN client to connect to the corporate network.
- Advice end-users what anti-virus client to use.
- Handheld firewalls are configured to log security events and send alerts to security_manager@company.com.
- Handheld groups and net groups are set up to restrict access privileges only to services and systems required.

Platform Security

The convergence between the fixed Internet and the mobile Internet has raised the question of security of mobile devices. The introduction of packet-based services, such as GPRS and 3G, has opened the wireless industry to new services, but also to new mobile device vulnerabilities. The handsets are actually changing from a formerly closed nature to a more open system. Incidents involving spamming, denial-of-services, virus attacks, content piracy, and malicious attacks have become a growing problem in the wireless world. With this openness a new level of security is required to protect wireless networks and handsets (Sundaresan, 2003).

Security is an integral part of a mobile handset because security affects all parts of the mobile device. Security issues can therefore not be treated as separate add-on features. Instead, security needs to be built into the platform (Trusted Computing Group, 2004).

Mobile Device Forensics

Computer forensics is the process of investigating data processing equipment—typically a PC computer but also mobile device like a PDA computer or a mobile smartphone—to determine if the equipment has been used for illegal, unauthorized, or unusual activities. Computer forensics is especially needed when a mobile device is found or returned after being lost or stolen in order to estimate if and what damage has occurred. By using specially designed forensics software, computer experts are able to identify suspects and sources of evidence, to preserve and analyze evidence, and to present findings (Phillips, Nelson, Enfinger, & Steuart, 2005; Robbins, 2005).

Physical Protection

Physical security of a mobile device means protection against physical and electromagnetic damage, and protection of stored content against power failures and functional failures. There must be possibilities to recover stored content after damage, after a functional failure, or after

an intrusion attack. Furthermore, protection of stored content in case of theft or other loss of the mobile device must be included.

Physical Damage

Most common physical damage is caused by a mechanical shock when the mobile device is dropped or by water when the mobile device has fallen into water or has been exposed to rain.

In case of a mechanical shock the content interface, the display and/or the keyboard, is usually damaged and the stored content is accessible after the content interface is repaired or the electronics and memory modules are transferred to another device. Water will cause damage to the power supply battery and also to electronic components.

Electromagnetic Damage and Unwanted Wireless Communication

The electronic and/or the stored content of a mobile device can be damaged by strong electromagnetic radiation. A switched-on mobile device may also have an always open wireless network connection. For prevention of unwanted electromagnetic radiation and unwanted wireless communication, an electromagnetic shielding bag (MobileCloak™ Web Portal, 2005) can be used.

Power and Functional Failures

The operating system should store configuration changes and sufficiently often, for example every five minutes, backup the working spaces of open applications in non-volatile memory in order to minimize the data losses in case of sudden power failure or other functional failure.

Backup and Recovery

Recovery after power failure or other functional failures is usually managed by the operating system from stored configuration data and from timed backups of the working spaces of open applications. A usual backup function of mobile devices is synchronization with a desktop computer. Examples of synchronization are HotSync in Palm OS, ActiveSync in Windows CE and Windows Mobile, and SyncML in Symbian OS. Recovery from a synchronization backup or from a traditional backup is needed after physical and electromagnetic damage and after an intrusion attack.

Loss and Theft

To minimize damage caused by a lost or stolen device, the following precaution measures are suggested:

- Confidentiality level classification of stored content;
- Content encryption of sensitive data;
- Use of bit wiping software, for stored content with the highest confidentiality level classification; and
- Visible ownership information, for example a phone number to call if a mobile device is found, for return of a lost or stolen mobile device.

Activated bit wiping software will permanently delete data and program code according to user settings. Good bit wiping software will also delete all related information in working memory (RAM) and in plugged external memory cards. Bit wiping software is typically activated when a wrong device access password is entered a preset number of times and/or the mobile device is not synchronized within a preset timeframe. Because bit wiping software can also be triggered by accident, a fresh synchronization backup or a fresh traditional backup of the stored content should always be available.

Device Access Control

Physical access control mechanisms are ineffective for PDAs and smartphones since such devices are small, portable, and thus easily lost and stolen. Access control mechanisms on the device itself are thus important for protecting the data stored in the device. Currently, there are no widely adopted standards for access control services in mobile devices. Manufacturers are concentrating more on securing the communication protocols than securing stored data in the device. However, this trend is expected to change in the near future. The use of handheld devices is constantly growing, and thus the need for secure access control mechanisms is also increasing.

Access control on a mobile device can be implemented with a combination of security services and features (Perelson & Botha, 2004):

- Authentication service
- Confidentiality service
- Non-repudiation service
- Authorization

Primarily, the security services for access control are authentication and authorization services. The principle of access control on a mobile device is shown in Figure 1.

Authentication

An authentication service is a system for confirming a claimed user identity. There are many methods in which a user is able to authenticate to a handheld device. These methods include:

- Passwords/PINs
- Visual and graphical login
- Biometrics

Figure 1. The principle of access control

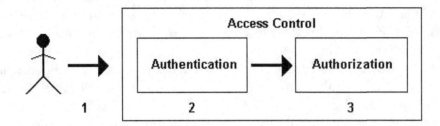

1. The user presents an identity (e.g. password or biometric)
2. The user's identity is confirmed
3. An authenticated user is allowd access to a resource

PIN and Password Authentication

Many handheld devices use a PIN code for user authentication. The PIN is four digits of length and is entered by the user from a ten-digit (0-9) numerical keypad. PINs are suitable for personal handheld devices, not containing critical information. However, PINs do not provide sufficient security for handheld devices in corporate use. PINs may be susceptible to shoulder surfing or to systematic trial-and-error attacks due to their limited length and alphabet. Passwords are more secure than PINs since they support a larger alphabet and increase the number of digits in the password string (Jansen, 2003).

Most PDA and smartphone operating systems provide inherent support for traditional alpha-numeric passwords. Strong password authentication solutions can also be implemented by installing additional security software, see section, *Available Security Solutions*.

Visual and Graphical Login

Visual authentication allows a user to select a sequence of icons or photo images as a password value. Instead of alpha-numeric characters a user must remember image sequences from a list of pictures shown on the display of the device. A visual authentication method is more user-friendly and more secure than standard password authentication. In order for a password to be secure, it must consist of many digits including both upper and lower case characters as well as both characters and numbers. Picture combinations are for a user easier to remember than complex password strings. Furthermore, a picture password system can be designed to require a sequence of pictures or objects matching a certain criteria and not exactly the same pictures (Duncan, Akhtari, & Bradford, 2004).

Graphical login relies on the creation of graphical images to produce a password value. A user enrolls a password value by drawing a picture on a display grid including block text or graphical symbols. Login process strokes can start anywhere and go in any direction on the touch screen, but they must occur in the same sequence as in the enrolled picture. The system maps each continuous stroke to a sequence of coordinate pairs by listing all cells through which the stroke passes and the order in which the stroke

Figure 2. A five-stroke graphical password

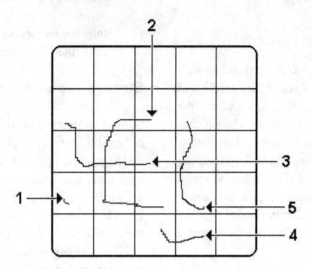

crosses the cell boundaries. An example with a five-stroke password entry, which can be drawn in eight different ways, is shown in Figure 2. The authentication system controls the ordering of the strokes and the beginning and end point of each stroke. The numbered items in Figure 2 indicate in which order the strokes were drawn and point to the starting point of each stroke in the enrolled picture (Jansen, 2003).

Currently, there are no PDAs or smartphones providing inherent support for visual or graphical login. To implement such an authentication method, an additional security application must be installed. Examples of such applications are presented in section, *Available Security Solutions*.

Biometrics

Biometric user authentication is a hardware solution for examining one or more physical attributes of an authorized user. Biometric controls are becoming more and more common in handheld devices (Perelson & Botha, 2004).

Fingerprint Verification

Fingerprint verification provides both user friendliness and security. A user only needs to touch a certain point of the handheld device in order to authenticate to it. The user does not need to remember a complex password string or an image sequence and the security level is still high since every fingerprint is unique.

Signature Verification

Signature verification is a technology where a number of dynamic characteristics from a physical signing process are captured and compared. Dynamic characteristics are speed, acceleration, direction, pressure, stroke length, sequential stroke pattern, and the time and distance when the pen is lifted away from the surface. The physical signature is made with an electronic pen on the touch screen (Jansen, 2003).

Voice Verification

In voice verification identifying and authenticating is based on the user's voice. The enrollment process is normally performed in a way that the user speaks a set of specific words. Usually this process is repeated several times. A template is then extracted from this voice input. This template defines the characteristics of the recorded voice. During authentication, the system prompts the user to pronounce a set of randomly chosen digits as they appear on the display of the handheld device (Jansen, 2003).

Voice verification for smartphones and PDAs has not yet broken through, but a number of companies have recently developed new biometric software and devices (Kharif, 2005).

Authorization

Handheld devices are typically personal and the authentication process infers that the user is authorized. It is often assumed that all data stored on a device is owned by the user. Today, there are large gaps in the features of the authorization services in both smartphone and PDA devices (Perelson & Botha, 2004).

It is becoming more common that handheld devices replace desktop and notebook computers in companies. A single device may be used by several employees and may contain confidential company information. Thus, the need for proper user authorization services is becoming more important. Such services are Perelson and Botha (2004):

- **File Masking:** Certain protected records are prevented from being viewed by unauthorized users.

- **Access Control Lists:** A list defines permissions for a particular object associated with a user.
- **Role-Based Access Control:** Permissions are defined in association with user roles.

Storage Protection

Storage protection of a mobile device includes online integrity control of all stored program code and data, optional confidentiality of stored user data, and protection against unauthorized tampering of stored content. Protection should include all removable storage modules used by the mobile device.

The integrity of the operating system code, the program code of installed applications, and system and user data can be verified by using traditional tools like checksums, cyclic redundancy codes (CRC), hashes, message authentication codes (MAC, HMAC), and cryptographic signatures. Only hardware-based security solutions for protection of verification keys needed by MACs, HMACs, and signatures provide strong protection against tampering attacks, since a checksum, a CRC, and a hash of a tampered file can easily be updated by an attacker. Online integrity control of program and data files must be combined with online integrity control of the configuration of a mobile device. This is needed to give sufficient protection against attempts to enter malicious software like viruses, worms, and Trojans.

Required user data confidentiality can be granted by file encryption software. Such software also protects the integrity of the stored encrypted files, since successful decryption of an encrypted file is also an integrity proof.

Network and Network Service Access Control

Once a user is authenticated and granted access to the handheld device, the device can be connected to and used in several types of networks and network services. This section presents access control mechanisms and user authentication issues related to some of these systems.

Hardware Authentication Tokens

Hardware tokens are used in handheld devices to authenticate mobile end users to mobile networks and network services. Such hardware tokens are SIM (subscriber identity module), PKI SIM (public key infrastructure SIM), USIM (universal SIM) and ISIM (IP multimedia services identity module) chips.

SIM

A basic SIM card is a smartcard securely storing an authentication key identifying a GSM network user. The SIM card is technically a microcomputer, consisting of a CPU, ROM, RAM, EEPROM, and I/O (input/output) circuits. This microcomputer is able to perform operations, such as cryptographic calculations with the individual authentication key needed for authenticating the subscriber. The SIM card also contains storage space for SMS (short message services) messages, MMS (multimedia messaging system) messages, and a phone book. The use and content of a SIM card is protected by using PIN codes.

PKI SIM

A PKI SIM card is a basic SIM card with added PKI functionality. A co-processor is added to take care of decryption and signing with private keys. A PKI SIM card contains space for storing private keys for digital signatures and for decryption (Setec Portal, 2005).

PKI SIM cards open the possibility for handheld users to generate digital signatures and authenticate to online services. Currently, the use of PKI SIM cards is not widely supported.

However, systems for utilizing this technique are being developed.

USIM

A USIM card is a SIM used in 3G mobile telephony networks, such as UMTS. USIM is based on a different type of hardware; it is actually an application running on a UICC (universal integrated circuit card). The USIM stores a pre-shared secret key as the basic SIM.

ISIM

An ISIM card consists of an application (ISIM) residing on a UICC. The ISIM application provides secure authentication of handheld users to IMS (IP multimedia systems) services (Dietze, 2005).

Cellular Networks

GSM/GPRS

User authentication in GSM networks is handled by a challenge-response based protocol. Every MS (mobile station) shares a secret key Ki with its home network. This key is stored in the SIM card of the MS and the AuC (Authentication Centre) of the home network. Ki is used to authenticate the MS to the visited GSM network and for generating session keys needed for encrypting the mobile communication. The authentication process, shown in Figure 3, is started by the MSC (Mobile Switching Centre) which requests an authentication vector from the AuC of the home network of the MS. The authentication vector, generated by the AuC, consists of a challenge response pair (RAND, RES) and an encryption key Kc. The MSC of the visited network sends the 128-bit RAND to the MS. Upon receiving the RAND, the MS computes a 32-bit response (RES) and an encryption key Kc using the received RAND and the Ki stored in the SIM. The MS sends the RES

Figure 3. GSM authentication and key agreement

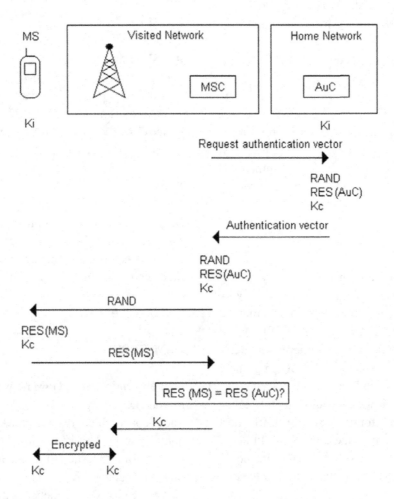

back to the MSC. MSC verifies the identity of the MS by comparing the received RES from the MS with the received RES from the AuC. If they match, authentication is successful and the MSC sends the encryption key Kc to the base station serving the MS. Then the MS is granted access to the GSM network service and the communication between the MS and the base station is encrypted using Kc (Meyer & Wetzel, 2004).

GSM networks provide reasonably secure access control mechanisms. The most serious concern is the lack of mutual authentication. This opens a possibility for an attacker to setup a false base station and imitate a legitimate GSM network. As a result, the key Ki can be cracked and the attacker can then impersonate a legitimate user (GSM Security Portal, 2005).

UMTS

The authentication and key management technique used in UMTS networks is based on the same principles as in GSM networks (see Figure 4). A secret authentication key is shared between the network and the MS. This key is stored on the MS USIM and in the AuC of the home network. Unlike in

Figure 4. UMTS authentication and key agreement

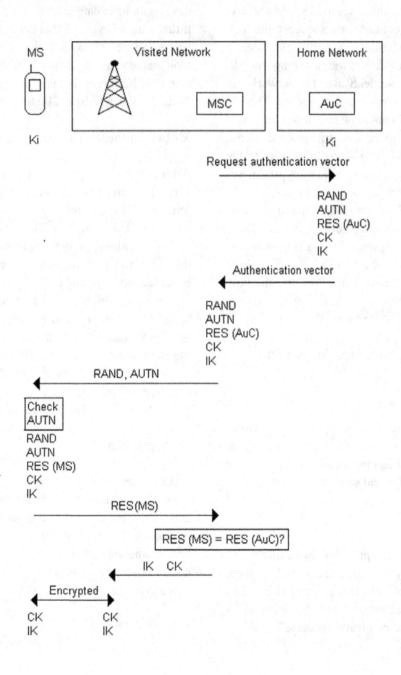

GSM networks, UMTS networks provide mutual authentication. Not only MS is authenticated to the UMTS network but also the UMTS network is authenticated to MS. This protects MS from attackers trying to impersonate a valid network to the MS. Network authentication is provided by a so-called authentication token AUTN.

The MSC (Mobile Switching Centre) of the visited network sends the AUTN together with the authentication challenge to MS. Upon receiving the AUTN, containing a sequence number, the MS checks whether it is in the right range. If the sequence number is in the right range the MS has successfully authenticated the network and the authentication process can proceed. The MS computes an authentication response, here called RES, and encryption and integrity protection keys, called CK and IK, and sends these back to the MSC. The MSC verifies the identity of the MS by checking the correctness of the received RES. Upon successful authentication, the MSC sends the encryption key CK and integrity key IK to the UMTS base station. The MS is now able to communicate with the UMTS network and the communication between the MS and the base station is encrypted with CK and the integrity is protected with IK (Meyer & Wetzel, 2004).

Wireless Personal Area Networks

IrDA

The IrDA standard does not specify any security measures. However, since IrDA is a short line of sight-based connection, security threats can be eliminated by physical security measures.

Bluetooth

Bluetooth technology provides device authentication, not actually user authentication. The whole authentication process can be divided into two phases, an initial process called pairing and a mutual device authentication process.

Pairing

Two Bluetooth devices, that want to set up a connection, share a PIN code which is entered on both devices in the beginning of the pairing process. The PIN code, the address of the Bluetooth device, and a 128-bit random number is used as inputs to an algorithm (E1) for creating an initialization key. A new random value is generated on both devices and exchanged after XORing it with the initialization key. With this new 128-bit random value, and the Bluetooth device address, the common shared secret key, called link key, is generated using the E21 algorithm (Gehrmann et al., 2004; Xydis & Blake-Wilson, 2002).

Mutual Authentication

After pairing, the actual device authentication is performed through the use of a challenge-response scheme. One of the devices is the verifier and the other is called a claimant. The verifier generates a 128-bit challenge value. The claimant then uses the algorithm E1 with the challenge, its 48-bit Bluetooth address, and the link key as inputs, for creating a 128-bit value. The 32 most significant bits of this value are returned to the verifier. The verifier verifies the response word by performing the same calculations. If the response value is successful, then the verifier and the claimant change roles and repeat the authentication process (Gehrmann et al., 2004; Xydis et al., 2002).

Authorization

Bluetooth technology also offers a way of performing user authorization. Devices have two levels of trust. They are divided into trusted or untrusted devices. A device is trusted only when it first has been authenticated. Interplay of authentication and authorization defines three service levels (see Table 1) (Gehrmann et al., 2004).

Table 1. Bluetooth service levels

	Authorization	Authentication	Encryption
Service Level 1	Yes	Yes	Yes
Service Level 2	No	Yes	Yes
Service Level 3	No	No	Yes

At service level 3, any device is granted access to any service. At the next level, only authenticated devices get access to all services, and at service level 1, only authenticated devices are granted access to one or more certain services.

Security Analysis

One of the major weaknesses in Bluetooth access control is the lack of support for user authentication. A malicious user can easily access network resources and services with a stolen device. Furthermore, PIN codes are often allowed to be short and thus susceptible to attacks. However, the coverage range of a Bluetooth network is very short. Malicious access to a Bluetooth network can therefore mostly be prevented by the use of physical access control measures.

Wireless Local Area Networks

Implementation and use of secure access control mechanisms is essential in order to protect WLANs from unauthorized network access. WLANs became, during their first years, known for their serious security vulnerabilities. One of the most significant concerns has been the lack of proper user authentication methods. Today, WLANs provide acceptable security through the recently ratified security standard IEEE 802.11i.

Access Control Based on IEEE 802.11

The authentication mechanisms defined in the original WLAN standard IEEE 802.11 are weak and not recommended. The standard only provides device authentication in form of the use of static shared secret keys, called WEP (wired equivalent privacy) keys. The same WEP key is shared between the WLAN access point and all authorized clients. WEP keys have turned out to be easily cracked. If a WEP key is cracked by an intruder, the intruder gets full access to the WLAN.

WEP authentication can be strengthened by using MAC filters and by disabling SSID broadcasting on the access point. However, SSIDs are easily determined by sniffing probe response frames from an access point and MAC address are easily captured and spoofed.

Access Control Based on IEEE 802.11i

The recently ratified WLAN security standards WPA and WPA2 address the vulnerabilities of WEP. WPA is a subset of the IEEE 802.11i standard, and WPA2 provides full 802.11i support. The difference between WPA and WPA2 is the way how the communication is encrypted. Furthermore, WPA2 provides support for ad-hoc networks which is missing in WPA. User authentication in WPA and WPA2 are based on the same techniques. Currently, there are already a number of PDAs and smartphones supporting WPA. WPA2, however, is still not supported in any handheld device (Wi-Fi Alliance Portal, 2005).

Pre-Shared Key. 802.11i provides two security modes: home mode and enterprise mode. The home mode is based on a shared secret string, called PSK (pre-shared key). Compared to WEP, PSK is never used directly as an input for data encryption algorithms. For large WLANs,

the enterprise mode is recommended. The home mode is suitable for small WLAN environments, such as small office and home WLANs, where the number of users is small.

IEEE 802.1X. The enterprise security mode utilizes the IEEE 802.1X standard for user authentication. The 802.1X standard is designed to address open network access. Three different components are involved: supplicant (client),

Figure 5. 802.1X authentication in unauthorized state

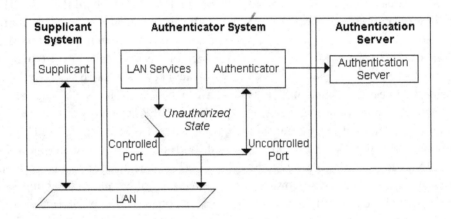

Figure 6. 802.1X authentication in authorized state

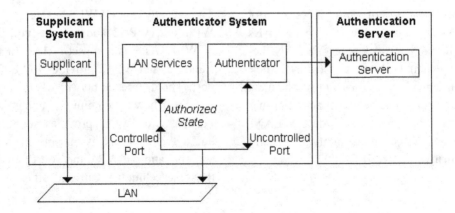

authenticator (WLAN access point) and AAA (authentication, authorization, and accounting) server. The supplicant wants to be authenticated and accesses the network via the authenticator. The AAA server, typically a RADIUS (remote authentication dial-in user service) server, works as a back-end server providing authentication service to an authenticator. The authentication server validates the identity and determines, from the credentials provided by the supplicant, whether the supplicant is authorized to access the WLAN or not.

The authenticator works as an intermediary between the supplicant and the authentication server passing authentication information messages between these entities. Until the supplicant is successfully authenticated on the authentication server, only authentication messages are permitted between the supplicant and the authentication server through the authenticator's uncontrolled port. The controlled port, through which the supplicant can access the network services, remains in *unauthorized state* (see Figure 5). As a result of successful authentication, the controlled port switches to *authorized state*, and the supplicant is permitted access to the network services (see Figure 6).

EAP (Extensible Authentication Protocol). 802.1X binds the EAP protocol which handles the exchange of authentication messages between the

Figure 7. EAP authentication exchange messages

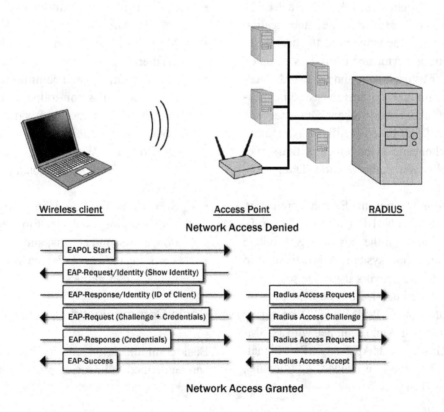

supplicant and the AAA server. The exchange is performed over the link layer, using device MAC addresses as destination addresses. A typical EAP authentication conversation between a supplicant and an AAA server is shown in Figure 7.

EAP supports the use of a number of authentication protocols, usually called EAP types. The following EAP types are WPA and WPA2 certified (Wi-Fi Alliance Portal, 2005):

- EAP-TLS (EAP-Transport Layer Security);
- EAP-TTLS (EAP-Tunneled Transport Layer Security);
- PEAPv0/EAP-MSCHAPv2 (Protected EAP version 0/EAP-Microsoft Challenge Authentication Protocol version 2);
- PEAPv1/EAP-GTC (PEAPv1/EAP-Generic Token Card); and
- EAP-SIM.

EAP-TLS, EAP-TTLS, and EAP-PEAP are based on PKI authentication. EAP-TTLS and EAP-PEAP however only use certificate authentication for authenticating the network to the user. User authentication is performed using less complex methods, such as user name and password. EAP-TLS provides mutual certificate based authentication between clients and authentication servers. Therefore, a X.509 based certificate is required both on the client and on the authentication server for user and server authentication (Domenech, 2003).

EAP-SIM is an emerging EAP authentication protocol. This standard is still an IETF draft. EAP-SIM is based on the existing GSM mobile phone authentication system. A user is able to authenticate to the network using the secret key and algorithms embedded on the SIM card. In order to implement EAP-SIM, a RADIUS server supporting EAP-SIM and equipped with a GSM/MAP/SS7 (GSM/Mobile Application Part/Signalling System 7) gateway is needed. Additionally,

the client software must support the EAP-SIM authentication protocol. The RADIUS server contacts the user's home GSM operator through the GSM/MAP/SS7 gateway and retrieves the GSM triplets. The triplets are sent to the client via the access point, and if the supplicant and the user's SIM card are able to validate the GSM triplets, then the RADIUS server requests the access point to grant the client network access.

Network Connection Security

Connection security means:

- Integrity of communication;
- SIM, USIM, PKI SIM, ISIM cards in cellular mobile networks (GSM, GPRS, UMTS);
- Confidentiality of communication (SSH, VPN, SSL, voice, video);
- Intrusion prevention
  - traffic filtering (firewall)
  - firewall log of connection attempts;
- Malware rejection: Anti-virus software and anti-spyware;
- Mutual authentication of communicating partners;
- Security settings and commitment to security policy rules controlled by centralized security management software;
- Control of remote synchronization (Palm OS/HotSync, Windows CE & Windows Mobile/ActiveSync, Symbian OS/SyncML based);
- Security audits based on communication event logging, analysis of logged information, alerts, and alarms; and
- Shielding the mobile device from unwanted wireless communication.

Basic Communication Security

Basic communication security of a mobile device can be defined as intrusion prevention and malware rejection. The basic intrusion prevention tool is a configurable firewall with communication event logging and alert messaging. The basic malware rejection tools are anti-virus and anti-spyware with suspicious event alarming features. The core of a malware rejection tool is a malware recognition database. Malware rejection tool providers constantly update this database and the updated malware recognition database is available to users through some network connection. An installed malware rejection tool should always use the latest update of the malware recognition database. Anti-virus software for mobile devices is delivered for example by Symantec Corporation (2005) and Kaspersky (2007).

Authentic Data Communication

Authentic data communication is based on mutual authentication of communicating parties. In 2G cellular networks (GSM Data) authentication is uni-directional. The mobile device is authenticated to the cellular network by use of the shared secret key in the SIM card. Mutual authentication, for example, based on a public key certificate, is however possible for packet data communication in GSM networks (GPRS) in addition to PIN-based GSM authentication. In 3G cellular networks, like UMTS, authentication is mutual. The mobile device and the network are authenticated to each other by the authentication and key agreement (AKA) mechanism (Cremonini, Damiani, De Capitani di Vimercati, Samarati, Corallo, & Elia, 2005).

In a WLAN authentication is mutual for WPA and IEEE 802.11i (WPA2). The authentication of a mobile client is based on presented credentials and information registered in an AAA server. The authentication protocol, EAP, also requires authentication of the AAA server to the mobile client (Pulkkis, Grahn, Karlsson, Martikainen, & Daniel, 2005). Also a Bluetooth connection can be configured for mutual authentication. The default security level of a Bluetooth service is:

- **Incoming Connection:** Authorisation and authentication required.
- **Outgoing Connection:** Authentication required (Muller, 1999).

Integrity and Confidentiality of Data Communication

Confidentiality and integrity of all data communication to and from cellular mobile networks (GSM, GPRS, UMTS) is provided by the security hardware of SIM/USIM/PKI SIM/ISIM cards in mobile devices.

For data communication through other network types (WLAN, Bluetooth, IrDA) connection specific security solutions must be installed, configured, and activated. Alternatively, end-to-end security software like VPN and SSH must be used. Available PDA VPN products are listed in (Taylor, 2004-2005, Part IV).

For WLAN connections, available solutions for confidentiality and integrity of all data communication are WEP, WPA, and IEEE 802.11i (WPA2). WEP security is however weak, since WEP protection can be cracked from recorded WEP protected data communication (WEPCrack, 2007).

For Bluetooth connections link-level security corresponding to security mode 3 should be used (Sun, Howie, Koivisto, & Sauvola, 2001a).

Roaming Security

Widely accepted Internet protocol standards supporting roaming for mobile devices are Mobile IP (Perkins, 2002) and SIP (Rosenberg et al., 2002). Mobile IP supports network level roaming, which is transparent for all IP-based network applications. SIP supports application level roaming, mainly for multimedia streaming services. Roaming security means:

- mutual authentication of a mobile device and the access network;

- integrity protection and confidentiality of all data communication, including roaming control signaling in handover situations.

A mobile device with Mobile IP support has an IP address registered by a home agent in the home network. A roaming mobile device visiting a new foreign network is registered with an additional IP address, the care-of-address by a foreign agent in the visited network and also by the home agent. The home agent can then redirect IP packets with the home address of the mobile node to the foreign network, which is presently visited by the mobile node. Home and foreign agent functionality is usually implemented on the gateway router of a network.

Authentication services for the Mobile IP protocol are:

- Security associations for authentication of mobile device registration messages (Rosenberg et al., 2002);
- AAA server based authentication of the home agent and the foreign agent of a mobile device (Calhoun, Johansson, Perkins, Hiller, & McCann, 2005; Glass, Hiller, Jacobs, & Perkins, 2000); and
- Public key cryptography based authentication of a mobile device (Hwu, Chen, & Lin, 2006; Lee, Choi, Kim, Sohn, & Park, 2003).

Secure Mobile IP communication in a handover situation can be achieved:

- on data link level by the WLAN security protocol WPA or IEEE 802.11i, when the new access network is a WLAN;
- as end-to-end protection by the security protocols IPSec, SSL, SSH, or by mobile node identity based public key cryptography (Barun & Danzeisen, 2001; Hwu, Chen, & Lin, 2006; Islam, 2005).

Security of SIP supported mobile device roaming is described in the section, *Mobile Multimedia Security*.

Connection Security Management

Security settings and commitment to security policy rules should be controlled by centralized security management software. Security audits based on communication event logging, analysis of logged information, and on alerts and alarms should be performed with timed and manual options.

Special attention should be paid to control of remote synchronization (Palm OS/HotSync, Windows CE & Windows Mobile/ActiveSync, Symbian OS/SyncML based). Remote synchronization should be disabled when not needed. Desktop and laptop computers should have basic communication security features like personal firewalls, anti-malware protection software with updated malware recognition data and latest software patches installed.

Use and attempts to use remote synchronization ports (see Table 2) should be logged

Table 2. TCP and UDP ports used by synchronization software

	TCP Ports Used	UDP Ports Used
ActiveSync	990, 999, 5678, 5679	
HotSync	14237, 14238	14237
SyncML based	80 (http)	

and alerts and alarms should be triggered by unauthorized use or usage attempts. Passwords used by synchronization software in desktop and laptop computers should resist dictionary attacks and the PC/Windows option to save connection passwords should not be used.

SECURITY OF MULTIMEDIA NETWORK APPLICATIONS

Security requirements for multimedia network applications and approaches to fulfill these requirements are surveyed in Sun, Howie, Koivisto, and Sauvola (2001b). The requirements include authentication, availability, protection, and performance.

Authentication Requirements

Authentication requirements for secure multimedia network applications are source authentication, receiver authentication, and authentication of delivered content.

Receiver authentication means that a user of a multimedia service must identify himself/herself with a password, by proving the possession of a security token, or biometrically before he/she is authorized for the service. Source authentication usually means that a user of a multimedia service checks the possession of a security token for the server or the provider before using the service. Digital signing and watermarking techniques can be used for authentication of delivered content.

A digital watermark is an imperceptible signal, which can be inserted into a digital content—a document, an image, an audio file/stream, a video file/stream, and so forth—for purposes like copyright control and binding digital content to content owners. Watermarks resist tampering and are robust to signal transformations like file filtering and compression. Content copyright and ownership can be checked with special watermark reading software. Digital pirates can be forensi-

cally revealed by watermark checks. A special type of watermarks, called annotation watermarks, can be used for access control based on binding usage rights to digital content. However, the current state of watermarking technology is imperfect. A hacker knowing the used watermark technique can often tamper watermarks beyond recognition (Cox & Miller, 1997; Dittmann, Wohlmacher, & Ackermann, 2001; Liu, Reihaneh, & Sheppard, 2003; Stamp; 2002).

Availability Requirements

Delivery of multimedia content to authorized receivers must be granted. Granted delivery of streaming multimedia implies quality of service (QoS) requirements, which include sufficient bandwidth from source to receiver for successful real-time content transport.

Requirements on IP for granted content delivery to authorized receivers are secure routing and DNS security. A misconfigured, faulty, or tampered router misdirects multimedia content transport and a fake DNS gives an incorrect IP address of the receiver.

Protection Requirements

Delivered multimedia content must protected against tampering, capture, unauthorized copying, and denial-of-service (DoS) attacks. With methods based on encryption, digital signing, and watermarking, some protection can be achieved.

Performance Requirements

Multimedia network applications usually generate and transmit a waste amount of data. Some multimedia applications, like interactive multipoint video streams, must also meet real-time constraints. Cryptographic operations needed by security services are therefore associated with strict execution time constraints, which can be met only by hardware implemented cryptographic algorithms. Also the complexity and specific characteristics of video coding technologies require careful choice as well as special design and application of cryptographic algorithms in necessary security services.

Secure Multimedia Streaming Services

For streaming service sessions both the session setup SIP signaling and the media transport should be protected. Security mechanisms in SIP for authentication of SIP user agent clients and confidential end-to-end and hop-to-hop communication are surveyed in Salsano, Veltri, and Papalilo (2002).

Streaming content delivery can be protected with the Secure Real-time Transport Protocol (SRTP) (Baugher, McGrew, Naslund, Carrara, & Norrman, 2004), which can provide confidentiality, message authentication, and reply protection to the RTP traffic and to the control traffic for RTP, the Real-time Transport Control Protocol (RTCP), which provides out-of-band control information for an RTP flow (Schulzrinne et al., 2003). RTPC cooperates with RTP in the delivery and packaging of multimedia data, but the protocol does not actually transport any data itself. The function of RTCP is to gather statistics on a media connection and also information such as bytes sent, packets sent, lost packets, jitter, feedback, and round trip delay. An application may then use this information to increase the QoS. SRTCP (secure real-time transport control protocol) is a security protocol designed to provide the same security-related features to RTCP, as the ones provided by SRTP to RTP.

RTP does not provide any QoS guarantee. QoS can, however, be achieved using resource reservation techniques in the network. An example of such a technique is defined in RSVP (resource reservation protocol) (Braden, Zhang,

Berson, Herzog, & Jamin, 1997; Featherstone & Zhang, 2004).

The RTP protocol itself is not suitable for integrated heterogeneous network environments consisting of both wired and wireless networks. As a solution for this weakness, mixers have been added to the RTP solution. The mixers are using a process called transcoding in order to deal with different levels of bandwidth. The transcoding process translates an incoming data stream into a different one with possible lower data-rate.

The use of mixers and translators to perform transcoding processes to achieve mobility support compromises security for example in video conferencing systems. Mixers and translators must decrypt data in order to be able to manipulate the data. After manipulation, the data streams are again re-encrypted. Thus, end-to-end confidentiality is compromised and furthermore since the data has been altered, it is impossible to perform source authentication.

The RSVP provides an authentication technique called hop-by-hop authentication (Baker, Lindell, & Talwar, 2000). This technique has however a high overhead, and is thus not suited for use in wireless networks with limited bandwidth. The hop-by-hop authentication in RSVP does furthermore not provide any key management or confidentiality services. This means that a separate key distribution algorithm is needed prior to the start of a conference and an attacker, who is eavesdropping the resource reservation messages could infer useful and sensitive information.

A solution to address security concerns with RSVP is proposed in RSVP-SQoS (RSVP with scalable QoS protection) (Talwar & Nahrstedt, 2000). This solution makes the design scalable by dividing a network into different sub-networks and it assumes that security requirements within one sub-network are weaker than the requirements for inter-sub-networks. Immediately, when a security violation is detected, the resource reservation is removed. However, considering mobile multimedia streaming services, this solution is not suitable since user mobility is not taken into account.

An end-to-end authentication scheme for resource reservation designed to eliminate DoS attacks, where an attacker alters the amount of resource requests, is worked out in Wu, Wu, Huang, and Gong (1999). In this solution, a digitally signed request is sent to the receiver by the source of the data stream. The digital signature is generated with the source's private key. If the receiver is able to verify the signed request, then it sends out a resource reservation message, which is a piggybacked version of the original request from the source, and a resource reservation request. These are also digitally signed before sent. As this message is sent back along the network, intermediate routers check if the piggybacked message is equivalent to the original one and when the message arrives back to the sender, the sender checks the signature of the packet. If any of these checks fail, the resource reservation is rejected. This protocol has, however, two major weaknesses. The first one is that it does not provide confidentiality to the reservation message. A potential attacker can thus eavesdrop on the messages. The second major weakness is that it lacks any support for mobility.

MOBILE MULTIMEDIA SECURITY

A mobile device for multimedia applications must fulfill the specific security requirements of mobile multimedia in addition to:

- all basic security requirements of mobile data communication; and
- all security requirements of multimedia network applications.

Security requirements of multimedia processing are included in the framework model of mobile security in Zeng, Zhuang, and Lan (2004).

A mobile device for multimedia applications should support:

- SIP signaling based application level mobility for multimedia streaming services; and
- Mobile IP-based network level mobility for file download services and carousel services for text and images.

Mobil IP functionality with security mechanisms described in the section, *Network Connection Security*, as well as SIP user agent functionality with security mechanisms surveyed in Salsano et al. (2002) should thus be integrated in secure mobile devices for multimedia applications. SIP signaling can during inter-domain roaming be separated from other network traffic by domain edge routers (Nasir & Mah-Rukh, 2006).

SRTP (Baugher et al., 2004) is used to encrypt the video, audio, or data stream for securing a multimedia streaming service, since real-time media transport is based on RTP and UDP. Setting up a secured session with SIP signaling involves exchange of separate SRTP key between the mobile device and the host delivering the stream for the specific media type (Dutta et al., 2004a).

Mobile Multimedia Device Security Based on IPSec VPN Technology

Mobile device security based on a IPSec VPN architecture is proposed for mobile multimedia services in a MAN network in Zhang, Lambertsen, and Yamada (2003). Bandwidth requirements of multimedia services are provided by a PON (passive optical network) backbone access network, which allows IP multicasting through optical fibres to a group of adjacent wireless base stations implemented by optical network units (ONU). The PON gateway to the Internet is an optical line terminal (OLT). An IPSec VPN agent with a VPN tunnel to the home network of the mobile devices is installed on this OLT. A mobile device entering the MAN is registered by the VPN agent, which also records the multicast address a roaming domain consisting of some adjacent ONU base stations. The location information of the mobile device is also updated in the VPN server in the home network of the mobile device. Smooth handover of IP multicast communication is now possible even if the mobile device roams rapidly between the ONU base stations. The privacy of the mobile device and the communication between the mobile device and the VPN agent can be protected by another IPSec VPN tunnel. The SAs and the VPN tunnel must not be renegotiated when the mobile device roams to another ONU base station, but the radio channel to the new ONU must be recreated and the location information of the mobile device must be updated at the VPN agent and at two ONUs. A correspondent C can now use the IP address in the home network of a mobile device. The VPN server in the home network can redirect the IP packets to the visited MAN network through the VPN tunnel to the VPN agent, which forwards the packets—unencrypted, pre-shared key encrypted, or ESP encrypted in another VPN tunnel—to the mobile device through the registered ONU.

Secure Mobile Video Conferencing

Video conferencing is one of the proposed future multimedia applications for mobile handheld devices. In a video conferencing system one or more data streams must be delivered to a client device in real time. The data streams may consist of video, audio, or other types of multimedia data. Two or more participants are taking part in a conference and the participants may want to have security provisions to protect their communication. For a mobile video conferencing system, where one or more of the participants are mobile, there are many challenging security issues:

- Implementing security with minimal loss of processing power;

- Implementing integrity and authentication algorithms, where authentication will not fail because of single bit errors caused by disturbances or disconnections in a wireless network;
- Roaming support;
- Protecting video streams from interception and interference; and
- Providing security for multicast multimedia data.

Security solutions for mobile multimedia video conferencing systems are under ongoing research. An example of such a research project is presented in Featherstone and Zhang (2003). The project started with studies of limitations and requirements for implementing security in mobile video conferencing environments. Related work was investigated to find out how current solutions can be improved to meet the requirements.

The requirements for mobile video conferencing can, according to the research project presented in Featherstone and Zhang (2003), be divided into three categories:

- Performance requirements
- Functional requirements
- Security requirements

Performance Requirements

Performance requirements of a mobile video conferencing system include QoS (quality of service) provision, minimal overheads, and accommodating heterogeneous network technologies. Real-time data needs bounded response times. Thus, bandwidth always needs to be guaranteed, also when a user is roaming between two mobile networks. A video conferencing system with implemented security and QoS provision introduces extra overhead to the system and also consumes extra bandwidth. These parameters need to be minimized in order to maintain QoS provision. Current and also future communication systems integrate both wired and wireless access networks of different standards. Thus, a video conferencing system design must be independent of underlying network technologies.

Functional Requirements

A video conferencing system should include at least the following functions:

- Initiation of a conference;
- Finishing of a conference;
- Admitting people to a conference;
- Allow someone to leave a conference;
- Deal with unexpected departure of a member from a conference;
- Split a conference;
- Merge two or more conferences; and
- Temporarily suspend a member from a conference.

Security Requirements

Security requirements in a mobile video conferencing system include:

- Confidentiality of the multiple multimedia data streams generated by a conference system;
- Integrity of multimedia data;
- Authentication of each member of a conference;
- Anonymity service for preventing outsiders from knowing that a conference is taking or has taken place;
- Non-repudiation service for providing evidence that a particular conference has actually taken place;
- Efficient and effective key exchange process for providing key distribution in a multicast environment;
- End-to-end protection; and

- Flexible security services providing users the possibility to choose different levels of protection.

CURRENT MOBILE MULTIMEDIA RESEARCH

The purpose of this section is to provide a picture of current and future trends in mobile multimedia security research. The research area is wide, and there is a large amount of ongoing research projects. Therefore, this section will only concentrate on a few important areas, such as secure roaming and video conferencing. A few examples of related research projects are presented. A testbed for developing and testing secure mobile multimedia systems is also presented.

Proposals for Secure Roaming

The current trend in mobile networking is towards mobile devices having multiple network interfaces such as WLAN, UMTS, and GPRS. A mobile user may want to securely and continuously access an enterprise network as well as securely receive real-time data such as multimedia data streams regardless of whether the user is physically located in the enterprise or not. The user should also be able to seamlessly roam between different types of wireless networks and still maintain an one-going secure application session. In order for a user to access an enterprise network from an external network it is typically required to use VPN technique. Using VPN, the enterprise network is able to authenticate the user and to determine whether the user should be permitted to use enterprise network applications or not. A key issue in developing a mobile multimedia system supporting seamless and secure roaming across heterogeneous radio systems is how to fulfill the security requirements and at the same time maintain QoS.

Current VPN technologies, such as IPSec, do not have sufficient capabilities to support seamless mobility. For instance, an IPSec tunnel will break when a mobile device changes its IP address as a result of roaming between two networks. Researchers are currently developing a new version of IKE (Internet Key Exchange), IKEv2, also known as MOBIKE. This version adds a mobility extension to the IKE protocol. The current MOBIKE specification is available as an IETF Internet-draft (Kivinen & Tschofenig, 2006). However, the work is still in a quite early stage. Another problem with VPN and mobility is that the VPN establishment requires manual actions from users when a time-variant password is used to set up the VPN. Furthermore, a VPN may cause significant overhead.

Roaming Based on Mobile IP and SIP

A secure universal mobility (SUM) network has been proposed (Dutta et al., 2004a). The network is able to support seamless mobility and the mobile user does not need to maintain an always-on VPN session. The SIP protocol can be used to achieve dynamic VPN establishment and Mobile IP for mobility support. A dynamic VPN is achieved by using a double tunneled Mobile IP. This Mobile IP system needs two home agents, an internal home agent, and an external home agent. One tunnel is set up from the mobile device to the external home agent and another tunnel is set up from the external home agent to the internal home agent. When both tunnels are up, a corresponding node or a mobile user is able to initiate a communication by sending a SIP INVITE message. The receiver then checks if there is an existing VPN session. If not, IKE is used to negotiate security parameters and establish a new VPN session.

A testbed realization of secure universal mobility is also presented. Testbed experiments proved that smooth handoff is achieved during roaming between heterogeneous networks. However, an additional delay appeared while moving from a 802.11 network to a cellular network. VoIP and video streaming traffic was also tested.

Host Identity Protocol

When the current Internet architecture was designed, functionalities such as mobility and multi-homing were not taken into account. An IP address represents both an identity and a topological location of the host. This overloading has led to several security problems, such as the so-called address ownership problem (Nikander, 2001). This makes IP mobility and multi-homing hard from the security point of view.

A new protocol, HIP (host identity protocol) is currently under development and the current specifications are available as an IETF draft (Moskowitz & Nikander, 2006). In the HIP architecture, the locators and end-point identifiers are separated from each other at the network layer of the TCP/IP stack. A new cryptographic name space, the HI (host identity) is introduced, which is typically a cryptographic public key. This key serves as the end-point identifier of a network node. The IP address retains the role of a locator. Each host has at least one HI assigned to its networking kernel or stack. The HIs can be either public or anonymous. Public HIs may be stored in directories, such as DNS, allowing the host to be contacted by other hosts. In other words, HIP provides a binding between the HIs and the IP addresses using DNS.

The major advantages of the HIP architecture are that the problems of dynamic readdressing, anonymity, and authentication are solved. In HIP, the IP address has no longer the function of an end-point identifier. This significantly simplifies mobility since the node may easily change its HI and IP address bindings while moving. Furthermore, the security is improved since the name space is cryptographically based and it is thus possible to perform public key-based authentication using the HIs.

For further reading about the HIP protocol and HIP-related research projects, see Lundberg and Candolin (2003), Nikander, Ylitalo, and Wall (2003), and Helsinki University of Technology (2006).

Research Issues for Mobile Video Conferencing

Based on the analysis of current work in Featherstone and Zhang (2003), three major research issues for developing security and QoS provision in video conferencing systems for mobile computers were identified:

- **Mobility:** A satisfactory solution should provide a seamless QoS with mobility support and addressing of both wireless and wired networks.
- **Integrated Wired and Wireless Network:** Most mobile resource reservation schemes do not consider how QoS provision for wireless networks integrates with QoS provision on wired networks.
- **Security:** Systems providing both secure resource reservations as well as support for mobility are missing.

Testbed for Secure Mobile Multimedia

In Dutta et al. (2004b), a wireless Internet telephony and streaming multimedia testbed is set up based on a testbed framework discussed in Dutta, Chen, Madhani, McAuley, NakaJima, and Schulzrinne (2001). Two types of mobility approaches are discussed: network-layer mobility by Mobile IP and application-layer mobility using SIP mobility. A moving mobile device is a target of potential attacks in different parts of the access and core network. Thus, a multi-layer security scheme is needed for providing comprehensive solutions that can possibly prevent any

such attack and for supporting a dependable and secured multimedia application.

The testbed presented in Dutta et al. (2004b) is using an AAA protocol running on NAS (network access servers) and AAA servers to provide profile-based verification services. Additionally, a new protocol, PANA (protocol for carrying authentication for network access) (Yegin et al. 2005), is used for providing user-to-network access control. PANA provides access control mechanism to the mobile client and works as a front-end for the AAA server. It is implemented as a user level protocol to enable a flexible access control that works independently of any IPv4 or IPv6 layer 2 technology.

SIP-AAA Security Model

In the testbed two different security models were implemented. The first model is based on SIP and AAA and it was implemented in order to realize how SIP signaling can interact with an AAA infrastructure in a mobile environment. When the SIP server receives an SIP register message from the mobile device, it consults with the home AAA server for authentication and authorization. The profile verification database of the user is located in the home AAA server. In other words, in this model SIP registration is authenticated only after consulting with the AAA server. Normally, SIP registration is performed at the SIP server after the client has obtained a new address. The intention of the interaction between the SIP and the AAA server is to provide secure monitoring of user activities for accounting and auditing purposes.

IPSec-PANA-AAA Security Model

The second model, based on PANA and AAA, helps access control on the edge routers and interacts with IPSec to provide packet encryption. This model is a proposed multi-layered security scheme which provides packet-based encryption and application-layer authentication based on user NAI (network access identifier). In this model, PANA registration with the PANA server in the domain is performed when a user is roaming to a new network domain. In order to authenticate the user, the PANA server consults with the home AAA server either directly or indirectly through a local AAA server. As a result of successful PANA registration, an LSA (local security association) is established between the mobile device and the PANA server. Hereby, any further authentication required for intra-domain handoff is performed locally and quickly at the PANA server without the need to contact the home AAA server. The local authentication is also performed periodically with the intention to detect the event that the user disappears from the domain due to, for example, bad radio conditions.

The PANA server at the ERC (edge router & controller) maintains an association between the user identity, such as a NAI, and the lower-layer identity, such as an IP-address, for each user. The ERC has firewall functionality preventing packets sent from/to a mobile host not belonging to an authorized user to pass through the firewall. As a result of handoff, the association between the user identity and lower-layer identity dynamically changes. The ERC updates the access control list of the firewall if there is a change in the association and the resulting PANA registration or local authentication is successful. This prevents SIP register or SIP re-invite messages to pass through the firewall until the access control list is updated in the edge router.

In other words, PANA is used in the testbed to provide dynamic control of a router with firewall functionalities. Full network access is authorized only for hosts associated with authenticated PANA clients. An open firewall is closed immediately as a result of a failure of a periodical PANA re-authentication.

Packet Encryption and End-to-End Security

An IPSec-based encryption mechanism is used to secure data packets on the last hop in wireless networks. An IPSec tunnel is set up between the mobile client and the edge router. The PANA protocol is used for distributing IKE credentials to an authorized host. At first, the mobile client is authorized by PANA-based authentication. Then, the IKE credentials are carried in a PANA message and are transferred from the PANA authentication agent to the mobile client. These credentials are then used for setting up an IPSec tunnel between the mobile client and the access router. This provides a secure uni-cast communication channel in the access network including a wireless LAN segment. Since there is no need for a host to pre-configure the IKE credentials due to the dynamic distribution of the IKE credentials, mobile clients are enabled to smoothly roam among different domains.

In a mobile multimedia system it is essential to provide end-to-end security for both data and signaling. A combination of Mobile IP and IPSec solutions can be used to provide such an architecture, but such a combination is however not suitable for mobile systems since it will cause large overhead.

Real-time traffic is based on RTP/UDP. Thus, SRTP is used to provide encryption to different types of multimedia data. A secured RTP session requires the exchange of a separate RTP key between the mobile client and the correspondent host for a specific real-time media type. While registering with the SIP servers, SIP clients use PGP based authentication. The RTP key exchange is performed by an INVITE-exchange method using SIP signaling at the time of setting up calls. The RTP key is protected using a S/MIME mechanism. By these procedures, both data and signaling can be end-to-end secured without suffering from extra overhead.

CONCLUSIONS

Mobility and wireless communications introduce for multimedia applications similar problems with respect to security as for other applications. Advanced mobile terminals face new threats due to openness. Platforms are open for external software and content. Malicious software, like Trojan horses, viruses, and worms, has started to emerge. Fine-grained software authorisation has been proposed. Downloaded software may then access particular resources only through user authorisation. Vulnerabilities in OS implementation still remain a challenge because of difficulties in minimizing OS code running in privileged mode. Integrated hardware may be the solution.

Security aspects must be taken into account from the very beginning of the design of a mobile device for multimedia applications, not added on later. Security standards (SIP, secure RTP), technologies (IPSec) and testbed solutions for mobile multimedia are available, but significant improvements are still needed before acceptable security for mobile multimedia devices is achieved. Especially secure mobile video conferencing is still a research challenge.

REFERENCES

Baker, F., Lindell, B., & Talwar, M. (2000). *RSVP cryptographic authentication.* RFC 2747, IETF. Retrieved May 10, 2006, from http://www.ietf. org/rfc/rfc2747.txt

Barun, T., & Danzeisen, M. (2001). Secure mobile IP communication. In *Proceedings of 26th Annual IEEE Conference on Local Computer Networks (LCN'01),* Tampa, FL, November 14-16 (pp. 586-593). Washington DC, USA: IEEE Computer Society.

Baugher, M., McGrew, S., Naslund, M., Carrara, E., & Norrman, K. (2004). *RTP: The secure real-time transport protocol (SRTP).* RFC 3711, IETF. Retrieved May 9, 2006, from http://www. ietf.org/rfc/rfc3711.txt

Braden, R., Zhang, L., Berson, S., Herzog, S., & Jamin, S. (1997). *Resource ReSerVation Protocol (RSVP).* RFC 2205, IETF. Retrieved May 10, 2006, from http://www.ietf.org/rfc/rfc2205.txt

Calhoun, P., Johansson, T., Perkins, C., Hiller,T., & McCann, Ed. P. (2005). *Diameter mobile IPv4 application.* IETF, RFC 4004. Retrieved May 8, 2006, from http://www.ietf.org/rfc/rfc4004.txt

Cremonini, M., Damiani, E., De Capitani di Vimercati, S., Samarati, P., Corallo, A., & Elia, G. (2005). Security, privacy, and trust in mobile systems and applications. In M. Pagani (Ed.), *Mobile and wireless systems beyond 3G: Managing new business opportunities* (pp. 312-340). Hershey, PA: IRM Press.

Dietze, C. (2005). The smart card in mobile communication: Enabler of next-generation (NG) services. In M. Pagani (Ed.), *Mobile and wireless systems beyond 3G: Managing new business opportunities* (pp. 221-253). Hershey, PA: IRM Press.

Dittmann, J., Wohlmacher, P., & Ackermann, R. (2001). Conditional and user specific access to services and resources using annotation watermarks. In R. Steinmetz, J. Dittman, & M. Steinebach (Eds.), *Communications and multimedia security issues of the new century* (pp. 137-142). Norwell, MA: Kluwer Academic Publishers.

Domenech, A. L. (2003). *Port-based authentication for wireless LAN access control.* Report of Graduation Degree. Department of Electrical Engineering Eindhoven University of Technology (TU/e), Eindhoven, Netherlands. Retrieved September 24, 2007, from http://people.spacelabs. nl/~alex/Port_Based_Authentication_for_Wireless_LAN_Access_Control.pdf

Duncan, M. V., Akhtari, M. S., & Bradford, P. G. (2004). Visual security for wireless handheld devices. In *JOSHUA Journal of Science & Health at The University of Alabama, 2/2004*. The University of Alabama.

Dutta, A., Chen, J., Madhani, S., McAuley, A., NakaJima, N., & Schulzrinne, H. (2001). Implementing a testbed for mobile multimedia. In *Proceedings of the IEEE Conference on Global Communications (GLOBECOM)*, San Antonio, Texas, November 25-29 (pp. 1944-1949). IEEE.

Dutta, A., Das, S., Li, P., McAuley, A., Ohba, Y., Baba, S., & Schulzrinne, H. (2004a). *Secured mobile multimedia communication for wireless Internet*. ICNSC 2004, Taipei.

Dutta, A., Zhang, T., Madhani, S., Taniuchi, K., Fujimoto, K., Katsube, Y., Ohba, Y., & Schulzrinne, H. (2004b). Secure universal mobility for wireless Internet. In *Proceedings of the 2nd ACM International Workshop on Wireless Mobile Applications and Services on WLAN Hotspots (WMASH)*, Philadelphia, PA, October 1 (pp. 71-80). New York: ACM Press.

ETSI EN 302 304. (2004). *Digital video broadcasting (DVB): Transmission system for handheld terminals (DVB-H)*. Retrieved September 26, 2007, from http://pda.etsi.org/pda/queryform.asp

Featherstone, I., & Zhang, N. (2003). Towards a secure videoconferencing system for mobile users. In the *5th European Personal Mobile Communications Conference*, Glasgow, Scotland, April 22-25 (pp. 477-481). London, UK: IEEE.

Gehrmann, C., Persson, J., & Smeet, B. (2004). *Bluetooth security*. Norwood, MA: Artech House, Inc.

Glass, S., Hiller, T., Jacobs, S., & Perkins, C. (2000). *Mobile IP authentication, authorization, and accounting requirements*. IETF, RFC 2977.

Retrieved May 8, 2006, from http://www.ietf.org/rfc/rfc2977.txt

GSM Security Portal. (2005). Retrieved July 11, 2005, from http://www.gsm-security.net/

Havinga, P. J. M. (2000). *Mobile multimedia systems*. Ph.D. thesis, University of Twente, Netherlands, ISBN 90-365-1406-1.

Helsinki University of Technology. (2006). *InfraHIP project portal*. Retrieved May 10, 2006, from http://infrahip.hiit.fi

Hwu, J.-S., Chen, R.-J., & Lin, Y.-B. (2006). An efficient identity-based cryptosystem for end-to-end mobile security. *IEEE Transaction on Wireless Communications, 5*(9), 2586-2593.

Islam, R. (2005). *Enhanced security in mobile IP communication*. MSc Thesis, Department of Computer and Systems Sciences, Royal Institute of Technology, Stockholm, Sweden.

Jansen, W. A. (2003). Authenticating users on handheld devices. In the *15th Annual Canadian Information Technology Security Symposium (CITSS)*, Ottawa, Canada, May 12-15. Ottawa, Canada: Communications Security Establishment. Retrieved September 24, 2007, from http://csrc.nist.gov/mobilesecurity/publications.html#MD

Kaspersky Lab Products & Services. (2007). *Kaspersky anti-virus mobile*. Retrieved September 24, 2007, from http://www.kaspersky.com/antivirus_mobile

Kharif, O. (2005). May I see your voice, please? In *BusinessWeek online*, April 20. Retrieved July 11, from http://www.businessweek.com/technology/content/apr2005/tc20050420_1036_tc024.htm?campaign_id=rss_techn

Kivinen, T., & Tschofenig, H. (2006). *Design of the IKEv2 Mobility and Multihoming (MOBIKE) Protocol, draft-ietf-mobike-design-08.txt*. RFC

4621, IETF. Retrieved March 3, 2006, from http://www.ietf.org/internet-drafts/draft-ietf-mobike-design-08.txt

Lee, B.-G., Choi, D.-H., Kim, H.-G., Sohn, S.-W., & Park, K.-H. (2003). Mobile IP and WLAN with AAA authentication protocol using identity-based cryptography. *10th International Conference on Telecommunications ICT 2003, 1,* Colmar, France, February 23-March 1 (pp. 597-603). USA: IEEE.

Liu, Q., Reihaneh, S.-N., & Sheppard, N. P. (2003). Digital rights management for content distribution. In *Proceedings of the Australasian Information Security Workshop Conference on ACSW Frontiers 2003, 21,* Adelaide, Australia, February 1 (pp. 49-58). Darlinhurst, Australia: Australian Computer Society, Inc.

Lundberg, J., & Candolin, C. (2003). Mobility in the host identity protocol (HIP). In *Proceedings of the International Symposium on Telecommunications (IST2003),* Isfahan, Iran, August (pp. 754-757). Iran: Iran Telecom Research Center.

Meyer, U., & Wetzel, S. (2004). On the impact of GSM encryption and man-in-the-middle attacks on the security of interoperating GSM/UMTS networks. In the *15th IEEE International Symposium on Personal, Indoor and Mobile Radio Communications (PIMRC 2004), 4,* Barcelona, Spain, September 5-8 (pp. 2876-2883). USA: IEEE. Retrieved July 12, 2005, from http://www.cdc.informatik.tu-darmstadt.de/~umeyer/UliP-IMRC04.pdf

MITA. (2002). *Mobile Internet technical architecture: The complete package.* Nokia, Finland, ISBN 951-826-669-7.

Mobile Broadcast/Multicast Service (MBMS). (2004, August). White Paper, TeliaSonera. Retrieved July 11, 2005, from http://www.medialab.sonera.fi/workspace/MBMSWhitePaper.pdf

MobileCloak™ Web Portal. (2005). Retrieved July 11, 2005, from http://www.mobilecloak.com

Moskowitz, R., & Nikander, P. (2006). *Host Identity Protocol (HIP) Architecture.* RFC 4423, IETF.

Muller, T. (1999). *Bluetooth security architecture: Version 1.0.* Bluetooth White Paper, Document # 1.C.116/1.0. Retrieved September 26, 2007, from http://www.bluetooth.com/NR/rdonlyres/C222A81E-D9F9-48CA-91DE-9C81F5C8B94F/0/Security_Architecture.pdf

Nasir, A., & Mah-Rukh, M.-R. (2006). Internet mobility using SIP and MIP. *Proc. Third International Conference on Information Technology: New Generations ITNG'06,* Las Vegas, NV, April 10-12 (pp. 334-339). USA: IEEE Computer Society.

Nikander, P. (2001). *Denial-of-service, address ownership, and early authentication in the IPv6 world.* Retrieved May 10, 2006, from http://www.tml.tkk.fi/~pnr/publications/cam2001.pdf

Nikander, P., Ylitalo, J., & Wall, J. (2003). Integrating security, mobility, and multi-homing in a HIP way. *Proceedings of Network and Distributed Systems Security Symposium (NDSS'03),* San Diego, USA, February (pp. 87-99). Reston, USA: Internet Society.

Olzak, T. (2005). *Wireless handheld device security.* Retrieved July 11, 2003, from http://www.securitydocs.com/pdf/3188.PDF

Ong, C. S., Nahrstedt, K., & Yuan, W. (2003). Quality of protection for mobile multimedia applications. In *Proceedings of IEEE International Conference on Multimedia and Expo (ICME2003),* Baltimore, MD, July 7-9 (pp. 137-140). IEEE. Retrieved July 11, 2005, from http://cairo.cs.uiuc.edu/publications/papers/ICME03-chui.pdf

Perelson, S., & Botha, R. (2004). An investigation into access control for mobile devices. In

H. S. Venter, J. H. P. Eloff, L. Labuschagne, & M.M. Eloff (Eds.), *ISSA 2004 Enabling Tomorrow Conference, Peer-Reviewed Proceedings of the ISSA 2004 Enabling Tomorrow Conference, Information Security South Africa (ISSA),* Gallagher Estate, Johannesburg, South Africa, June 30-July 2. CDROM, ISBN 1-86854-522-9. ISSA. Retrieved July 11, 2005, from http://www.nmmu.ac.za/rbotha/Pubs/docs/LC_017.pdf

Perkins, C. (2002). *IP mobility support for IPv4, IETF.* RFC 3344. Retrieved May 8, 2006, from http://www.ietf.org/rfc/rfc3344.txt

Phillips, A., Nelson, B., Enfinger, F., & Steuart, C. (2005). *Guide to computer forensics and investigations* (2nd ed.). USA: Course Technology Press.

Pulkkis, G., Grahn, K., Karlsson, J., Martikainen, M., & Daniel, D. E. (2005). Recent developments in WLAN security. In M. Pagani (Ed.), *Mobile and wireless systems beyond 3G: Managing new business opportunities* (pp. 254-311). Hershey, PA: IRM Press

Robbins, J. (2005). *An explanation of computer forensics.* Retrieved July 11, 2005, from http://www.computerforensics.net/forensics.htm

Rosenberg, J., Schulzrinne, H., Camarillo, G, Johnston, A., Peterson, J., Sparks, R., Handley, M., & Schooler, E. (2002). *SIP: Session initation protocol.* RFC 3261, IETF. Retrieved May 9, 2006 from http://www.ietf.org/rfc/rfc3261.txt

Salsano, S., Veltri, L., & Papalilo, D. (2002). SIP security issues: The SIP authentication procedure and its processing load. *IEEE Network, 16*(6), 38-44.

Schulzrinne, H., Casner, S., Frederick, R., & Jacobson, V. (2003). *RTP: A transport protocol for real-time applications.* RFC 3550, IETF. Retrieved May 8, 2006, from http://www.ietf.org/rfc/rfc3550.txt

Setec Portal. (2005). Retrieved July 11, 2005, from http://www.setec.fi

Stamp, M. (2003). Digital rights management: The technology behind the hype. Journal of Electronic Commerce Research, 4(3), 202-212. Long Beach, CA: California State University Long Beach. Retrieved May 12, 2006, from http://home.earthlink.net/~mstamp1/DRMpaper.pdf

Sun, J., Howie, D., Koivisto, A., & Sauvola, J. (2001a). A hierarchical framework model of mobile security. *Proc. 12th IEEE International Symposium on Personal, Indoor and Mobile Radio Communication, 1,* San Diego, CA, September 30-October 3 (pp. 56-60). USA: IEEE.

Sun, J., Howie, D., Koivisto, A., & Sauvola, J. (2001b). Design, implementation, and evaluation of Bluetooth security. *Proc. IEEE International Conference on Wireless LANs and Home Networks*, Singapore, December 5-7 (pp. 121-130). USA: IEEE. Retrieved July 11, 2005, from http://www.mediateam.oulu.fi/publications/pdf/87.pdf

Sundaresan, H. (2003). *OMAP™ platform security features.* White Paper. Retrieved July 1, 2005, from http://focus.ti.com/pdfs/wtbu/omapplatformsecuritywp.pdf

Symantec. (2003). *Wireless handheld and smartphone security.* Retrieved September 26, 2007, from http://whitepapers.zdnet.co.uk/0,1000000651,260085794p,00.htm?r=1

Symantec Corporation. (2005). *Symantec AntiVirus™ for handhelds annual service edition, Symantec antiVirus for handhelds safeguards Palm and Pocket PC mobile users.* Retrieved July 9, 2005, from http://www.symantec.com/sav/handhelds/ and http://www.symantec.com/press/2003/n030825.html

Talwar, V., & Nahrstedt, K. (2000). Securing RSVP for multimedia applications. *Multimedia Security Workshop ACM Multimedia*, Los Angeles, CA, October 30-November 3 (pp. 153-156). USA: ACM Press.

Taylor, L. (2004-2005). *Handheld security, part I-V.* Retrieved September 25, 2007, from http://www.pdastreet.com/articles/2004/12/2004-12-6-Handheld-Security-Part.html

Trusted Computing Group. (2004). *Security in mobile phones.* Retrieved July 1, 2005, from http://www.trustedcomputinggroup.org/downloads/whitepapers/TCG-SP-mobile-sec_final_10-14-03_V2.pdf

WEPCrack. (2007). Retrieved September 26, 2007, from http://wepcrack.sourceforge.net

Wi-Fi Alliance Portal. (2005). Retrieved July 11, 2005, from http://www.wi-fi.org

Wikipedia. (2005). *Secure sockets layer (SSL) and transport layer security (TLS).* Retrieved July 10, 2005, from http://en.wikipedia.org/wiki/Secure_Sockets_Layer

Wikipedia, the free encyclopedia. Retrieved July 11, 2005, from http://www.wikipedia.org

Wikipedia. *Computer forensics.* Retrieved July 11, 2005, from http://en.wikipedia.org/wiki/Computer_forensics

Wu, T., Wu, S., Huang, H., & Gong, F. (1999). Securing QoS: Threats to RSVP messages and their countermeasures. *Proceedings of IWQoS'99 – Seventh International Workshop on Quality of Service*, UCL, London, May 31-June 4 (pp. 62-64). USA: IEEE.

Xydis, T. G., & Blake-Wilson, S. (2002). *Security comparision: Bluetooth™ Communications vs. 802.11.* Retrieved July 11, 2005, from http://ccss.isi.edu/papers/xydis_bluetooth.pdf

Yegin, A., Ohba, Y., Penno, R., Tsirtsis, G., & Wang, C. (2005). *Protocol for carrying authentication for network access (PANA) requirements.* RFC 4058, IETF.

Zeng, W., Zhuang, X., & Lan, J. (2004). Network friendly media security: Rationales, solutions, and open issues. *2004 International Conference on Image Processing ICIP '04, 1,* Singapore, October 24-27 (pp. 565-568). USA: IEEE.

Zhang, L., Lambertsen, G., & Yamada, T. (2003). Security scenarios within IP-based mobile multimedia metropolitan area network. In *Proceedings of Global Telecommunications Conference, GLOBECOM '03, 3,* San Francisco, CA, December 1-5 (pp. 1522-1526). Piscataway, NJ: IEEE Operations Center.

APPENDIX: AVAILABLE SECURITY SOLUTIONS

Market offers several products as a solution to the increasing security problems of mobile devices. These products are designed to solve individual or more comprehensive security problems like unauthorized access to data or device, viruses, and unencrypted data transfers. The following list of the product groups reveal versatility of the mobile device security solutions (Douglas, 2004; Taylor, 2004, part III):

- Platform security products
- Authentication products
- Encryption products
- Anti-virus products
- Bit wiping products
- Firewall products
- VPN products
- Wireless security products
- Forensic products
- Database security products
- Centralized security management products
- Backup/restore products

In this section, available software products for platform security, authentication, encryption, anti-virus, VPNs (virtual private networks), firewall, and computer forensics are presented. The specific features of some products from each category are briefly described. In addition, characteristics and benefits of three well-known multifunctional security products for mobile devices, Pointsec, Bluefire and PDASecure, will be handled. Product information was searched during spring 2005.

Platform Security Solutions

There are different embedded on-chip security solutions, but mostly the security solution relies on a combination of hardware and software components. In the following, the two new security solutions, Texas Instruments OMAPTM Platform (Sundaresan, 2003) and Intel Wireless Trusted Platform (Intel, 2004), are shortly described.

The TI solution relies on three layers of security: application layer security, operating system layer security, and on-chip hardware security. The platform includes four main security features:

- **Secure Environment:** This feature enables secure execution of critical code and data. Sensitive information is hidden from the outside world with the use of four security components. These are *secure mode, secure keys, secure ROM,* and *secure RAM.* Secure mode allows secure execution of "trusted" code via on-chip authentication. Secure mode can be seen as a 3rd privilege level. Secure keys, ROM, and RAM are accessible only in secure mode. The keys are OEM specific one-time programmable keys used for authentication and encryption. Secure ROM is a storage feature using secure keys, key management and cryptographic libraries. Secure RAM is used for running OEM specific authenticated code.
- **Secure Boot/Flash:** The secure boot/flash process prevents security attacks during device flashing and booting. The authentication process must guarantee the origin and the integrity of the software stored on the platform. The process also prevents execution of any spoofed software code.
- **Run-Time Security:** The secure environment has been integrated into the OS. Thus, OS applications for short security critical tasks can be performed. Such tasks include encryption/decryption, authentication, and secure data management.
- **Crypto Engine:** Hardware crypto engines are used to enhance performance and security. Available crypto engines are DES/3DES,

SHA1/MD5, and RNG. Two configuration modes are available: secure mode for short and/or high security level applications, and user mode for long and/or low-level security applications.

The building blocks of the Intel platform consists of performance primitives (hardware) and cryptographic primitives (optimized software). The solution enables services such as trusted boot and safe processing of secrets, protected key storage, and attestation measuring the security status of the platform during trusted boot.

The components of the platform include:

- **Trusted Boot ROM:** Validates the integrity of the platform and boots it into the right configuration.
- **Wireless Trusted Module:** This is the module where the secrets are processed. This suite of cryptographic engines include random number generation, symmetric and asymmetric cryptography, key creation, key exchange, digital signature operations, hashing, binding, and a monotonic counter.
- **Security Software:** A security software stack enables the OS and applications to access the platform resources through standard cryptographic APIs.
- **Protected Storage:** Protected storage in system flash allows secure non-volatile storage of secrets.
- **Physical Protection:** Physical protection is supported by integration of the security hardware in a single device and by packaging the discrete components into a single physical package.

Authentication Products

Unauthorized access has not always been recognized as a security risk especially among private users, even though mobile devices such as PDAs and smartphones are small, portable, and easily lost or stolen. These features lead to a high risk of vital data loss. From this point of view, it is easy to understand the necessity of authentication.

There are several authentication methods (Douglas, 2004, p.13):

- Electronic signature authentication
- Picture-based password authentication
- Fingerprint authentication
- Card-based authentication
- Storage-card-based certificate authentication
- Legacy host access

This section concentrates on products for electronic signature, picture-based password, and fingerprint authentication. A summary of authentication product examples for Pocket PCs and smartphones is presented in Tables A1-A3.

Electronic Signature Authentication Products

Electronic signature authentication has several benefits. It provides a high level of security and an electronic signature is, from the user's point of view, a simple password which cannot be forgotten. The main problem with this kind of technology is that the biographic signature is varying from time to time, which causes the possibility of access denial.

To mention a couple of these kinds of solutions, Romsey Associates Ltd. offers PDALok with dynamic signature recognition and Transaction Security Inc. has PDA Protect with crypto-sign pattern recognition.

When logging on or synchronizing of data, PDALok forces users to set PIN (personal identification number) or signature. PDALok's biometric technology, dynamic signature recognition, monitors different characteristics of writing as rhythm, speed, pressure, flow, and acceleration.

Table A1. Signature authentication products

Company	Product Name	Feature / Function
Certicom Corporation	Security Builder	Cross-platform cryptographic module, which allow developers to handle all encryption, decryption, digital signatures and message authentication codes.
Communication Intelligence Corporation	InkTools	Developers' kit with electronic ink capture and display, signature verification, encryption, ink compression and other algorithms.
Romsey Associates Ltd.	PDALok	Uses biometric technology, Dynamic Signature Recognition to prevent unauthorized access to Pocket PC.
Transaction Security Inc.	PDA Protect	Crypto-Sign™ is a biometric pattern recognition technology based upon the submission of a secret sign.
VASCO	Digipass	Digipass offers strong user authentication and electronic signatures.

The software is compatible with almost every mobile device running on Window CE and Palm OS (Romsey Associates Ltd., 2005).

Topaz Systems Inc. (2005) provides a more traditional signature authentication alternative. The set with separate writing pads includes software and inkless pen such as SignatureGem. Features of signature authentication products are shown in Table A1.

Picture-Based Password Authentication

Instead of the traditional alpha-numeric login passwords and PINs, Pointsec's PicturePIN and sfr Gesellschaft für Datenverarbeitung mbH's visKey Palm OS picture-based authentication are alternative solutions. Picture-based login style is easier to remember and more user-friendly for mobile devices with small screens and mini keyboards.

Figure A1. Pointsec's PicturePIN picture-based authentication

Pointsec allows users to select a password consisting of a combination of icons (Figure A1). The position of the Pointsec's icons changes place each time the device is activated (Pointsec Mobile Technologies AB., 2005).

Alternatively, visKey Palm OS users select several desired spots in an image (Figure A2). Before creating the password, visKey Palm OS splits up the selected image into cells. Each cell represents a single character (sfr Gesellschaft für

Figure A2. Gesellschaft für Datenverarbeitung mbH's visKey Palm OS picture-based authentication

Figure A2a

Figure A2b

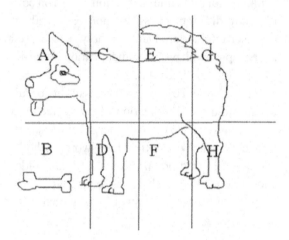
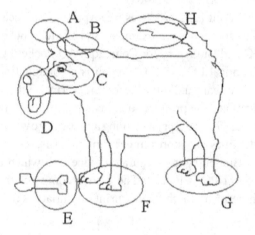

Table A2. Picture-based password authentication

Company	Product Name	Feature / Function
Pointsec Mobile Technologies	Pointsec for Smartphones, Pointsec for Pocket PC, Pointsec for Palm OS, Pointsec for Symbian OS,	The picture-based authentication software enables users to select a password consisting of a combination of icons.
sfr Gesellschaft für Datenverarbeitu ng mbH	visKey Palm OS	The picture-based authentication software enables users to select several desired spot in an image.
Softava	PicturePassword	User authentication with one click on the desired point of an image.

Datenverarbeitung mbH, 2005). The user identifies to the device by choosing the same order of clicks. Features of picture-based password authentication products are shown in Table A2.

Fingerprint Authentication

Fingerprint-based identification is the oldest method of biometric techniques. The uniqueness of a human fingerprint prevents effectively forgery attempts. The security level of fingerprint authentication is depending on such factors as the quality of scanning and the visual image recognition (ROSISTEM, 2005).

Veridt LLC produces fingerprint authentication solutions, for example BioHub for Pocket PC and BioSentry for Compaq iPAQ Pocket PC (Veridt, LLC., 2005). BioHub is a combination of hardware and software for fingerprints authentication. The product contains a plug-in capture device for fingerprint imaging and software which identifies and stores fingerprint patterns.

BioSentry is a separate device into which a Pocket PC is placed. The device compares the fingerprint to the fingerprint minutiaes saved in the Pocket PC's memory. The minutiaes of a fingerprint are defined from a scanned image and the coordinates of these minutiaes are calculated. Finally these coordinates are encrypted and saved. The device uses an encryption algorithm which meets the requirements of AES-128, AES-192, and AES-256 standards. Features of fingerprint authentication products are shown in Table A3.

Encryption Products

One of the simplest methods to protect sensitive data is encryption. Pointsec for Pocket PC provides a solution for real-time encryption of data on both PDAs and different types of memory cards without any user interaction. Data can only be accessed or decrypted with proper authentication (Pointsec Mobile Technologies AB, 2005).

Trust Digital 2005 offers the possibility of full background synchronization of encrypted data (Trust Digital, 2005). Airscanner Mobile Encrypter secures data with optional password methods. Instead of the traditional individual file and folder encryption, where each file can be encrypted/decrypted with its own password, one global pass-

Table A3. Fingerprint authentication products

Company	Product Name	Feature / Function
Hewlett-Packard Development Company, L.P.	HP iPAQ 5450, 5455 and 5555 Pocket PC.	The HP iPAQ series Pocket PC with inbuilt biometric fingerprint reader.
Veridt, LLC	BioHub	BioHub is a combination of hardware and software for fingerprints authentication solutions.
Veridt, LLC	BioSentry	A fingerprint verification "jacket" for Pocket PC.

Table A4. Storage encryption products

Company	Product Name	Feature / Function
Airscanner Corp.	Airscanner Mobile Encrypter	File and folder encryption for Pocket PC
Asynchrony	PDA Defense	Encrypts all databases, files and memory cards on Pocket PC.
Certicom Corporation	movianCrypt	Encrypts and locks all information without impeding performance or usability on Pocket PC.
Glück & Kanja Group	CryptoEx Pocket	Secure email communication and file storage encryption on Pocket PC.
PC Guardian Technologies, Inc.	PDASecure	Provides data encryption and access control. The software encrypts data based on administrator and user preferences on Pocket PC, Palm OS, smartphone, Symbian and BlackBerry.
Pointsec Mobile Technologies	Pointsec for Pocket PC, Pointsec for Smartphone, Pointsec for Symbian OS	Provides real-time encryption of all data on both their PDAs and all removable media on Pocket PC, smart phone, Symbian OS.
SoftWinter	Sentry 2020 for Pocket PC	Provides encrypted virtual volumes. Data stored on a Sentry volume is transparently encrypted/decrypted.
Cranite Systems, Inc.	WirelessWall	provides FIPS 140-2 certified AES data encryption for Pocket PCs.

word can be applied. After sign-on, transparent and automatic volume encryption takes place within a user-defined time (Airscanner® Corp., 2005). Features of storage encryption products are shown in Table A4.

Anti-Virus Products

Major anti-virus software companies started to develop anti-virus products since the first Symbian OS mobile phone virus, the Cabir worm, was found in 2004 (Kaspersky Lab, 2005).

F-Secure manufactures anti-virus security software, especially for Nokia's mobile phones, for example, Series 60/90 and Nokia 9300/9500 Communicator Series. Symantec offers anti-virus products for Palm OS- and Microsoft Pocket PC-compatible devices. To prevent infection, F-Secure Mobile Anti-Virus and Symantec AntiVirus for Handhelds scan all files automatically and trans-

Table A5. Anti-virus products

Company	Product Name	Feature / Function
ALWIL Software	avast! 4 PDA Edition (WinCE/Palm)	Provides virus protection and updates of the virus database.
F-Secure	F-Secure Mobile Anti-Virus for Series 60/90, F-Secure Mobile Anti-Virus for Nokia 9300 / 9500 Communicator Series	An antivirus service with invisible virus signature updates. It offers fully automatic protection against malware such as viruses.
Kaspersky Labs	Kaspersky Security for PDAs	Virus protection for Pocket PC scans both data storage locations and memory extension cards.
McAfee Corporate	McAfee® VirusScan® PDA Enterprise 2.0	Virus protection and Automatic updating for Pocket PC.
Symantec	Symantec AntiVirus for Handhelds	Symantec AntiVirus for Handhelds provides real-time defense against viruses on a Pocket PC and Palm OS.
SOFTWIN	BitDefender for Windows CE, BitDefender for Palm OS	Provides on-demand virus scanning of all files stored on the mobile device's memory and automatic updates without user intervention. (freeware).

parently during modification and transference of data, without user intervention. When an infected file is detected, the file is immediately quarantined by the system to protect all other data.

F-Secure Mobile Anti-Virus for Nokia 9300/9500 Communicator Series provides automatic updates for the virus signature database over a secure HTTPS data connection or incrementally with SMS messages.

With LiveUpdate Wireless feature from Symantec AntiVirus for Handhelds, users are en-

abled to download virus definitions and Symantec product updates directly to their mobile device with a wireless Internet connection (F-Secure Corporation, 2005; Symantec Corporation, 2005).

Instead of commercial products, BitDefender offers free anti-virus software, BitDefender Antivirus—Free Edition for Windows CE and Palm OS. They are freeware, which means that no license is required for using the products. Features of anti-virus products are shown in Table A5 (BitDefender, 2007).

VPN (Virtual Private Network) and Firewall Products

Wireless networks cause a remarkable security risk because the transmitted data over the air can be easily exploited by outsiders. Secure VPNs use cryptographic tunneling protocols to ensure sender authentication, confidentiality, and integrity of data.

Columbitech's Wireless VPN is a client/server software architecture in which a secure encrypted virtual tunnel between the VPN client (mobile device) and the VPN Server is created to secure the data traffic between the devices. Authentication, validation, and the transmitted data is encrypted with WTLS (Wireless Transport Layer Security). WTLS is applied because it has better performance than IPSec (IP Security) over wireless networks. Wireless VPN uses the AES algorithm, with up to 256 bit encryption keys.

The software uses PKI (public key infrastructure), smart cards, biometric identification, and one-time passwords as authentication mechanisms. In order to prevent unauthorized access Columbitech's Wireless VPN has firewall functionality on both client and server. The client integrity is ensured with security status monitoring before the VPN tunnel is created. See Figure A3 (Columbitech AB, 2005).

Compared to the more common VPN, that uses IPSec technology, the newer VPN with SSL (secure sockets layer) cryptographic protocol is easier for administrators and users to set up and manage. The benefit of using SSL VPN instead of IPSec VPN is that SSL does not require any client software to be installed on each remote device because SSL is widely supported on most Web browsers (Ferraro, 2003).

Intoto's iGateway SSL-VPN allows users to securely access to key corporate resources at remote locations and create a secure encrypted virtual tunnel through any standard Web browser. iGateway SSL-VPN offers alternative authentication methods: RADIUS (Remote Authentication Dial-In User Service), LDAP (Lightweight Directory Access Protocol), Active Directory, Windows NTLN (NT LAN Manager), and digital certificates (Intoto Inc., 2005).

Figure A3. Columbitech's Wireless VPN solution

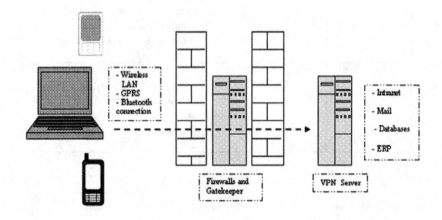

Table A6. VPN and firewall products

Company	Product Name	Feature / Function
Certicom Corporation	MovianVPN	The client for Palm and Pocket PCs operates with VPN gateways from Check Point, Cisco, and Nortel.
Check Point Software Technologies Ltd.	VPN-1 Secure Client	Allow users to securely access resources protected by VPN-1 gateways and provides a personal firewall on Pocket PCs.
Columbitech	Wireless VPN	A client/server software architecture, which create a secure WTLS encrypted virtual tunnel between the VPN client and the VPN Server, to secure the data traffic between the devices.
Ecutel Inc.	Viatores Mobile IP VPN	A client-server VPN solution. The software is based on Mobile IP and IPSec.
Funk Software Inc.	AdmitOne VPN Client for Pocket PC	offers advanced IPsec and IKE technologies to secure the transmission of data between Pocket PCs and the VPN server.
Intoto	iGateway SSL-VPN	enable users to create a secure encrypted virtual tunnel through at any standard web browser. The software offers alternative authentication methods: RADIUS, LDAP, Active Directory, Windows NTLN and digital certificates.
Symbol Technologies Inc.	AirBEAM Safe	A wireless VPN solution for full "end-to-end" security for Pocket PCs. It delivers strong, end-to-end encryption between the corporate firewall and the application server, using industry-standard authentication methodology.
the Familiar Project	Familiar v0.8.2	The firewall program "iptables" is included on Familiar's Linux distribution. Familiar v0.82 supports the iPAQ h series, Siemens Simpad and Sharp Zauri.

There are also Linux-based firewalls for PDAs on the market. Firewall programs, such as iptables or netfilter, are included on Familiar's Linux distribution list. Features of VPN and firewall products are shown in Table A6 (The Familiar Project, 2005; Villamil, 2005).

Forensic Analysis Products

While a large variety of forensic analysis software is available for personal computers, the range of solutions is much more limited for mobile devices. The problem is not only fewer software products for PDAs. These products operate only in most common families of PDAs.

The forensic analysis products have three main functionalities: acquisition, examination, and reporting. Only few products have all these mentioned functionalities. In such cases, several software products have to be purchased in order to accomplish the forensic examination process. The forensic analysis products need full access to the devices, in order to start acquisition of data. If the examined device is protected with some authentication method, cracking software is needed.

Paraben's PDA Seizure and Guidance Software Inc's EnCase Forensic Version 5 are forensic analysis tools that allow forensic examiners to acquire, examine, and analyze data. The PDA Seizure tool supports Palm OS, Windows-based OS, and BlackBerry devices, while Encase Forensic currently only supports Palm OS. Both of these mentioned tools have possibility for physical and logical acquisition of data. Physical acquisition means complete bit-by-bit copying of physical storage, for example a disk drive. Logical acquisition also means exact copying of logical storage objects, such as files and folders (Ayers & Jansen, 2004, p.14).

The EnCase Evidence file, created by EnCase, is suitable for the computer security industry and even for the law enforcement. EnCase allows users to determine how or what kind of data will be presented. This depends on the examination purposes. More Forensic analysis products are shown in Table A7 (Guidance Software, Inc., 2005; Paraben Corporation, 2007).

Table A7. Forensic analysis products

Company	Product name	Feature
Paraben Corporation	PDA SEIZURE version 3.0.2.43	Forensic acquisition, examination and reporting of data for Palm OS and Pocket PC.
Guidance Software Inc.	ENCASE Version 5	Forensic acquisition, examination and reporting of data for Palm OS.
Symantec Corporation	PDD	Memory imaging and forensic acquisition for Palm OS.
PalmSource Inc.	Palm OS Emulator (POSE)	Forensic examination and reporting for Palm OS.
Compelson Laboratories	MOBILedit!	Forensic gathering and reporting of data for mobile phones.

Multi-Functional Products

Multi-functional products are developed to solve comprehensively all security needs of mobile devices. From an administrator's point of view, these kind of products are appealing while there can be saved a lot of resources in terms of effective central administration. In this part of the chapter, three well-known commercial multi-functional solutions from Pointsec Mobile Technologies, PC Guardian Technologies, and Bluefire Security technologies will be presented.

Pointsec has solutions for the following devices and purposes: smartphone, Palm OS, Symbian

Table A8. Multi-functional products

Features	Company: Pointsec Product: Pointsec for Pocket PC, Smartphone, Symbian OS and PalmOS	Company: PC Guardian Technologies Product: PDASecure	Company: Bluefire Security Technologies Product: Bluefire Mobile Security Suite 3,5
Security policy enforcement from one central location	X	X	X
Remote Help and 3rd party distribution	X	X	X
FIPS 140-2 certified, AES algorithm with 128 bit encryption key	X	X	X
Automatic and immediate encryption	X	X	X
Memory card encryption	X	X	X
Picture-based passwords authentication	X		
Alphanumeric password authentication	X	X	X
Authenticated activeSync	X	X	
Antivirus software / virus protection		X	X
Firewall			X
VPN			X

OS, Pocket PC, and media encryption. PC Guardian Technologies software PDASecure is applicable to the smartphones, Pocket PCs, Palm OS, BlackBerry, and Symbian OS. Bluefire Security Technologies software Bluefire Security Suite 3.5 operates on Palm OS and Windows-based mobile platforms.

One of the most important features of all these multi-functional solutions is the third party management possibility. Administrators can enforce, update, and deploy the company security policy, which is applied also on desktops and laptops. Wished security level is ensured when a user does not adjust the policy according to their preferences. This means that the security system is user transparent and does not contain or demand any user intervention. Pointsec's software is also compatible with all major third-party management tools, which means that administrators can easily integrate their Pointsec's software into their existing system.

Available third-party management system is used to give remote help such as password reset for users. Bluefire highlights their real-time logging function, which enables the administrator to log and monitor all suspicious activities such as password attempts and changes in firewall security level.

The key function of PDASecure, Bluefire, and Pointsec software is encryption. This feature ensures high security level because all the data can be automatically and immediately encrypted before they are stored or transferred and decrypted automatically by an authenticated user. All of the three example products use the AES (Advanced Encryption Standard) algorithm, the U.S. government approved cryptographic standard with strong 128-bit encryption keys. PDASecure even allows users or administrators to choose from six alternative encryption algorithms.

Authenticated synchronization has been mentioned as an essential part of PDASecure's and Pointsec's mobile device software. Pointsec's ActiveSync system requires the users to authenticate themselves to the mobile device before synchronization with a PC can be started.

None of the example software emphasizes the anti-virus or VPN part of their solution. In this case only Bluefire offers anti-virus and VPN software as an additional feature (Bluefire Security Technologies, 2005; PC GuardianTechnologies, 2005; Check Point Software Technologies LTD., 2007). In Table A8, features of Pointsec, PDASecure, and Bluefire Security Suite 3.5 are shown.

Case Study: Encryption Software Performance Measurements

This case study presents performance measurements on the security software Pointsec for Symbian OS. The purpose was to measure the influence of Pointsec on data communication performance of Symbian OS. Pointsec is presented in more detail in the section, *Add-on Security Software*. According to Pointsec Mobile Technologies, the Pointsec security software should not reduce speed or other performance measures even when the strong 128-bit AES encryption is used to protect the information in the device and in memory cards.

Measurements

However, security solutions may reduce data communication performance measures of mobile operating systems, such as download speed and connection times. These performance measures were measured for a Pointsec security software installation in a Nokia Communicator 9500 for:

- downloading a 4.92 MB file;
- connection to an e-mail server (penti. arcada.fi) with imaps based e-mail client software;

Table A9. Download speed measurements (download times for a 4.92 MB file)

With "Pointsec for Symbian OS"		Without "Pointsec for Symbian OS"	
WLAN (11 Mbit/s)	GPRS (56 Kbit/s)	WLAN (11 Mbit/s)	GPRS (56 Kbit/s)
1 min	16 min 33 s	59.62 s	21 min 1 s
57 s	20 min 54 s	58.28 s	19 min 52 s
59 s	18 min 16 s	59.14 s	18 min 2 s
1 min 2 s	19 min 25 s	1.0 min	18 min 42 s
1min 3 s	20 min 40 s	59.61 s	18 min 14 s
59.2 s	19 min 48 s	1 min 1 s	19 min 3 s
Average 60.03 s	Average 19 min 16 s	Average 59.61 s	Average 19 min 9 s
Standard Deviation 2.174093 s	Standard Deviation 97.912206 s	Standard Deviation 3.253739 s	Standard Deviation 67.337954 s
		Average +0.42 s with Pointsec	Average +7 s with Pointsec

Table A10. Connection time measurements (to mailbox on e-mail server)

With "Pointsec for Symbian OS"		Without "Pointsec for Symbian OS"	
WLAN (11 Mbit/s)	GPRS (56 Kbit/s)	WLAN (11 Mbit/s)	GPRS (56 Kbit/s)
31.62 s	36.96 s	24.29 s	35.91 s
32.23 s	40.70 s	25.16 s	32.07 s
32.17 s	37.11 s	26.21 s	33.88 s
31.86 s	38.42 s	25.34 s	31.30 s
31.19 s	39.71 s	25.45 s	43.76 s
32.42 s	42.51 s	25.46 s	30.14 s
Average 31.915 s	Average 39.235 s	Average 25.31833 s	Average 34.51 s
Standard deviation 0.454962 s	Standard deviation 2.165777 s	Standard deviation 0.618948 s	Standard deviation 4.965360 s
		Average +6.60 s with Pointsec	Average +4.72 s with Pointsec

Table A11. Connection time measurements (to the Web site www.nokia.fi)

With "Pointsec for Symbian OS"		Without "Pointsec for Symbian OS"	
WLAN (11 Mbit/s)	GPRS (56 Kbit/s)	WLAN (11 Mbit/s)	GPRS (56 Kbit/s)
22.68 s	58.68 s	21.55 s	27.17 s
26.70 s	57.65 s	22.27 s	28.98 s
27.62 s	59.06 s	23.24 s	29.20 s
31.47 s	56.06 s	24.58 s	29.76 s
27.87 s	59.03 s	22.94 s	30.43 s
25.14 s	58.23 s	23.31 s	30.11 s
Average 26.91333 s	Average 58.11833 s	Average 22.98167 s	Average 29.275 s
Standard deviation 2.942419 s	Standard deviation 1.140341 s	Standard deviation 1.028308 s	Standard deviation 1.165345 s
		Average + 4.93 s with Pointsec	Average +28.84 s with Pointsec

Table A12. Connection time measurements (to a SSH server)

With "Pointsec for Symbian OS"		Without "Pointsec for Symbian OS"	
WLAN (11 Mbit/s)	GPRS (56 Kbit/s)	WLAN (11 Mbit/s)	GPRS (56 Kbit/s)
<1 s	4.85 s	<1 s	4.9 s
<1 s	4.65 s	<1 s	4.71 s
<1 s	4.62 s	<1 s	4.13 s
<1 s	4.79 s	<1 s	4.61 s
<1 s	4.86 s	<1 s	4.42 s
<1 s	4.80 s	<1 s	4.52 s
Average <1 s	Average 4.761667 s	Average <1 s	Average 4.548333 s
Standard deviation < 1 s	Standard deviation 0.102258s	Standard deviation <1 s	Standard deviation 0.263015 s
			Average +0.21 s with Pointsec

- connection to a www site (www.nokia. com);
- connection to a ssh server (penti.arcada.fi) with a putty ssh client.

All four performance measures were measured six times with and without installed Pointsec security software for two different access network types, WLAN and GPRS. The network bandwidths were 11 Mbit/s for the WLAN and 56 kbit/s for GPRS. Measurement results are presented in Tables A9-A12.

Usefulness of Measurement Results

Conditions cannot be assumed to be equal for different measurements, since the download speed and connection times were measured for data communication through the public Internet. The utilization of Internet during a measurement session is not deterministic. Measurement results have been considered to be useful if standard deviation is less than 10% of the calculated average for measurements with the same mobile device configuration. Standard deviation exceeded 10% of calculated average only in one measurement case, GPRS connection to an e-mail server without Pointsec security software installed, being about 15% of calculated average (see Table A10).

Measured Degradation of Data Communication Performance Caused by Pointsec

The influence of Pointsec was considered to be noticeable if the intervals defined by measured average and standard deviation do not overlap with and without Pointsec for otherwise the same mobile device configuration. Noticeable performance degradation was measured only for connection time to a Web site, about twice as long as for a GPRS connection and about 17%

longer for a WLAN connection (see Table A11). However, the influence of:

- the traffic load on the Internet, and
- the load on the selected Web server

during carried out performance measurements is unfortunately unknown.

The measurements can thus be considered to support the view of the provider of Pointsec security software, that the performance degradation from this security software is insignificant on a Symbian device.

REFERENCES

Airscanner® Corp. (2005). *Airscanner® Mobile Encrypter V2.2b (BETA) user's manual.* Retrieved July 8, 2005, from http://airscanner.com/downloads/encrypter/EncrypterManual.htm

Ayers, R., & Jansen, W. (2004). *PDA forensic tools: An overview and analysis.* Retrieved July 6, 2005, from http://csrc.nist.gov/publications/nistir/nistir-7100-PDAForensics.pdf

BitDefender. (2007). *BitDefender free edition for Windows CE—AntiVirus Freeware, BitDefender Free Edition for PALM OS—AntiVirus Freeware.* Retrieved September 27, 2007, from http://www.johannrain-softwareentwicklung.de/bitdefender_antivirus_free_edition_pour_windows_ce.htm or http://www.johannrain-softwareentwicklung.de/bitdefender_antivirus_free_edition_pour_palm_os.htm

Bluefire Security Technologies. (2005). *Wireless device security products.* Retrieved July 9, 2005, from http://www.bluefiresecurity.com/products.html

Check Point Software Technologies LTD. (2007). *Welcome Pointsec and Reflex Magnetics custom-*

ers! Retrieved September 27, 2007, from http://www.checkpoint.com/pointsec/

Columbitech AB. (2005). *Columbitech wireless VPN, technical description.* Retrieved September 26, 2007, from http://www.columbitech.com/img/2007/5/7/5300.pdf

Douglas, D. (2004). *Windows mobile-based devices and security: Protecting sensitive business information.* Retrieved July 1, 2005, from http://download.microsoft.com/download/4/7/c/47c9d8ec-94d4-472b-887d-4a9ccf194160/6.%20WM_Security_Final_print.pdf

Ferraro, C. I. (2003). *Choosing between IPsec and SSL VPNs.* Retrieved July 10, 2005, from http://searchsecurity.techtarget.com/qna/0,289202,sid14_gci940324,00.html

Guidance Software, Inc. (2005). *EnCase Forensic Version 5.* Retrieved July 8, 2005, from http://www.guidancesoftware.com/products/ef_index.asp

Intel. (2004). *Intel wireless trusted platform: Security for mobile devices.* White Paper. Retrieved September 26, 2007, from http://whitepapers.zdnet.co.uk/0,1000000651,260091578p,00.htm

Intoto Inc. (2005). *iGateway SSL-VPN.* Retrieved July 9, 2005, from http://www.intoto.com/product_briefs/iGateway%20SSL%20VPN.pdf

Kaspersky Lab. (2005). *Worm.SymbOS.Cabir.a.* Retrieved July 12, 2005, from http://www.viruslist.com/en/viruslist.html?id=1689517

Paraben Corporation. (2007). *Device seizure v1.2: Cell phone & PDA forensic software.* Retrieved September 27, 2007, from http://www.paraben-forensics.com/catalog/product_info.php?cPath=25&products_id=405

PC Guardian Technologies. (2005). *PDASecure—Powerful security. Simple to use. Superior service & support.* Retrieved July 9, 2005, from http://www.pcguardiantechnologies.com/PDASecure/PDASecure_Brochure.pdf

Romsey Associates Ltd. (2005). *Details—PDALok V1.0..., PDALok—The technology behind the security software.* Retrieved July 3, 2005, from http://www.pdalok.com/pda_security_products/PDALok_details.htm

ROSISTEM. (2005). *Biometric education » Fingerprint.* Retrieved July 9, 2005, from http://www.barcode.ro/tutorials/biometrics/fingerprint.html

sfr Gesellschaft für Datenverarbeitung mbH. (2005). *The patented technology in all of our visual key products.* Retrieved July 3, 2005, from http://www.viskey.com/viskeypalm/index.html and http://www.viskey.com/tech.html

Sundaresan, H. (2003). *OMAP™ platform security features.* White Paper. Retrieved July 1, 2005, from http://focus.ti.com/pdfs/wtbu/omapplatformsecuritywp.pdf

Symantec Corporation. (2005). *Symantec Anti-Virus™ for handhelds annual service edition, Symantec antiVirus for handhelds safeguards Palm and Pocket PC mobile users.* Retrieved July 9, 2005, from http://www.symantec.com/sav/handhelds/ and http://www.symantec.com/press/2003/n030825.html

Taylor, L. (2004). *Handheld security, part III: Evaluating security products.* Retrieved July 1, 2005, from http://www.firewallguide.com/pda.htm

The Familiar Project. (2005). *The familiar project.* Retrieved July 10, 2005, from http://familiar.handhelds.org/

Topaz Systems Inc. (2005). *Electronic signature pad with interactive LCD display.* Retrieved July 7, 2005, from http://www.topazsystems.com/products/specs/TL462.pdf

Trust Digital. (2005). *Trust digital 2005™*. Retrieved July 5, 2005, from http://www.trustdigital. com/downloads/productsheet_trust2005.pdf

Veridt, LLC. (2005). *Veridt verification and identification, product information.* Retrieved

September 27, 2007, from http://www.veridt. com/Products.html

Villamil, F. (2005). *Firewall wizards: Looking for PDA firewall.* Retrieved July 10, 2005, from http://seclists.org/lists/firewall-wizards/2004/ Jan/0031.html

Chapter XIII
Security in Ad-Hoc Networks

Muhammad Mahmudul Islam
Monash University, Clayton, Australia

Ronald Pose
Monash University, Clayton, Australia

Carlo Kopp
Monash University, Clayton, Australia

ABSTRACT

Due to the nature of wireless media, dynamic network topology, resource constraints, and lack of any base station or access point, security in ad-hoc networks is more challenging than with cabled networks. In this chapter, we discuss various attacks on the network layer of ad-hoc networks. We also review security protocols that protect network layer operations from various attacks.

INTRODUCTION

An ad-hoc network consists of a set of nodes that communicate using a wireless medium over single or multiple hops and do not need any pre-existing infrastructure such as access points or base stations. Ad-hoc networks can comprise of mobile, static, or both types of nodes. Ad-hoc networks containing mobile nodes are known as MANETs (mobile ad-hoc networks). An example of ad-hoc networks with static nodes is SAHN

(suburban ad-Hoc network) (Kopp & Pose, 1998). Since ad-hoc networks can be rapidly deployed, they are attractive for digital communication in battlefields, rescue operations after a disaster, and so forth. Ad-hoc networks are also useful in civilian forums for running demanding multimedia applications such as video conferencing.

Due to the lack of a clear physical boundary, a node in an ad-hoc network is very likely to hear the transmissions of a neighbouring node operating in the same frequency channel. If the node cannot distinguish the packets transmitted by an authorised neighbour from the ones transmitted by a malicious node, then the malicious node can: (1) cause the node to accept misleading information, and (2) propagate unnecessary traffic and misleading information to other parts of the network. As a result normal network operation could be disrupted.

The wireless medium makes eavesdropping easier than with a cabled network. If the packets are not encrypted properly, eavesdroppers can make unauthorised use of the received information and cause trouble. For example, an eavesdropper can forward unencrypted routing information to an accomplice to disrupt the normal operation of the network.

For the aforementioned reasons security is a primary concern in ad-hoc networks in order to provide secure communication among the nodes in a potentially hostile environment (Yang, Luo, Ye, Lu, & Zhang, 2004). Resource constraints (e.g., battery or computational power), dynamic network topology, and lack of infrastructure (e.g., fixed trusted nodes, base stations, or access points) make the security issue more challenging.

Existing ad-hoc routing protocols, such as DSR (dynamic source routing) (Johnson & Maltz, 1996) or AODV (ad-hoc on-demand distance vector) (Perkins & Royer, 1999) assume a trusted and cooperative environment. These routing protocols have to be protected from malicious nodes that can disrupt the network operation by intentionally disobeying of the protocol specifications (Yang et al., 2004).

There are two main approaches to secure a network: (1) pro-active and (2) reactive. A pro-active approach tries to prevent any attacks happening in the first place. On the other hand, a security protocol using a reactive approach detects any anomaly in network operation and

Figure 1. Classification of network layer security protocols

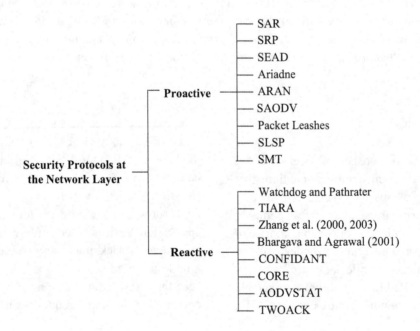

reacts accordingly. The latter group of protocols is also referred to as the intrusion detection system (IDS). See Figure 1 for the classification and list of network layer security protocols that we review in this chapter.

The rest of the chapter is organised as follows. In the next section, we discuss possible malicious attacks in the network layer of ad-hoc networks. Then we give an overview of the various authentication primitives used in the security protocols and review various key management protocols required for distributing and updating keys in ad-hoc networks. Next we present various pro-active approaches based secure routing protocols. Following this section, we review other security protocols that use the reactive approach to secure ad-hoc networks. We also outline open research challenges in the security domain of ad-hoc networks.

POSSIBLE ATTACKS

This section gives an overview of the attacks pertaining to the network layer.

Drop Packets

A malicious node may disrupt the normal operation of a network by dropping packets (Gupte & Singhal, 2003). This type of attack can be classified into two groups:

- **Black-Hole Attack:** In this type of attack, an attacker drops all types of packets.
- **Grey-Hole Attacks:** Here an attacker selectively drops packets (e.g., only data packets).

Delay Transmissions

A malicious node can give preference in transmitting its own or friends' packets by delaying packets in its output queue. If the packets belong to any time sensitive application, such as real-time video, they may become useless if not received at the receiver in due time.

Protocol Field Modification

A router is supposed to modify various protocol fields in the messages passing through itself according to the associated protocol specification. Existing resource allocation, transmission control, or routing protocols are based on these assumptions. Here are some of the attacks that can be launched by a malicious node exploiting these assumptions to cause disruption in the network.

- **Divert Traffic through Malicious Node:** The path length to a destination in protocols like AODV (Perkins & Royer, 1999) is measured by the hop count field of the route reply packets. A malicious node can advertise a shorter hop count to a destination by decreasing the actual hop count value in the route reply packet. Thus it succeeds in diverting all the traffic for that destination through itself (Sanzgiri, Dahill, Levine, Shields, & Belding-Royer, 2002).
  Since the malicious node has become part of the routing path, it can do anything with the packets diverted to it, such as dropping or delaying packets.
- **Divert Traffic through Endless Loop:** Protocols like DSR (Johnson & Maltz, 1996) require the inclusion of the list of intermediate nodes in each packet, known as the source route, for routing packets. A malicious node can become part of the routing process and modify the source route to create loops (Sanzgiri et al., 2002). Consequently it can succeed in launching a DoS attack against the destination of the route since the destination will not receive the packets.
- **Divert Traffic through Bottleneck Node:** A QoS protocol may send probe packets

along a set of possible paths for a particular destination to gather information about how much bandwidth is available along each path. The path having the most bandwidth available may be selected as a route for the new flow. A QoS protocol usually assumes that the nodes, which are on the routing path, set the fields in the probe packets. Moreover, a node is not supposed to modify a value that has been set by another node.

If a malicious node becomes an intermediate node of any of the paths that do not have enough bandwidth available for the new flow, it can increase the available bandwidth values set by other nodes in such a way that the path will ultimately be selected as the routing path for the new flow. When traffic starts to flow along the selected path, it is very likely that all flows passing through the bottleneck node/nodes will be impeded and hence the overall network performance will degrade.

Message Fabrication

Sanzgiri et al. (2002) have defined message fabrication as the generation of false routing messages.

However, we have redefined this term to expand its scope. That is, in our work message fabrication would imply the generation of any (except for the MAC (medium access control) control messages) false control message. Message fabrication can take any of the following forms:

- **Divert Traffic through Malicious Node:** We have already discussed this type of attack previously that occurs when a malicious node becomes part of the route reply paths and makes changes in the routing protocol fields. Here we present other ways that a malicious node may divert traffic through itself. We will explain these mechanisms in terms of DSR.

    Nodes in DSR learn new routes by overhearing transmissions from neighbouring nodes. A malicious node can use this vulnerability to announce false routes that are shorter than existing routes to redirect traffic through itself. For example (see Figure 2), a malicious node M can transmit spoofed packets with a shorter source route to E via itself. B may add this route to its route cache and divert all traffic, destined for E, to M. Thus M becomes part of the routing path and can

Figure 2. The malicious node M is becoming a part of the routing path from A to E

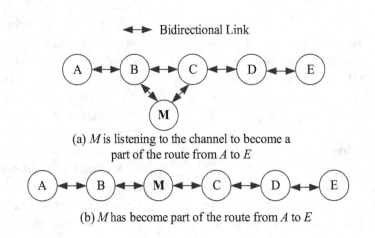

(a) M is listening to the channel to become a part of the route from A to E

(b) M has become part of the route from A to E

drop or delay subsequent data packets. This particular attack is also known as "route cache poisoning" (Sanzgiri et al., 2002).

- **Force Termination of Ongoing Flows:** Some routing protocols, such as AODV and DSR, facilitate route maintenance by sending error messages from the nodes preceding broken links. Since these messages are not authenticated in traditional AODV or DSR, a malicious node can send a false error message against a working node W so that the source node, sending packets to a destination through W, deletes its corresponding route entry (Sanzgiri et al., 2002). If the source node does not have any other route to the destination, the malicious node can succeed in launching a DoS attack by stopping the ongoing traffic towards that destination.
- **Routing Table Overflow:** A node can contain a finite number of route entries in its route table. In reactive routing protocols, a malicious node can try to overflow route tables of other nodes by initiating route discovery for non-existent nodes (Sanzgiri et al., 2002). In pro-active routing, the same purpose can be achieved by broadcasting route information of non-existent nodes (Sanzgiri et al., 2002). In this way, the malicious node can prevent other nodes adding legitimate route entries in their route tables.
- **Denouncing a Benign Node:** A malicious node can report to other nodes of the network that a benign node is misbehaving. If the report is not authenticated, the benign node may not get any legitimate service from other members of the network.

Replay Attack

In this attack a malicious node intercepts a message and retransmits it at a later time, modified or unmodified. By replaying stale information the malicious node can cause inconsistency in the network and consequently can prevent the network

from operating properly. A repetitive replay attack can blind a target node with a storm of traffic so that other nodes are prevented from getting any service from the attacked node.

Broadcast Storm Attack

A malicious node can send network-wide broadcast packets to a node within its transmission range. If the packets are not authenticated, the node will rebroadcast the packet to its neighbourhood. If this process continues, the whole network can be flooded with so much traffic that transmission contention (if the network uses a contention-based MAC protocol) and network congestion may become inevitable. Consequently legitimate flows may not be able to get their required QoS.

Piggyback Attack

If a malicious node knows how packets are routed in a network, it can piggyback its messages on network-wide broadcast packets to send to its accomplice located on another part of the network. If the malicious node sends a large number of packets within a short-time period, the network may face problems similar to the broadcast storm attack. Figure 3 illustrates a piggyback attack.

Tunnelling Attack

This is also known as the *wormhole attack* (Hu, Perrig, & Johnson, 2003) or *route hijacking* (Ramanujan, Ahamed, Bonney, Hagelstrom, & Thurber, 2000). In this attack one or more malicious nodes link two parts of a network through a path that may seem shorter in distance or duration than would otherwise be expected (Sanzgiri et al., 2002). This attack does not require the malicious nodes to have any knowledge of cryptographic keys (Gupte & Singhal, 2003). The results of this attack could be diverting traffic through malicious nodes or preventing nodes from discovering paths more than a certain length.

Figure 3. Piggyback attack where a malicious node M1 is using the network as a carrier for sending messages to another malicious node M2

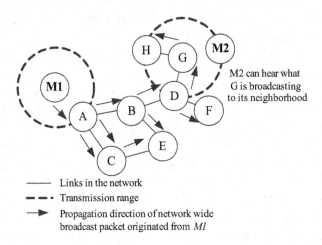

——— Links in the network

– – – Transmission range

——▶ Propagation direction of network wide broadcast packet originated from *M1*

Identity Theft

In the absence of any authentication mechanism a malicious node can easily masquerade as a legitimate node by changing its identity (e.g., IP and MAC addresses) to that of the legitimate node (Sanzgiri et al., 2002). In this way the malicious node can perform any of the attacks mentioned so far without being detected.

AUTHENTICATION PRIMITIVES

Existing network layer security protocols for ad-hoc networks depend on one or more of the following authentication primitives (Yang et al., 2004):

- Hashed message authentication code (HMAC),
- Digital signature, and
- One-way hash chain.

Hashed Message Authentication Code (HMAC)

HMAC requires the communicating nodes to negotiate a symmetric key prior to sending any packet. This key and some parts of the transmitted packet are fed into a one-way hash function to generate a message authentication code. This code is then attached to the transmitted packet. The intended receiver recomputes a message authentication code of the received packet with the same key and hash function. If the recomputed code and the attached code match, the receiver can be assured that the packet originated from an authentic node. The computation involved in this process is very low due to the use of a hash function and symmetric cryptosystems.

Digital Signature

A digital signature may be based on asymmetric cryptography such as RSA (Rivest Shamir Adleman) (Rivest, Shamir, & Adleman, 1978). In this scheme the sender of a packet signs the packet with its private key and the receivers verify the authenticity of the packet using the associated public key. To make the receivers believe that the public key belongs to an authentic sender, the public key needs to be signed by trusted entities. Digital signatures involve more computational overhead for authentication and encryption than a symmetric crypto-system (Yang et al., 2004).

One-Way Hash Chain

In a one-way hash chain a one-way hash function H maps an input of arbitrary length to a fixed length bit string. The function of H is expected to be simple yet robust enough not to be invertible by finite computational capability. To create a hash chain of length $n + 1$, a node chooses a random x where x is a variable length bit string. H is applied to x repetitively to compute a chain $h_0, h_1...h_n$ where $h_0 = x$ and $h_i = H(h_{i-1})$ for $0 \leq i \leq n$. For a given element in a hash chain, it is possible to verify the elements having lower subscript values of that chain. For example, given h_i, a node can authenticate h_{i-3} by computing the value of $H(H(H(h_{i-3})))$ and verifying that the computed value equals h_i. The computational requirement for hash chain-based authentication is lower than for digital signatures. Coppersmith and Jakobsson (2002) describe efficient mechanisms for storing and generating hash chains.

KEY MANAGEMENT IN AD-HOC NETWORKS

Keys are needed for authentication and encryption of transmitted messages. The primary goals of a key management system are to distribute the keys to nodes securely, and to update them periodically or on-demand. Due to the lack of infrastructure, key management in ad-hoc networks is a challenging task. Below some well-known key management protocols suitable for ad-hoc networks have been described.

Threshold Cryptography

Public key infrastructure can be used for easy distribution and certification of keys. A certification authority (CA) certifies the public key of a node by signing the public key and identity of the node with the CA's private key. The CA has to be online all times, which may not be feasible for ad-hoc networks. A CA consisting of a single node is vulnerable to a single point of failure and compromise. If the CA is replicated to multiple nodes, it will not be vulnerable to a single point of failure any more. However it would still be susceptible to a single point of compromise.

Zhou and Haas (1999) have proposed to distribute the CA functionality to several nodes using threshold cryptography. In a *(n, m)* threshold cryptography scheme, the public key of the CA is known to all nodes, the private key of the CA is shared among n nodes called servers according to the secret sharing model of Shamir (1979), and a minimum of m shares of the private key are needed to create a certificate. Hence this scheme can withstand up to n-m compromised nodes.

Yi and Kravets (2002) have extended the work of the *(n, m)* threshold cryptography scheme. They have discussed which nodes to select as servers and how to make the servers available to other nodes efficiently without requiring too much network overhead. They have suggested that physically more secure and computationally more powerful nodes should be selected as servers. If a node has enough cached routes towards the servers, it should send requests to the servers for certificates without employing any broadcast-based route discovery mechanism. Using unicast traffic may reduce network overhead that would otherwise increase if a broadcast-based route discovery mechanism were used.

Luo and Lu (2000) also use a threshold cryptography scheme but allow any m nodes to carry a share of the private key of the CA. Hence this scheme provides a fairer distribution of the burden of creating certificates than the schemes discussed earlier. Since any *m* nodes in the local neighbourhood of a requesting node can create a certificate, communication overhead is minimised. This scheme allows any node not possessing a share of the private key of the CA to obtain a share from any group of *m* nodes that jointly possess the private key. This feature can make the scheme vulnerable to Sybil attack (Douceur, 2002) where a malicious node can take as many identities as necessary to collect enough shares and reconstruct the CA's private key (Capkun, Buttyan, & Hubaux, 2003).

Self-Organised Public Key Management for MANETs

Capkun et al. (2003) have proposed a self-organised public key system for ad-hoc networks that does not need any CA or special nodes. Each node issues certificates to others based on personal acquaintances.

Each node stores the certificates it has issued to others and others have given it. Periodically neighbours exchange new certificates. Two certificates are said to be conflicting if they contain the same public key or if they belong to the same node. A conflict may arise because a malicious node has issued a false certificate. If a node finds any of the certificates is conflicting, it labels the certificate as conflicting. To resolve the dispute, the node obtains certificates from other nodes and takes a majority decision on the collected information.

When a node X wants to obtain the public key of another node Y, X acquires a chain of valid public key certificates such that: (1) the first certificate can be directly verified by X using a public key that X holds and trusts; (2) each remaining certificates can be verified using the public key contained in the previous certificate of the chain; and (3) the last certificate contains the public key of Y.

Since the key authentication is performed via chains of pubic key certificates, the protocol assumes the trust is transitive, that is, if A trusts B and B trusts C, then A trusts C.

SECURE ROUTING IN AD-HOC NETWORKS

Secure routing protocols try to prevent malicious nodes from taking part in any routing process (e.g., sending updated information) and becoming a part of the routing path. Next we review some of the well-known secure routing protocols.

Secure Aware Ad-Hoc Routing (SAR)

SAR (Yi, Naldurg, & Kravets, 2001) outlines various alternatives to authenticate and encrypt routing packets. It incorporates a trust hierarchy in the network in order to associate different privileges or "trust levels" with different nodes.

In SAR, the headers of route request and reply packets are signed and encrypted. Since only nodes with the same level of trust share the same key, any node that does not belong to the same trust level cannot decrypt the packets and hence is forced to drop the packets. The discovered route may not be the shortest for a particular source and destination pair but may be the most secure in terms of the desired trust level.

An implementation of SAR in AODV has been provided. The headers of route request and reply packets are signed using digital signatures and encrypted using the key shared among all the nodes having the same level of trust.

SAR relies on certification authorities for distributing keys in the network. If a centralised certification authority is used, it has to be online all the time and could be a single point of failure that would make the whole system ineffective. If a distributed certification authority is employed then more than one trusted node would be required to construct a certificate. Depending on the distance in the number of hops, the neighbourhood density and the number of the trusted nodes constituting the certification authority, and the update frequency of the keys, the associated communication overhead and response time of getting a reply from the certification authority may not be negligible.

Secure Routing Protocol (SRP)

SRP (Papadimitratos & Haas, 2002) ensures that each pair of end nodes can differentiate between legitimate requests and replies. This is realised through a security association between the two end nodes. A security association can be achieved by establishing a shared key (K_{ST}) between the end nodes. Both the requests and replies are broadcast to thwart potentially malicious nodes from controlling them to reach their destinations.

The SRP adds a header, called the SRP header, to the underlying routing protocol. Three fields $(Q_{ID}, Q_{SEQ},$ and $SRP_{HMAC})$ in the SRP header are used to authenticate the requests and replies at the end nodes. The Q_{ID} is a 32-bit query identifier generated by the source and used by the intermediate nodes to identify duplicate requests. The Q_{SEQ} is a monotonically increasing 32-bit sequence number. It is negotiated during the security association and maintained by the source node for each destination. The SRP_{HMAC} is a 96-bit HMAC calculated by the source using K_{ST} over the addresses of the source and destination nodes, Q_{ID}, and Q_{SEQ}.

When the destination receives a request, it checks Q_{SEQ} for outdated or replayed requests. It recomputes the HMAC using K_{ST} over the message it receives. If the recomputed value and the received SRP_{HMAC} are the same, the destination can be assured the route request originated from the source and has not been tampered with. Otherwise the destination drops the request.

If the destination accepts the packet as a valid route request, it puts the accumulated route into a reply packet. It puts the received values of Q_{ID} and Q_{SEQ} into the reply packet. It calculates a message authentication code using K_{ST} in the same way the source did and puts this value in the reply packet. This enables the source of the route request to verify the integrity and authenticity the reply packet.

Each node monitors the rate at which each of its neighbours forwards route request packets. It gives priority to the neighbours that less frequently forward request packets. This mechanism attempts to limit the flooding of route request packets from malicious nodes.

SRP does not have intermediate nodes checking the authenticity of route requests and route replies. This means that a malicious node can become part of a routing path and later disrupt the operation of the routing protocol. To handle the problem of malicious nodes flooding the network with unnecessary route requests and replies, SRP employs a system at each node that monitors the rate at which each of the neighbours forwards route request packets. Each node forwards packets, with lower priorities for the neighbours that are found sending excessive route requests or route replies. This monitoring system may exacerbate greedy behavior of selfish nodes. A selfish node may misbehave by not forwarding route request packets originated from other nodes. Its neighbours will consider it a low frequency route request packet forwarder and give it priority. Consequently this selfish node may achieve better performance than benign nodes. Moreover SRP does not protect route error messages. Therefore a malicious node can send false route error messages to isolate a node from the network.

Secure Efficient Ad-Hoc Distance (SEAD) Vector Routing Protocol

SEAD (Hu, Johnson, & Perrig, 2002a) aims to protect the route updates of DSDV (destination sequenced distance vector) (Perkins & Bhagwat, 1994) from malicious attacks. DSDV is a table driven or pro-active routing protocol.

A node's shortest distance, usually expressed in hops, is referred to as the metric in DSDV. A node accepts a route update packet containing updated route information for a particular destination if the sequence number included in the packet is greater than the corresponding entry currently available in the node's routing table. The node also accepts the packet if the sequence number is equal but the new metric is lower than those entries in its routing table. A malicious node can send route updates with higher sequence numbers, or with the same sequence number and lower metric for the destination. Neighbours of the malicious node receiving these updates may believe that the malicious node has the shortest or the updated path to that destination. This will make the malicious node the first hop towards that destination from those neighbours. Consequently traffic destined for that destination will be forwarded to that malicious node. The malicious node can then disrupt the flow of the traffic by dropping or delaying packets. SEAD uses a one-way hash chain to prevent malicious nodes from providing incorrect route updates containing a higher sequence number or the same sequence number with lower metrics.

In SEAD each node generates its one-way hash chain of length n and divides the hash chain into groups of m elements each. The value of m is chosen in such a way that the maximum diameter of the network is m-1. Each group corresponds to a sequence number, that is, for a new sequence number the previously used elements are replaced by a new group of elements. Before a node can use a group of elements for authentication, it has to distribute the element with the highest subscript value of that group securely to all other nodes in the network.

Here is an example that illustrates how SEAD works. A destination node X wants to send its updated route information to its neighbours in a route update packet. X sets the value of the metric in the route update packet to 0. The sequence number field is set to a new value, say i. Assume

the generated hash chain at X consists of the sequence $h_0, h_1...h_n$ and $k = n/m - i$. SEAD requires X to send $h_{mk+(m-1)}$ to all other nodes in the network before using this value to authenticate h_{mk}, h_{mk+1}... $h_{mk+(m-2)}$. X puts the first value h_{mk} from the selected group in the route update packet.

A node receiving a route update has to apply the hash function on the received hash value *(m-1-(value of the metric))* times. If the recomputed hash value matches the previously sent $h_{mk+(m-1)}$, the receiving node can assume that a malicious node has not increased the sequence number or lowered the value of the metric keeping the sequence number unchanged. For example when a neighbour A receives the packet it applies the hash function *(m-1-0)* times, where 0 is the value of the metric in the packet, on the received h_{mk} to get $h'_{mk+(m-1)}$. If $h'_{mk+(m-1)} = h_{mk+(m-1)}$, A accepts the sequence number and the metric value of the route update packet as valid. Otherwise A ignores the packet.

When A rebroadcasts the route update packet, it increases the metric value by 1, keeps the sequence number field as it is and replaces the hash value with $h_{mk+1} = H(h_{mk})$.

When a neighbouring node B receives this packet, it applies the hash function on the received h_{mk+1}, m-1-1 times to get $h''_{mk+(m-1)}$. If $h''_{mk+(m-1)}$ matches $h_{mk+(m-1)}$, B accepts the information contained in the route update packet as valid and considers the packet for further processing. Before rebroadcasting, B increases the received metric value by 1, keeps the sequence number field unchanged and replaces h_{mk+1} with $h_{mk+2} = H(h_{mk+1})$. This process of validating and rebroadcasting is repeated at each node that receives the route update packet.

If a node decreases the metric value of a route update packet, it also has to replace the hash value of the packet with an earlier hash value. Due to the one-way nature of the hash function, this is not possible. On the other hand, if a node increases the sequence number in a route update packet, it must replace the hash element of the packet with a hash element that belongs to the next group of the corresponding hash chain. Though the node can apply H on the hash element of the packet an appropriate number of times to get the new element suitable for a higher sequence number, it will still fail to convince others to accept the packet. This is because the node, which initially sent the route update packet, has not yet distributed the authentic hash element from the next group to other nodes.

SEAD requires authenticated elements from the hash chain of each node to be distributed to all other nodes prior to sending update packets. It cannot prevent the "same distance attack" where a malicious node rebroadcasts a route update with the same sequence number and hop count, but with a different sender address (e.g., replacing the original sender address with its own). A secured layer that can identify any unauthorised packet modification is needed to identify such attacks.

Ariadne

Nodes using Ariadne (Hu, Perrig, & Johnson, 2002b) can authenticate routing messages using any of the three schemes: (a) shared secrets between each pair of nodes, (b) shared secrets between communicating nodes combined with broadcast authentication, or (c) digital signature. Here we discuss Ariadne based on shared secrets between communicating nodes combined with a broadcast authentication protocol called the timed efficient stream loss-tolerant authentication (TESLA) (Perrig, Canetti, Tygar, & Song, 2000).

A broadcast authentication protocol is used to authenticate broadcast packets sent to multiple nodes. In a traditional broadcast authentication protocol the sender distributes a public key to all the intended receivers from its generated public and private key pair. Then it includes a message authentication code with each packet. The message authentication code is computed using the private key. The receivers can verify the authenticity of the received packets by recomputing the

message authentication code using the public key and comparing this value with the one included in the packet. If they match, the receiver assumes that the packet has been sent from the sender. TESLA differs from this traditional asymmetric protocol in that it achieves this asymmetry using clock synchronisation and delayed key disclosure, rather than from computationally expensive encryption/decryption functions.

To use TESLA for authentication each sender generates a one-way hash chain and pre-determines a schedule at which it discloses each key of its hash chain to other nodes. The sender discloses keys in reverse order, that is, in order of $h_n, h_{n-1} ... h_0$. A simple key disclosure schedule, for example, would be to disclose key h_i at time $T_0 + i{\times}t$ where T_0 is the time at which h_n has been disclosed and t is the key disclosure interval.

TESLA assumes each receiver can determine which keys a sender might have already disclosed based on loose time synchronisation between nodes. Let Δ be the maximum time synchronisation error between any two nodes, which is known to all nodes. When a sender sends a packet, it adds an HMAC of the packet using a key h_i, which has not been used before.

When a receiver receives the packet it verifies that h_i cannot yet have been disclosed. For example, if the packet arrival time at the receiver is t_r and the receiver knows that the earliest time at which the sender will disclose h_i is $T_0 + i{\times}t$, the receiver needs to verify that $t_r \leq (T_0 + i{\times}t - \Delta)$ implying h_i has not yet been disclosed. If the check is not successful, the receiver discards the packet since the sender may already have disclosed h_i and a malicious node may have forged the packet. Otherwise the receiver buffers the packet and waits for the sender to disclose h_i. When the receiver receives h_i, it first authenticates h_i using the mechanism outlined in the *One-way Hash Chain* section. Then it authenticates buffered packets authenticated with h_j, where $j \geq i$.

Ariadne's basic idea of route discovery and maintenance is based on DSR. For end-to-end authentication Ariadne uses HMAC using shared keys. Let S denote the source that performs a route discovery for destination D, and K_{SD} and K_{DS} denote the keys they share for message authentication in each direction respectively. Setting up shared keys between the source and all intermediate nodes on a route discovery path may be expensive for some ad-hoc networks. Hence it has been proposed to use TESLA for authenticating the intermediate nodes.

A route request (RREQ) packet in Ariadne contains eight fields: *REQUEST, initiator, target, id, time interval, hash chain, node list,* and *HMAC list.* When S sends a RREQ, it sets the *initiator* and *target* to the addresses of S and D respectively. S sets the *id* to an identifier that can be used to distinguish between duplicate RREQs. S sets the *time interval* to the pessimistic expected arrival time t_i of the RREQ at D. S initialises the hash chain to h_0 where $h_0 = HMAC[REQUEST, S, D, id, t_i]K_{SD}$. Here $HMAC[...]K$ denotes the HMAC of $[...]$ computed using K. S does not initialise the *node list* and the *HMAC list*. The RREQ appears as follows:

$S \rightarrow *$: $\langle REQUEST, S, D, id, t_i, h_0, (), () \rangle$

When a neighbour A of S receives the RREQ, it checks the $\langle S, id \rangle$ to determine if it has already seen the same RREQ. If it has, it discards the packet. It also verifies t_i by checking that t_i is not too far in the future and that the corresponding hash chain value has not yet been disclosed. If the *time interval* value is not valid, A discards the RREQ. Otherwise A modifies the RREQ as follows. A appends its address to the *node list* field. It updates the hash chain field with $h_1 = H[A, h_0]$ and appends the $M_A = HMAC[REQUEST, S, D, id, t_i, h_1, (A), ()]h^A_i$ to the *HMAC list*. Here H is a one-way hash function known to all nodes and h^A_i is the TESLA key of the hash chain generated by A. Then A rebroadcasts the modified RREQ.

$A \rightarrow *$: $\langle REQUEST, S, D, id, t_i, h_1, (A), (M_A) \rangle$

When A's neighbour B receives the RREQ from A, it verifies the RREQ as discussed earlier. If the RREQ is valid, B calculates $h_2 = H[B, h_1]$ and $M_B = HMAC[REQUEST, S, D, id, t_i, h_2, (A,B), (M_A)]h^B_i$. Then it rebroadcasts a modified RREQ that looks as follows:

$B \rightarrow *$: $\langle REQUEST, S, D, id, t_i, h_2, (A,B), (M_B) \rangle$

Assume that this RREQ goes through node C before reaching destination D. The RREQ broadcast from C appears as follows:

$C \rightarrow *$: $\langle REQUEST, S, D, id, t_i, h_3, (A,B,C), (M_C) \rangle$

Here $h_3 = H[C, h_2]$ and $M_C = HMAC[REQUEST, S, D, id, t_i, h_3, (A,B,C), (M_A, M_B)]h^C_i$.

When D receives the RREQ, it checks the validity of packet determining that the keys from the *time interval* specified have not yet been disclosed and the hash chain field is equal to $H(C,(H(B,(H(A,HMAC[REQUEST, S, D, id, t_i]K_{SD})))))$.

If the RREQ is valid, D returns a route reply (RREP) packet to S containing eight fields: *REPLY, target, initiator, time interval, node list, HMAC list, target HMAC, key list.* The values of *target, initiator, time interval, node list,* and *HMAC list* are set to the values from the RREQ packet. The *target HMAC* is set to $M_D = HMAC[REPLY, D, S, t_i, (A,B,C), (M_A, M_B, M_C)]K_{DS}$. The key list is kept empty at D.

A node forwarding a RREP waits until it is able to disclose its key from the specified time interval. Then it appends its key to the *key list* field.

The RREPs forwarded by $D, C, B,$ and A contain the values as shown in the following:

$D \rightarrow C$: $\langle REPLY, D, S, t_i, (A,B,C), (M_A, M_B, M_C), M_D, () \rangle$

$C \rightarrow B$: $\langle REPLY, D, S, t_i, (A,B,C), (M_A, M_B, M_C), M_D, (h^C_i) \rangle$

$B{\rightarrow}A$: $\langle REPLY, D, S, t_i, (A,B,C), (M_A, M_B, M_C),$
$M_D, (h^C_i, h^B_i)\rangle$

$A{\rightarrow}S$: $\langle REPLY, D, S, t_i, (A,B,C), (M_A, M_B, M_C),$
$M_D, (h^C_i, h^B_i, h^A_i)\rangle$

When S receives the RREP, it verifies that each key in the *key list*, M_D and each HMAC in the *HMAC list* are valid. If tests are successful, S accepts the RREP. Otherwise S discards the packet.

Ariadne also authenticates route error packets using the TESLA protocol.

TESLA and Ariadne require a sender node to generate a one-way hash chain and attach a hash element from the hash chain to each transmitted packet. Then at a later time the sender discloses an element from its hash chain that can be used by the intended receivers to authenticate the previously attached element. The delayed discloser mechanism requires clock synchronisation. Receivers may have to buffer packets before the packets can be verified with the delayed release of the authentic hash value. The buffering of packets can increase the response time of the routing protocol (Yang et al., 2004). Moreover the release of the authentic hash values involves a second round of communication, increasing network traffic.

Authenticated Routing for Ad-Hoc Network (ARAN)

ARAN (Sanzgiri et al., 2002) is a secure on-demand routing protocol based on AODV (Perkins & Royer, 1999). It uses certificates from trusted entities and public key cryptography for authenticating route request, reply and error packets.

In ARAN each node gets a certificate (*CERT*) from a trusted certificate server T before joining the network. For a node A, the certificate looks as follows: $CERT_A = [IP_A, PBK_A, t, e]PVK_T$. Here IP_A is the IP address of A, PBK_A is the public key of A, t and e are the times indicating when the certificate was created and will expire respectively, and PVK_T

is the private key of T. The term $[...]PVK_T$ denotes that all the values inside the square bracket are concatenated and signed with PVK_T. All nodes are required to contain fresh certificates and must have the public key of T.

To discover routes from A to X, A broadcasts a route request packet with $[RDP, IP_X, CERT_A, N_A, t]PVK_A$. The route request consists of a RDP (route discover packet) marker, the destination's IP address IP_X, certificate of the initiator $CERT_A$, a nonce N_A and the current time t. The nonce is monotonically increased each time a route request is sent. Nodes record the nonce they have seen with the related time-stamp. Therefore they ensure the freshness of the request.

Upon receiving a route request, each intermediate node sets up a reverse path back to the previous node so that it can be used later to unicast the corresponding reply packet back to the initiator.

When an intermediate node B, neighbour of the initiator A, receives the route request with the pair $\langle N_A, IP_A\rangle$ for the first time it extracts A's public key from $CERT_A$ and validates the signature. If the signature appears valid B rebroadcasts a route request which contains $[[RDP, IP_X, CERT_A, N_A, t]PVK_A]PVK_B$, and $CERT_B$.

When an intermediate node C receives the route request from B for the first time, it verifies the authenticity of the request by validating the signature with the B's public key. B's public key is retrieved from $CERT_B$. Then C removes B's signature and certificate from the route request and rebroadcasts $[[RDP, IP_X, CERT_A, N_A, t]PVK_A]PVK_C, CERT_C$.

Eventually single or multiple route requests with the same $<N_A, IP_A>$ pair reach the destination X. The destination replies to the first route request with a given source and nonce pair. The route reply with a REP (Reply) marker from X for A consists of $[REP, IP_A, CERT_X, N_A, t]PVK_X$.

Each intermediate node processes a route reply in the same manner as outlined for route requests except that each reply is unicast rather than broadcast. The packet formats of route reply packets, while being unicast by C and B, are as follows:

$$C \rightarrow B : [[REP, IP_A, CERT_X, N_A, t]PVK_X]PVK_C, CERT_C.$$

$$B \rightarrow A : [[REP, IP_A, CERT_X, N_A, t]PVK_X]PVK_B, CERT_B.$$

In ARAN a node generates a signed route error packet if it finds out one of the links of an active route is broken. Here is an example. Assume the route from A to X consists of intermediate nodes B, C, and D. If the link $C \rightarrow D$ gets broken, C unicasts $[ERR, IP_A, IP_X, CERT_C, N_C, t]PVK_B$ to B. B verifies the authenticity of the error packet from the signature and the public key of C. Then it forwards the unmodified error packet to A. The nonce and the time-stamp together ensure the freshness of the message. Since the error packet is signed, A can be assured that it was not sent from a malicious node.

If a certificate $CERT_R$ needs to be revoked, the trusted entity T broadcasts a revocation packet with packet type *revoke*. It appears as $[revoke, CERT_R]PVK_T$. Any node receiving this message rebroadcasts it to its neighbours and avoids using routes passing through the untrusted node.

To reduce the processing overhead using asymmetric cryptography at each node at each time, the authors have suggested neighbouring nodes exchange secret keys using their public keys and certificates, and use the secret keys to sign routing packets (Sanzgiri, Dahill, Levine, Shields, & Belding-Royer, 2005). However the end nodes are still required to include full public key signatures.

If a node, whose certificate is being revoked, is the only means to connect two parts of a network, it may not propagate the certificate revocation message from one part to another and consequently may cause the network to be partitioned (Gupte & Singhal, 2003). Signing each route request or reply packet by intermediate nodes increases the size of the corresponding packet at each hop. Since ARAN uses a time-stamp together a nonce to ensure the freshness of a message, it may become less effective if clocks of all nodes in the network are not synchronized.

ARAN requires each node to communicate with a certification authority when the node needs to certify its updated public key. Therefore ARAN can suffer from the problems as stated previously for SAR.

Secure AODV (SAODV)

Zapata and Asokan (2002) have proposed SA-ODV to secure AODV. SAODV uses hash chains similar to SEAD. Unlike SEAD, SAODV does not require each node to disseminate some elements from a chain prior to authenticating earlier values of the hash chain. Unlike ARAN, SAODV uses a hash chain to authenticate the hop count field of route request and route reply packets at each intermediate node. Moreover, it uses a digital signature to sign all fields of route error, and most of the fields (except for the hop count field) of route request and route reply packets at each intermediate node.

To discover routes from S to D, S chooses a maximum hop count MH based on the expected diameter of the network. Then it generates a one-way hash chain $h_0, h_1 ... h_{MH}$, similar to SEAD, of length $MH+1$. Here h_0 corresponds to a random number generated by S and $h_i = H(h_{i-1})$ where H is a hash function known to all nodes in the network. Before broadcasting a RREQ, S includes MH, h_{MH}, the current value of hop count HC ($HC = 0$ for S), and an element h_i from the hash chain that corresponds to $HC = 0$ *(i.e. $h_i = h_0$)*. It also includes a digital signature computed using its private key PVK_s and all fields (except for h_i and HC) of the route request packet. The route request packet after adding the fields is referred to as the RREQ-SSE (route request with single signature extension).

When a node receives a copy of the RREQ-SSE, it calculates $h'_{MH} = H^{MH-HC}(h_i)$. $H^{MH-HC}(h_i)$ means applying the hash function H to the hash value enclosed within the brackets $MH\text{-}HC$ times. The node also recomputes the digital signature using S's public key PBK_s and the fields (except for h_i and HC) of the received RREQ-SSE. If $h'_{MH} = h_{MH}$ and the recomputed digital signature is valid, the node rebroadcasts the RREQ-SSE. Otherwise the node drops the packet. Before rebroadcasting the node updates previous the h_i with $H(h_i)$ and HC with $HC+1$ and performs other tasks required for the normal protocol operation of AODV.

When the RREQ-SSE reaches D, D checks the validity of the RREQ-SSE in a similar manner as described earlier. If the RREQ-SSE is valid, D replies with an RREP-SSE (Route Reply with SSE). The RREP-SSE contains the hash chain values and the signature generated by D.

When a node receives an RREP-SSE, it verifies the integrity and authenticity of the RREP-SSE in a similar manner to an RREQ-SSE. If the RREP-SSE is valid, the node updates the hash and hop count values in the same way as it would do for an RREQ-SSE and follows the normal protocol operation of AODV. Otherwise the node drops the RREP-SSE.

SAODV allows each intermediate node to send a route reply in response to a route request if the node contains a fresh route to the destination. This requires the intermediate node to sign the generated route reply on behalf of the final destination.

To solve this problem SAODV requires each intermediate node to store the last route reply it has forwarded for each destination. If an intermediate node receives a route request destined for D and contains a stored route reply originated from D with a sequence number greater than the sequence number contained in the route request, it can reply with a route reply denoted by RREP-DSE (RREP with double signature extension). The intermediate node puts the protocol fields (including the signature computed by D) of the

stored route reply within the RREP-DSE. The RREP-DSE must have a different value of route lifetime than the one stored in the stored route reply. To account for this change, the intermediate node includes a new lifetime value in the RREP-DSE and signs this new value with its public key. Thus the RREP-DSE contains two signature values, one generated by D and the other one generated by the intermediate node. This is why the term "double signature extension (DSE)" is used instead of the "single signature extension (SSE)". This RREP-DSE is verified and handled by other nodes in the same way as an RREP-SSE would have verified and handled.

SAODV also allows each intermediate node to create a reverse route to the originator S of a route request so that if another node asks for a route to S, the intermediate node can send a reply. This feature also requires the intermediate node to sign the generated route reply on behalf of S.

To address this issue, SAODV requires each source node to include the protocol fields of a route reply and a signature that signs the fields corresponding to the route reply in the route request packet. This extended route request is denoted by RREQ-DSE (RREQ with double signature extension). When a node receives this RREQ-DSE, it considers this packet containing a route request seeking a route to a destination and as well as a route reply sent by the originator of the route request. Therefore, it handles the packet in a similar way to handling RREQ-SSE and RREP-SSE.

SAODV uses digital signatures to identify the originator and forwarders of a route error packet. When a node generates or forwards a route error packet, it signs the whole message with its private key. A node receiving a route error packet verifies previously appended signatures to verify the originator and the forwarders of the route error packet.

SAODV assumes that there is a key management sub-system that enables each node to obtain public keys of other nodes of the same network.

Moreover, each node is assumed to be capable of verifying the association between the identity of a given node and the public key of that node. A general limitation of a hash chain-based approach is that it may be complicated and inefficient for continual metrics that take non-integer values (Yang et al., 2004).

Packet Leashes

Hu et al. (2003) have provided two approaches to handle the tunnelling attack using the notion of a *packet leash*. A leash is some information that is added to a packet to determine if the packet has traversed an unrealistic distance.

There are two types of packet leashes: *geographical* and *temporal*. A geographical leash ensures that the receiver of a packet is within a certain distance from the sender. A temporal leash puts an upper bound on the lifetime of a packet to restrict its maximum travel distance.

With a geographical leash each node knows its location and all nodes have loosely synchronised clocks. Let Δ be the maximum time synchronisation error between any two nodes and v be an upper bound on the velocity of any node. When sending a packet the sender includes its own location p_s and the time t_s at which it sent the packet.

When receiving the packet, the receiver computes an upper bound on the distance d_{sr} between the sender and itself based on Δ, received information, its current location p_r, local receive time t_r, and maximum relative error in location information δ:

$$d_{sr} \leq \| p_s - p_r \| + 2v \cdot (t_r - t_s + \Delta) + \delta$$

In temporal leashes all nodes are assumed to have tightly synchronised clocks such that the maximum difference between any two nodes' clocks is Δ. There are two approaches to construct temporal leashes:

1. When sending a packet the sender includes the time t_s at which it sent the packet. When the receiver receives the packet it records the receive time t_s. Based on the difference between t_r and t_s, and the speed of light the receiver can detect if the packet has traveled too far. If the sender's clock is faster than the receiver's, Δ is added to the difference. Conversely Δ is subtracted from the difference if the sender's clock is slower than the receiver's.

2. When the sender sends a packet it includes an expiration time t_e in the packet after which the receiver should not accept the packet. If t_s is the time at which the sender sent the packet, L is the maximum distance the packet is allowed to travel and c is the speed of light, then $t_e = t_s + L/c - \Delta$. When the receiver receives the packet it has to check if the receive time $t_r < t_e$. If not the receiver drops the packet.

HMAC, digital signature, or other authentication technique such as Schnorr's signature (Schnorr, 1991) or TESLA can be used to authenticate the information pertaining to packet leashes.

The main limitation for "packet leashes" is that the clocks have to be synchronised to handle the tunnelling attack. Moreover packet leashes have not been designed to guard against any other attacks except for the tunnelling attack.

Secure Link-State Protocol (SLSP)

SLSP (Papadimitratos & Haas, 2003a) secures the discovery of neighbours and distribution of link-state information of table driven or pro-active routing protocols using digital signatures and one-way hash chains.

During neighbour discovery in link-state routing protocols, each node broadcasts a HELLO message to its neighbours. The HELLO message contains a signed copy of the sender's IP and MAC (medium access control) addresses. Receiving nodes validate the signature and store the new neighbour's information. For signature verification SLSP assumes that each node has a public and private key pair where key certification is provided by a certification authority.

Each node contains a neighbour lookup protocol (NLP) which notifies SLSP of the following events: (1) Two MAC addresses are being used by the same IP address; (2) One MAC address is being used by two different IP addresses; and (3) Another node is using the same MAC address as the detecting node. Notifications from NLP enable SLSP to discard the packets having any of these aforementioned criteria.

Link-state updates in SLSP are signed and propagated for a limited number of hops. Similar to SEAD, a hash chain is used to authenticate the hop count value so that an SLSP update does not travel too many hops.

Like SRP, SLSP uses a monitoring system to limit the flooding of link-state updates from malicious nodes. For relaying link-state updates, a node gives priority to those nodes that relay or generate fewer link-state updates than other nodes.

SLSP is only concerned with securing the topology discovery. Therefore it does not guarantee that malicious nodes, which complied with its operation during route discovery, would still comply during the actual data transmission at a later time.

SLSP assumes that each node has a public/private key pair where certification is provided by a certification authority.

A malicious node masquerading as one of its neighbours with the same MAC address can flood the victim's neighbourhood with false link-state updates. Due to SLSP's "duplicate MAC address detection" functionality, other neighbours of the victim will reject all the link-state updates originating from both the malicious node and the victim. Thus the malicious node can succeed in launching a DoS attack against the victim.

Secure Message Transmission (SMT)

SMT (Papadimitratos & Haas, 2003b) protocol aims to safeguard the data transmission against malicious behavior of network nodes. It works in an end-to-end manner. It assumes that there exists a routing protocol that can find a set of routes to each destination.

With a set of routes to a destination, SMT disperses each outgoing message into several pieces based on Rabin's (1989) algorithm. Rabin's algorithm adds limited redundancy to each data packet to allow recovery from a certain number of faults. The message and the redundancy data are divided into a number of packets so that the destination node can reconstruct the whole message even if m out of n packets are received successfully.

SMT requires a security association, similar to SRP, to exchange the required information for successful reconstruction of the dispersed pieces. This requirement obviates the need for a separate secured layer that can perform the security association properly tolerating potential malicious activities.

INTRUSION DETECTION SYSTEMS FOR THE NETWORK LAYER

The security protocols discussed in the previous section have been designed to prevent attacks happening on the first place. In the course of time if any of the nodes in a network is compromised and consequently does not follow protocol specification properly, these protocols may not detect the misbehaving nodes. Though SRP and SLSP contain monitoring systems, these are not robust enough to detect various anomalies. Next we discuss some of the intrusion detection systems designed to detect node misbehaviour in ad-hoc networks.

Watchdog and Pathrater

Marti, Giuli, Lai, and Baker (2000) have proposed an intrusion detection technique, known as Watchdog, to detect nodes that agree to forward packets but fail to do so. They have used another module, known as Pathrater, which uses the information from Watchdog and helps the routing protocol to avoid misbehaving nodes. Watchdog and Pathrater are best implemented on top of source routing protocols, such as DSR.

When a node X forwards a packet, its Watchdog module verifies that the next node Y in the routing path also forwards the packet to another node Z. X's Watchdog module does this by listening to the transmissions of Y promiscuously. If Y does not forward the packet to Z, X's Watchdog module increments a failure tally. Once the failure tally exceeds a certain threshold, Y is considered to be misbehaving.

The Pathrater module of each node maintains a rating for every other node it knows about in the network. The Pathrater computes the rating, that is, reliability, of each node based on the information obtained from its Watchdog module. A path metric is calculated averaging the node ratings in the path and is used to pick the path most likely to be reliable among the paths between a source and destination pair.

The Watchdog module cannot detect misbehaving nodes in the presence of: (1) Ambiguous collisions, (2) Receiver collisions, (3) Limited transmission power, (4) Directional antennas, (5) False misbehavior, and (6) Partial dropping. Ambiguous collisions prevent a node from overhearing transmissions of its neighbours. In receiver collisions X's Watchdog module can tell if Y has forwarded a packet to Z but not if Z has received it or not. Thus if Y wants to circumvent the Watchdog module of X, it can purposefully cause a collision at Z by forwarding the packet to Z when Z is transmitting. A misbehaving node Y can limit its transmission power such that the transmitted signal is strong enough to reach the previous node X but not the actual recipient Z. A directional antenna prevents neighbouring nodes, which are not within its footprint, from overhearing its transmissions, thus making Watchdog modules ineffective. False misbehavior occurs when a node reports incorrect information about benign nodes. In partial dropping a malicious node can circumvent a Watchdog by dropping packets at a rate lower than the threshold set at that Watchdog.

Techniques for Intrusion Resistant Ad-Hoc Routing Algorithms (TIARA)

Ramanujan et al. (2000) have presented some general techniques, collectively called TIARA, to protect ad-hoc networks from resource depletion, packet dropping and delaying, tunnelling, and replay attacks. TIARA also aims to keep the network operational at an acceptable level during such attacks. It is independent of any specific ad-hoc routing algorithm.

TIARA consists of the following techniques: (1) flow-based route access control (FRAC), (2) flow monitoring, (3) source-initiated route switching, (4) fast authentication, (5) referral based resource allocation, (6) multi-path routing, and (7) incorporating sequence numbers. Table 1 shows the aforementioned design techniques with associated various attacks.

FRAC requires each routing node to maintain an access control rule base that defines the list of authorised flows that the node may forward. The router drops packets belonging to unauthorised flows. The incorporation of FRAC in existing routing protocols requires that flow identifiers instead of destination addresses index the routing tables. Moreover the routing protocol must forward packets based on flow identifiers as opposed to the destination address.

Multi-path routing allows routing nodes to discover and maintain multiple legitimate routes between the source and destination pair corresponding to a data flow. This enables seamless

Table 1. Countermeasure techniques with TIARA for various attacks

Attacks	TIARA countermeasures
Resource depletion	Flow-based route access control, fast authentication, referral-based resource allocation, and sequence numbers
Packet dropping or delaying	Multi-path routing, flow monitoring, source-initiated route switching
Tunnelling	Multi-path routing, flow monitoring, source-initiated route switching

flows of traffic along unharmed routes in the event of attacks on some of the routes.

Source-initiated route switching enables the source to choose a route or a set of routes among all the legitimate routes discovered with a multi-path routing protocol.

A flow monitoring mechanism detects the failure of a route and notifies the source. This provision enables the source-initiated route-switching module to switch the traffic flow to an alternate route and thus maintains seamless flow of traffic.

The flow monitoring system in the source node sends periodic flow status messages to the destination node. The flow status message contains the number of packets associated with the flow that has been transmitted by the source since the last flow status message was sent. This message is encrypted and protected by digital signature to ensure integrity of the message.

The flow monitoring function at the destination node keeps track of the number of packets successfully received from the source between consecutive flow status messages. If the destination does not receive a flow status message for a long time, or the number of packets received is less or more than a predefined threshold, it sends a route failure message to the source along all alternate routes that exist between the two nodes.

Fast authentication is a lightweight mechanism for authenticating data packets at each forwarding node. It relies on placing a path label, that is, an identification mark, of a packet at a node specific secret location within the packet that may differ for different nodes in the path between the source and destination. The information on the node specific secret location is sent to each node in a secured way during the route establishment process.

Sequence numbers, combined with FRAC and fast authentication, provides a counter measure for replay attacks. Similar to the fast authentication technique, the source embeds sequence numbers in node specific locations, corresponding to the nodes between the source and destination, within each data packet.

Referral-based resource allocation is claimed to prevent two intruders generating excessive traffic flow between themselves by collusion, hence depleting network resources of the legitimate intermediate nodes between them. In this scheme a node defines an initial threshold that it will allocate for a flow passing through it. Additional resources can only be granted if such a request is presented with referrals from a certain number of trusted nodes. These referrals from the trusted nodes are assumed to be digitally signed by them and have a time-bound validity.

Though the flow monitoring system may identify an attacked route, it cannot find the intruder since the system receives failure messages only from the destination node instead of each intermediate node. Since the thresholds are not dynamically adapted, a malicious node may fool the detection mechanism.

Intrusion Detection Techniques

Zhang and Lee (2000) and Zhang, Lee, and Huang (2003) have developed a distributed and cooperative IDS for detecting falsifying of route information and random packet dropping (i.e., grey-hole attack) in MANETs.

In the proposed architecture individual IDS agents are placed on each node. Each IDS agent runs independently, detects intrusion from local traces, and initiates a response. If the evidence of an anomaly from a local trace is inconclusive, neighbouring IDS agents cooperate to resolve the issue.

For multi-layer integrated systems, individual IDS agents are placed at each layer on each node. If a node detects a local intrusion at a higher layer, lower layers are notified of the anomaly. The detection module at the lower layer then further investigates the attack, responds by blocking access from the offending nodes, and notifies other nodes of the incident.

Each IDS agent can be structured into six parts: (1) data collection module, (2) local detection engine, (3) cooperative detection module, (4) local response module, (5) global response module, and (6) secure communication module.

The data collection module collects audit traces and activity logs within the radio range of the respective node. The local detection engine uses these data to detect anomalies locally. The cooperative detection module is used whenever the local detection module requires broader data sets or collaboration among neighbouring IDS agents. The local response module triggers actions local to the node. Some of the likely responses are: (1) re-initializing the communication channel between nodes, and (2) identifying compromised nodes and re-organizing the network topology to exclude the compromised node. The global response module is responsible for coordinating

the intrusion response actions of other nodes in the network. A secure communication module provides a secured communication channel among IDS agents.

To detect the aforementioned attacks, the anomaly detection module uses information-theoretic measures (Cover & Thomas, 1991), namely entropy and conditional entropy, to describe the characteristics of normal information flow and uses a classification algorithm to distinguish between normal and abnormal behaviors. The following steps are used for anomaly detection: (1) select or partition audit data into subsets so that each subset contains one class of data (i.e., low entropy or high information gain), (2) perform appropriate data transformation based on the entropy measures, for example, construct new features with low entropy, (3) form classifier using training data, (4) apply the classifier to classify the collected data, and (5) produce intrusion reports.

The authors have suggested using local data sources for anomaly detection since information obtained from external data sources can be compromised and hence cannot be trusted. The information provided by the local data sources are: (1) local routing information including cache entries and traffic statistics, and (2) position location with GPS to provide location and velocity information of nodes.

To evaluate the feasibility of the proposed IDS scheme, the authors have conducted a series of experiments with DSR, AODV, and DSDV in a simulation environment. Results show that the proposed intrusion detection scheme is best suited for on-demand ad-hoc routing protocols (e.g., DSR and AODV) instead of table driven protocols (e.g., DSDV).

The random packet dropping detection scheme relies on overhearing transmissions by listening in promiscuous mode. However, such dependence has short shortcomings as stated in the section, *Watchdog and Pathrater*. Collaboration with other nodes can be achieved properly if a secured communication layer acts as a shield to protect legitimate nodes from intruders.

Security Enhancements in the AODV Protocol

Bhargava and Agrawal (2001) have extended the Watchdog and Pathrater described previously to enhance the security in the AODV routing protocol. Their protocol consists of an intrusion detection model (IDM) and an intrusion response model (IRM).

In this scheme each node employs an IDM that uses neighbourhood information to detect misbehavior of its neighbours. When the misbehavior count (MalCount) for a particular neighbour reaches a predefined threshold, the information is sent to other nodes. When a node receives this information, it checks the local MalCount for the malicious node and adds its result to the initiator's response.

Each node also uses an IRM that identifies that another node has been compromised if the local MalCount corresponding to the malicious node increases beyond a threshold value. In such a case, the IRM module propagates this information to the entire network by transmitting a special type of packet called a Mal packet. If a node receiving the Mal packet also suspects the node indicated in the Mal packet to be malicious, it reports its suspicion to the entire network by transmitting a ReMal packet. If two or more nodes report about a particular node misbehaving, a special type of packet called the purge packet is transmitted to isolate the malicious node from the network. All nodes, that have routes through the identified compromised node, look for newer routes. All packets received from the compromised nodes are dropped as well.

The proposed scheme is claimed to identify false route request, DoS (denial of service), compromise of a destination and impersonation attacks in the following ways:

1. **Distributed False Route Request:** A malicious node may disrupt the normal operation of the network by sending frequent unnecessary route requests. If the nodes receive a number of route requests for a particular source and destination pair greater than a threshold count within a specific time interval, the source is identified as malicious.

2. **DoS:** A malicious node can transmit false control and data packets at an excessive rate and thus launch DoS attacks against legitimate nodes so that they cannot access the network resources (e.g., bandwidth) properly. This type of attack can be identified if a node generates packets at higher frequency than a threshold.

3. **Destination is Compromised:** This type of attack is detected if a source does not receive a reply from a particular destination within a given time interval. Furthermore, the neighbours generate probe or HELLO packets to determine connectivity. In this model if a node does not respond to a route request, it is identified as a malicious node.

4. **Impersonation:** A sender is identified as a malicious node if the receiver cannot decrypt the received packet sent by that sender.

If the threshold values are not adapted dynamically, a malicious node may be clever enough to work around the proposed detection system. The authors have not provided any mechanism to adjust the threshold values dynamically. Moreover it is not clear how a node can monitor ongoing traffic of its neighbours if the neighbours transmit packets with limited transmission power and directional antennas.

COoperative of Nodes, Fairness In Dynamic Ad-hoc NeTworks (CONFIDANT)

CONFIDANT (Buchegger & Boudec, 2002a, 2002b) is an extended version of Watchdog and

Pathrater. Unlike Watchdog and Pathrater, this scheme punishes the misbehaving nodes. Nodes with bad reputations are isolated in order to limit their activity in the network. Thus CONFIDANT, unlike Watchdog and Pathrater, stimulates misbehaving nodes to contribute to the normal operations of the network in order to be able to get services from other nodes.

CONFIDANT consists of four major components: (1) monitor, (2) reputation system, (3) path manager, and (4) trust manager. Each node is assumed to be equipped with all these components.

The monitor observes any routing or forwarding misbehavior of its neighbouring nodes by listening to the transmission channel in promiscuous mode. It then passes this information to the reputation system for further analysis.

The reputation system manages a table consisting of entries for nodes and their ratings. The rating of a node is updated when there is sufficient evidence that the node in question has misbehaved significantly. The rating is changed according to a rating function that assigns different weights to the type of behavior detection, namely the greatest weight for own experience, a smaller weight for observed activities and smallest weight for reported misconduct. There is a time limit for the reputation information in order to lessen the effect of false reporting.

The path manager takes input from the reputation system and performs the following functions: (1) path re-ranking according to the level of reputation of associated intermediate nodes, (2) deletion of paths containing malicious nodes, (3) actions on receiving a request for a route from a malicious node (e.g., ignore the request), and (4) actions on receiving a request for a route containing a malicious node from a benign node (e.g., alert the benign node).

The trust manager of a node issues ALARM messages to the node's friends. ALARM messages are generated if the node experiences, observes, or receives a report of malicious behavior. Upon receiving an ALARM message from a node, the trust manager assesses the trustworthiness of the received message according to the trust level of the reporting node.

CONFIDANT uses a mechanism similar to PGP (pretty good privacy) for expressing various levels of trust (e.g., unknown, none, marginal, and complete), key validation, and certification. CONFIDANT calculates the validity of a public key by examining the trust levels of all attached signatures and computes a weighted score of the validity. For example, two marginally trusted signatures could be accepted as a completely trusted signature.

If the trust manager considers the received report as trusted and has received similar trusted reports from other nodes, it passes this information to the reputation system so that the perception about the allegedly malicious node can be updated.

Since misbehaving nodes are isolated, reputable nodes may suffer from congestion as most of the routes are likely to pass through them. CONFIDANT allows negative ratings from other nodes, which may result in false accusation. Moreover, like Watchdog and Pathrater, CONFIDANT may not work properly in networks where nodes use limited transmission power and directional antennas.

COllaborative REputation (CORE)

CORE (Michiardi & Molva, 2002a) is a reputation-based system similar to CONFIDANT. Unlike CONFIDANT, it does not allow negative ratings to be broadcast by other nodes. This prevents false accusation. CORE is a generic mechanism that can be integrated with several network and application layer functions. Examples of the network functions include performing route discovery, forwarding data packets, and so forth.

A CORE consists of a set of reputation tables (RT) and a Watchdog (WD) module. There is one RT for each function that has to be monitored and a global RT that combines the different values of reputation calculated for different functions. CORE defines two types of reputations, namely subjective and indirect, each of which is calculated with respect to different functions. A subjective reputation is calculated directly from a node's own observation, and can take both negative and positive values. An indirect reputation is evaluated considering the observations of other nodes, and can take only positive values. Allowing only positive values for indirect reputation prevents DoS attacks based on malicious broadcasting of negative ratings for legitimate nodes.

The general formula to calculate subjective reputation is:

$$r_{s_i}(s_j \mid f) = \sum \rho(t, t_k) \cdot \sigma_k$$

Here $r_{s_i}(s_j \mid f) \in [-1, 1]$ denotes the subjective reputation of node s_j calculated by its neighbour s_i at time t with respect to the function f. $\sigma_k \in [-1, 1]$ is a rating factor given to the kth observation taken at time t_k and $\rho(t, t_k)$ is a time dependent function that gives higher relevance to past values of σ_k.

The indirect reputation is denoted by $ir_{s_i}(s_j \mid f)$. It can be calculated using a similar expression to that used for $r_{s_i}(s_j \mid f)$ provided that the corresponding rating factor takes only positive values. Now the aforementioned reputation information can be combined with the following formula:

$$r_{s_i} = \sum w_k \{ r_{s_i}(s_j \mid f) + ir_{s_i}(s_j \mid f) \} \qquad (1)$$

Here w_k represents the weight associated with the functional reputation value. For example packet forwarding can be given more weight than the routing function since the former has a more important impact on the overall performance compared to the latter (Michiardi & Molva, 2002b).

Every time a node needs to monitor the correct execution of a function by a neighbouring node, it triggers a WD specific to that function. The WD compares the observed result with an expected value to detect if the monitored function has been executed properly. If the WD detects any anomaly, a negative value is assigned to the rating factor corresponding to the monitored function. With this information a new reputation for the malicious neighbour is calculated using Equation (1).

In CORE the most reputed nodes may become congested as most of the routes are likely to pass through them. As in Watchdog and Pathrater, the limitations of the monitoring system in networks with limited transmission power and directional antennas have not been addressed.

AODV State Transition Analysis Technique (AODVSTAT)

AODVSTAT (Vigna, Gwalani, Srinivasan, Belding-Royer, & Kemmerer, 2004) is a state transition analysis technique (STAT) (Ilgun, Kemmerer, & Porras, 1995) based IDS designed for detecting attacks against AODV. The proposed system tries to reduce the number of false positives.

STAT was initially designed to model host-based and network-based intrusions in wired networks. In state transition analysis, computer penetrations (i.e., attacks) are described as a sequence of actions that leads an attacker from some initial state on a system to a target compromised state. Between the initial and the compromised states there are one or more intermediate state transitions that an attacker has to perform to achieve the desired compromised state.

When the initial and compromised states have been identified, the key actions (or signature actions) are determined. Here signature actions refer to those actions that would prevent an attack from completing successfully if one or more of those actions are omitted from the execution of an attack scenario. STAT can be a useful approach for intrusion analysis since it requires the analyst to identify a minimum number of signature actions for an attack and to organise them visually similar to a state transition diagram. In AODVSTAT the signature actions are either data or AODV control packets exchanged over a wireless network.

AODVSTAT sensors are deployed on a subset of nodes where each sensor performs state analysis of the packet streams to detect signs of intrusion. Figure 4 shows the architecture of an AODVSTAT sensor.

The detection process relies on an internal fact-base which contains information about neighbouring nodes. The internal fact-base is updated by analysing the observed data and AODV control packets transmitted from the neighbouring nodes. In the distributed mode, the fact base also contains information from other nodes obtained by means of UPDATE messages.

Each sensor monitors surrounding network traffic. The transmitted data packets are monitored to determine how much traffic has been generated, received, and forwarded by each node. On the other hand the control packets are monitored to extract the AODV sequence numbers of active nodes, the IEEE 802.11 header details and the MAC/IP pairs of the nodes residing within the range of the sensor. The packets retrieved are then matched against a number of state transition attack scenarios. When a match is found, an intrusion alert is initiated.

Here are some of the attacks that AODVSTAT can detect:

1. **Impersonation:** Many routing protocols use MAC/IP address pairs to uniquely identify a node. To detect impersonation attacks when a packet arrives with a new MAC/IP address pair, the pair is stored for the first time. If another packet arrives with the same IP but different MAC, or vice versa, the packet is detected as spoofed.

2. **Dropping of Packets:** To detect dropping of RREQ (route request) packets, the IDS maintains a list of all its neighbours. This list is refreshed dynamically to allow changes in network topology. The IDS detects a dropped RREQ packet if its neighbour fails to reply to that packet with another RREQ or RREP (route reply) packet. To detect dropped RREP packets, the IDS checks whether its neighbour forwards the received RREP packet towards the source node, provided that the neighbour is not the intended final destination for the RREP packet. Detecting

Figure 4. An AODVSTAT architecture

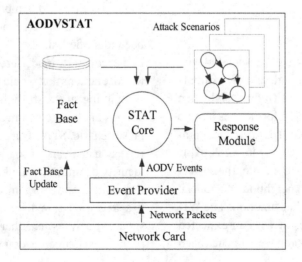

dropping of DATA packets is similar to the scenario for dropping RREP packets.

3. **Resource Depletion Attack:** To detect this attack the IDS maintains a count of the number of packets it receives from each node within a specific time window. If the count exceeds a certain threshold, then an alert is signaled.

4. **False Propagation of High Sequence Number:** This kind of attack can be detected by using a distributed set of sensor nodes that send neighbour information to every other sensor node. The sensor nodes as a group check if an RREP packet with a sequence number for a particular destination is received with a value greater than a certain threshold. If so the attack is identified.

5. **Diverting Traffic through Malicious Node:** It can be detected by a combination of impersonation and the packet drop attack.

The proposed scheme does not consider compromised nodes sending false intrusion information to other nodes. Moreover, like Watchdog and Pathrater AODVSTAT does not provide any solution for networks where nodes can transmit packets with limited transmission power, different frequency channels, or directional antennas.

TWOACK

Balakrishnan, Deng, and Varhney (2005) have proposed a different way to detect packet dropping in ad-hoc networks that addresses the problems of receiver collisions, limited transmission power, and directional antennas discussed in Watchdog and Pathrater. The scheme can be added on to any source routing protocol such as DSR.

Suppose node A has discovered a route to F with a source route $A{\to}B{\to}C{\to}D{\to}E{\to}F$. In TWOACK when B forwards a packet for A, C (the node two hops away from A) receives the packet and sends an acknowledgement to A indicating B has forwarded the packet properly. If

A does not get an acknowledgment for the packet within a certain timeout period it suspects B to be misbehaving. The same procedure is carried out by every set of three consecutive nodes along the source route.

In TWOACK each forwarded packet has to be acknowledged which may contribute to traffic congestion on the routing path. S-TWOACK (Selective TWOACK) reduces this extra traffic by sending a single acknowledgment for a certain number of packets instead of a single packet.

The proposed scheme may fail if two consecutive nodes along the route collude to misbehave. SAHN-IDS (Islam, Pose, & Kopp, 2005b) provides a mechanism to minimise this problem. Moreover, TWOACK/S-TWOACK cannot detect the misbehavior of a forwarding node if it delays packet transmissions for selective nodes.

OPEN RESEARCH CHALLENGES

Multimedia applications, such as video conferencing and interactive gaming, often have high bandwidth requirements and sometimes need very fast response times. Running such demanding applications over ad-hoc networks requires the network to be able to provide high bandwidth and real-time response. Security protocols often use computationally expensive encryption and authentication mechanisms, and need to exchange many messages for their correct operation. This computational and communications overhead make deploying secure multimedia applications on ad-hoc networks very challenging.

As the size and density of ad-hoc networks grow, the communication overhead of the security protocols increases. This can have an adverse effect of network performance in ad-hoc networks having limited bandwidth. In some cases it may be impractical to provide the required security, with its attendant overheads, since doing so may compromise the essential performance requirements of the multimedia applications.

Some ad-hoc networks, for example, sensor networks, consist of nodes having limited battery power and computational resources. Security protocols with high computational overheads, for example, computations for RSA, may not be feasible for this type of network.

There are tradeoffs among security strength, communication overhead, computational complexity, energy consumption, and scalability. Further security and performance analyses can be used to find a balance between security protocols and other network constraints so that the overall security strength and network performance are optimised to suit the requirements of network operations and users.

The security protocols discussed so far are mostly implemented in a single layer (except CORE) in the network protocol stack. One of the disadvantages of having a single layered security approach is if the security protocol in the single layer fails, other layers will be exposed to malicious attacks. Therefore a multi-layered security approach is required that can be integrated into possibly every component in the network protocol stack. The performance of the network should not critically depend on any single security component so that the overall security system is effective despite the breakdown of one or more security components.

The existing security protocols are designed with certain attacks in mind and hence may not work well in the presence of unanticipated attacks. One of the main reasons for this limitation is that the possible attacks are not well modelled. A comprehensive model of possible attacks needs to be developed in order to design robust security protocols that can handle a bigger problem space.

Malicious attacks and network faults due to node misconfigurations, extreme network overload, or operational failures share common symptoms (Yang et al., 2004). Security protocols should be able to cope with all types of anomalies.

To strengthen the level of security, a security protocol should have an access control management facility so that different nodes can be given different privileges to access the network services. SAR incorporates a trust of hierarchy in the network that provides a simple access control management in the network. Other concepts, such as the password-capability (Anderson, Pose, & Wallace, 1986; Castro, 1996; Kopp, 1997; Pose, 2001), can be incorporated in a security protocol to provide a fine grained access control management facility in the network. Islam, Pose, and Kopp (2004, 2005a, 2005c) have proposed a security protocol called SSP (SAHN security protocol) based on password-capability for static ad-hoc networks.

CONCLUSION

In this chapter we examined a number of possible classes of attacks on ad-hoc networks. These included drop packets, delay transmission, protocol field modification, message fabrication, replay attack, broadcast storm attack, piggyback attack, tunnelling attack, and identity theft.

For these attacks we outlined preventive strategies for dealing with them and discussed the merits of these approaches. SAR provides various alternatives to authenticate and encrypt routing packets. It incorporates a trust hierarchy in the network in order to associate different privileges or "trust levels" with different nodes. SRP ensures that each pair of end nodes can differentiate between legitimate route requests and route replies. SEAD attempts to protect the route updates of DSDV routing protocol from malicious attacks using one-way hash chains. Ariadne authenticates routing messages using any of the three schemes: (1) shared secrets between each pair of nodes, (2) shared secrets between communicating nodes combined with broadcast authentication called TESLA, or (3) digital signature. ARAN uses public key cryptography for authenticating route request, reply and error packets of AODV routing protocol. SAODV uses one-way hash chains to authenticate the hop count field of route request and route reply packets, and digital signature to sign all fields of route error, and most of the fields (except for the hop count field) of route request and route reply packets at each intermediate node for AODV routing protocol. Packet leashes handle the tunnelling attack by adding additional information in each packet to determine if the packet has traversed an unrealistic distance. SLSP secures the discovery of neighbours and distribution of link-state information of table driven or pro-active routing protocols using digital signatures and one-way hash chains. SMT aims to safeguard the data transmission against malicious behaviour of network nodes on end-to-end basis. All these protocols have been designed to plug various forms of attack, hence making the system more robust.

We also examined mechanisms and security protocols that can deal with various forms of anomalous behaviour of ad-hoc networks. Watchdog and Pathrater detect nodes that agree to forward packets but fail to do so in a DSR-like routing protocol and help the routing protocol to avoid misbehaving nodes. TIARA consists of some general techniques to protect ad-hoc networks from resource depletion, packet dropping and delaying, tunnelling, and replay attacks. The scheme proposed by Zhang et al. (2000, 2003) is a distributed and cooperative IDS that detects falsifying route information and random packet dropping. Bhargava and Agrawal (2001) have extended Watchdog and Pathrater to enhance the security in the AODV routing protocol. CONFIDANT is an extension of Watchdog and Pathrater that punishes the misbehaving nodes by limiting their activity in the network. Thus CONFIDANT encourages misbehaving nodes to contribute to the normal operations of the network in order to be able to get services from other nodes. CORE is a generic reputation-based IDS that can be integrated with several network and application

layer functions. AODVSTAT is a state transition analysis technique (STAT) based IDS designed for detecting attacks against AODV. TWOACK detects packet dropping in ad-hoc networks using DSR like routing protocols that addresses the problems of receiver collisions, limited transmission power, and directional antennas of Watchdog and Pathrater. All of these protocols deal with anomalous network behaviour and attempt to limit its spread and indeed preferably to isolate the offending network nodes. In so doing they assist in preserving the security of the system.

Our study of various security protocols suggests that no particular protocol or class of protocols is best suited for all scenarios. The field of security in ad-hoc networks is still an open research challenge, however significant security protocol developments as outlined in this chapter have led the way towards the elusive goal of a demonstrably secure system.

REFERENCES

Anderson, M., Pose, R., & Wallace, C. S. (1986). A password-capability system. *The Computer Journal, 29*(1), 1-8.

Balakrishnan, K., Deng, J., & Varhney, P. K. (2005). TWOACK: Preventing selfishness in mobile ad hoc networks. *Proceedings of the IEEE Wireless Communications and Networking Conference (WCNC), 4,* New Orleans, LA, March 13-17 (pp. 2137-2142). IEEE.

Bhargava S., & Agrawal, D. P. (2001). Security enhancements in AODV protocol for wireless ad-hoc networks. *Proceedings of the Vehicular Technology Conference, 4,* Atlantic City, NJ, October 7-11 (pp. 2143-2147). IEEE.

Buchegger, S., & Boudec, J.-Y. L. (2002a). Nodes bearing grudges: Towards routing security, fairness and robustness in mobile ad hoc networks. *Proceedings of the 10th Euromicro Workshop on*

Parallel, Distributed and Network-based Processing, Canary Island, Spain, January 9-11 (pp. 403-410). IEEE Computer Society.

Buchegger, S., & Boudec, J.-Y. L. (2002b). Performance analysis of the CONFIDANT protocol. *Proceedings of the 3rd ACM International Symposium on Mobile Ad Hoc Networking and Computing (MobiHoc)*, Lausanne, Switzerland, June 9-11 (pp. 455-465). ISBN 1-58113-501-7. ACM Press.

Capkun, S., Buttyan, L., & Hubaux, J.-P. (2003). Self-organizing public-key management for mobile ad hoc networks. *IEEE Transactions on Mobile Computing, 2*(1), 52-64.

Castro, M. D. (1996). *The walnut kernel: A password-capability based operating system.* Ph.D. Thesis, Monash University, Clayton, Australia.

Coppersmith, D., & Jakobsson, M. (2002). Almost optimal hash sequence traversal. *Conference on Financial Cryptography (FC)*, Southampton, Bermuda, March 11-14, Lecture Notes in Computer Science (pp. 102-119). ISBN: 978-3-540-00646-6. Springer-Verlag.

Cover, T. M., & Thomas, J. A. (1991). *Elements of information theory.* New York: Wiley.

Douceur, J. (2002). The Sybil attack. *1st International Workshop Peer-to-Peer Systems (IPTPS)*, Cambridge, MA, March 7-8, Lecture Notes in Computer Science (pp. 251-260). ISBN 3-540-44179-4. Springer-Verlag.

Gupte, S., & Singhal, M. (2003). Secure routing in mobile wireless ad hoc networks. *Ad Hoc Networks, 1*(1), 151-174.

Hu, Y.-C., Johnson, D. B., & Perrig, A. (2002a). SEAD: Secure efficient distance vector routing for mobile wireless ad hoc networks. *Proceedings of the 4th IEEE Workshop on Mobile Computing Systems and Applications (WMCSA)*, Callicoon, NY, June 20-21 (pp. 3-13). ISBN 0-7695-1647-5. IEEE.

Hu, Y.-C., Perrig, A., & Johnson, D. B. (2002b). Ariadne: A secure on-demand routing protocol for ad-hoc networks. *Proceedings of the 8th Annual International Conference on Mobile Computing and Networking (MobiCom)*, Atlanta, Georgia, September 23-28 (pp. 12-23). ISBN 1-58113-486-X. ACM Press.

Hu, Y.-C., Perrig, A., & Johnson, D. B. (2003). Packet leashes: A defense against wormhole attacks in wireless ad hoc networks. *Proceedings of the 22nd IEEE Computer and Communications Societies (InfoCom)*, San Francisco, CA, April 1-3 (pp. 1976-86). IEEE.

Ilgun, K., Kemmerer, R. A., & Porras, P. A. (1995). State transition analysis: A rule-based intrusion detection approach. *IEEE Transactions on Software Engineering, 21*(3), 181-199.

Islam, M. M., Pose, R., & Kopp, C. (2004). A link layer security protocol for suburban ad-hoc networks. *Australian Telecommunication Networks and Applications Conference (ATNAC)*, Sydney, Australia, December 8-10 (pp. 174-177). Australia.

Islam, M. M., Pose, R., & Kopp, C. (2005a). Link layer security for SAHN protocols. *3rd IEEE PerCom Workshops on PWN*, Kauai Island, Hawaii, March 8-12 (pp. 279-283). IEEE Computer Society.

Islam, M. M., Pose, R., & Kopp, C. (2005b). An intrusion detection system for suburban ad-hoc networks. *IEEE Tencon*, Melbourne, Australia, November 12-14. ISBN: 0-7803-9312-0. IEEE Press.

Islam, M. M., Pose, R., & Kopp, C. (2005c). Suburban ad-hoc networks in information warfare. *Proceedings 6th Australian InfoWar Conference*, Geelong, Australia, November 25-26 (pp. 71-80). ISBN: 1-74156-028-4. Australia: Deakin University.

Johnson, D., & Maltz, D. (1996). Dynamic source routing in ad hoc wireless networks. In T. Imielinski, & H. Korth (Eds.), *Mobile computing* (vol. 353, pp. 151-181). USA: Kluwer Academic Publishers.

Kopp, C. (1997). *An I/O and stream inter-process communications library for a password capability system.* Masters Thesis, Monash University, Clayton, Australia.

Kopp, C., & Pose, R. (1998). Bypassing the home computing bottleneck: The suburban area network. *3rd Australasian Computer Architecture Conference (ACAC), 20*(4), 87-100, Perth, Australia, February 2-3. ISBN: 981-3083-93-X. Springer-Verlag.

Luo, H., & Lu, S. (2000). *Ubiquitous and robust authentication services for ad hoc wireless networks.* Technical Report, TR-200030, Department of Computer Science, UCLA.

Marti, S., Giuli, T. J., Lai, K., & Baker, M. (2000). Mitigating routing misbehavior in mobile ad-hoc networks. *Proceedings of the 6th International Conference on Mobile Computing and Networking (MobiCom)*, Boston, MA, August 6-11 (pp. 255-265). ISBN 1-58113-197-6. ACM Press.

Michiardi, P., & Molva, R. (2002a). CORE: A collaborative reputation mechanism to enforce node cooperation in mobile ad hoc networks. *Proceedings of the IFIP TC6/TC11 Sixth Joint Working Conference on Communications and Multimedia Security*, Portoroz, Slovenia, September 26-27 (pp. 107-121). ISBN 1-4020-7206-6. Berlin: Springer.

Michiardi, P., & Molva, R. (2002b). Simulation-based analysis of security exposures in mobile ad-hoc networks. *Proceedings of the European Wireless Conference.*

Papadimitratos, P., & Haas, Z. J. (2002). Secure routing for mobile ad hoc networks. *Proceed-ings of the SCS Communication Networks and Distributed Systems Modelling and Simulation Conference (CNDS)*, San Antonio, Texas, January 27-31. USA: Society for Computer Simulation.

Papadimitratos, P., & Haas, Z. J. (2003a). Secure link state routing for mobile ad hoc networks. *Proceedings of the Symposium on Applications and the Internet Workshops (SAINT)*, Orlando, FL, January 27-31 (pp. 27-31). USA: IEEE Computer Society.

Papadimitratos, P., & Haas, Z. J. (2003b). Secure message transmission in mobile ad hoc networks. *Ad Hoc Networks, 1*(1), 193-209.

Perkins, C., & Bhagwat, P. (1994). Highly dynamic destination-sequenced distance-vector routing (DSDV) for mobile computers. *Proceedings of the ACM Conference on Communications Architectures, Protocols and Applications (SIGCOMM)*, London, UK, August 31-September 2 (pp. 234-244). ACM Press.

Perkins, C. E., & Royer, E. M. (1999). Ad-hoc on-demand distance vector routing. *Proceedings of the 2nd IEEE Workshop on Mobile Computing Systems and Applications (WMCSA)*, New Orleans, LA, February 25-26 (pp. 90-100). IEEE Computer Society.

Perrig, A., Canetti, R., Tygar, J. D., & Song, D. (2000). Efficient authentication and signing of multicast streams over lossy channels. *Proceedings of the IEEE Symposium on Security and Privacy*, Berkeley, CA, May 14-17 (pp. 56-73). IEEE Computer Society.

Pose, R. (2001). Password-capabilities: Their evolution from the password-capability system into walnut and beyond. In G. Heiser (Ed.), *6th Australasian Computer Systems Architecture Conference (ACSAC)*, 23, Gold Coast, Australia, January 29-30 (pp. 105-113). IEEE Computer Society.

Rabin, M. O. (1989). Efficient dispersal of information for security, load balancing, and fault tolerance. *Journal of ACM, 36*(2), 335-348.

Ramanujan, R., Ahamad, A., Bonney, J., Hagelstrom, R., & Thurber, K. (2000). Techniques for intrusion-resistant ad hoc routing algorithms (TIARA). *Proceedings of the 21st Century Military Communications Conference (MILCOM)*, Los Angeles, CA, October 22-25 (pp. 660-664). ISBN 0-7803-6521-6. Piscataway, NJ: IEEE.

Rivest, R. L., Shamir, A., & Adleman, L. (1978). A method for obtaining digital signatures and public-key cryptosystems. *Communication ACM, 21*(2), 120-126.

Sanzgiri, K., Dahill, B., Levine, B. N., Shields, C. & Belding-Royer, E. M. (2002). A secure routing protocol for ad hoc networks. *Proceedings of the 10th IEEE International Conference on Network Protocols (ICNP)*, Paris, France, November 12-15 (pp. 78-89). IEEE Computer Society.

Sanzgiri, K., Dahill, B., Levine, B. N., Shields, C., & Belding-Royer, E. M. (2005). Authenticated routing for ad hoc networks. *IEEE Journals on Selected Areas in Communications, 23*(3), 598-610.

Schnorr, C. P. (1991). Efficient signature generation by smart cards. *Journal of Cryptology, 4*(3), 161-174.

Shamir, A. (1979). How to share a secret. *Communications of ACM, 22*(11), 612-613.

Vigna, G., Gwalani, S., Srinivasan, K., Belding-Royer, E. M., & Kemmerer, R. A. (2004). An intrusion detection tool for AODV-based ad hoc wireless networks. *Proceedings of the Annual Computer Security Applications Conference (ACSAC)*, Miami Beach, FL, December 10-14 (pp. 16-27). IEEE Computer Society.

Yang, H., Luo, H., Ye, F., Lu, S., & Zhang, L. (2004). Security in mobile ad hoc networks: Challenges and solutions. *IEEE Wireless Communications, 11*(1), 38-47.

Yi, S., & Kravets, R. (2002). *Key management for heterogeneous ad hoc wireless networks.* Technical Report, UIUCDCS-R-2002-2290, Department of Computer Science, University of Illinois at Urbana-Champaign.

Yi, S., Naldurg, P., & Kravets, R. (2001). Security-aware ad hoc routing for wireless networks. *Proceedings of the 2nd ACM International Symposium on Mobile Ad Hoc Networking and Computing (MobiHoc)*, Long Beach, CA, October 4-5 (pp. 299-302). ISBN 1-58113-428-2. ACM Press.

Zapata, M. G., & Asokan, N. (2002). Securing ad hoc routing protocols. *Proceedings of the 3rd ACM Workshop on Wireless Security (WiSe)*, Atlanta, Georgia, September 28 (pp. 1-10). ISBN 1-58113-585-8. ACM Press.

Zhang, Y., & Lee, W. (2000). Intrusion detection in wireless ad-hoc networks. *Proceedings of the 6th International Conference on Mobile Computing and Networking (MobiCom)*, Boston, MA, August 6-11 (pp. 275-283). ISBN 1-58113-197-6. ACM Press.

Zhang, Y., Lee, W., & Huang, Y. (2003). Intrusion detection techniques for mobile wireless networks. *Kluwer Wireless Networks, 9*(5), 545-556.

Zhou, L., & Haas, Z. J. (1999). Securing ad hoc networks. *IEEE Networks, 13*(6), 24-30.

Chapter XIV
Security, Privacy, and Trust for Pervasive Computing Applications

Sheikh I. Ahamed
Marquette University, USA

Mohammad Zulkernine
Queen's University, Canada

Munirul M. Haque
Marquette University, USA

ABSTRACT

Pervasive computing has progressed significantly during this decade due to the developments and advances in portable, low-cost, and light-weight devices along with the emergence of short range and low-power wireless communication networks. Pervasive computing focuses on combining computing and communications with the surrounding physical environment to make computing and communication transparent to the users in day-to-day activities. In pervasive computing, numerous, casually accessible, often invisible, frequently mobile or embedded devices form an ad-hoc network that occasionally connects to fixed networks structure too. These pervasive computing devices often collect information about the surrounding environment using various sensors. Pervasive computing has the inherent disadvantages of slow, expensive connections, frequent line disconnections, limited host bandwidth, location dependent data, and so forth. These challenges make pervasive computing applications more vulnerable to various security-related threats. However, traditional security measures do not fit well in pervasive computing applications. Since location and context are key attributes of pervasive computing applications, privacy issues need to be handled in a sophisticated manner. The devices in a pervasive computing network leave and join in an ad-hoc manner. This device behavior creates a need for new trust models for pervasive computing applications. In this chapter, we address the challenges and requirements of security, privacy, and trust for pervasive applications. We also discuss the state-of-the-art of pervasive security, privacy, and trust along with some open issues.

INTRODUCTION

Pervasive computing embeds information communication and computation in an environment that overcomes the constraints of character, place, and time. Comparing virtual reality (VR), which builds an artificial world in the computer, to pervasive computing, which embeds computing in the real world, we observe two forms of computing on the opposite ends of the computing spectrum. Along with the forward march of wireless and sensor networks, the demand curve of PDAs, mobiles, smartphones, and other small handheld

devices is showing an exponential growth. Given that, pervasive computing is showing its potential in almost every aspect of our life including hospitals, emergency and critical situations, industry, education, hostile battle field, and so forth. Pervasive computing contains varieties of diversified devices varying from PDAs to the tiny, nearly invisible chips in watches, cordless phones, and other day-to-day items. Because these devices can communicate with one another wirelessly, forming an ad-hoc mobile network, the utility of the combined capability of these small devices can be great. These pervasive computing devices

Figure 1. A typical ad-hoc network scenario in pervasive computing environment

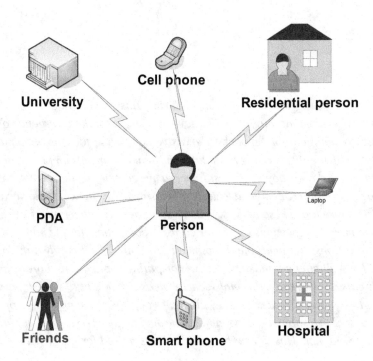

often collect information about the surrounding environment using various sensors. Figure 1 illustrates a typical ad-hoc network.

Pervasive computing has the inherent disadvantages of slow, expensive connections, frequent line disconnections, limited host bandwidth, location dependent data, and so forth. These challenges make the pervasive computing applications more vulnerable to various security-related threats. However, traditional security measures do not fit well in pervasive computing applications. Despite advancement in security, the requirement of unavoidable inter-device dependency along with common shared medium, fugacious connectivity, and the absence of fixed trust infrastructure make the pervasive systems susceptible and vulnerable to malicious active and passive snoopers. Therefore, the prospects of pervasive computing can be nipped in the bud. Moreover, the reluctance of knowledgeable users towards using this flourishing technology due to the fear of lack of privacy preservation hinder the usage of pervasive computing technologies. On the other hand, to provide the users with several unique aspects of pervasive computing like context awareness, invisibility, and reduced user interaction, it is required to feed nearly invisible pervasive devices with private information. For this reason, unlike distributed computing, pervasive computing often tries to explore user and resource location and other private information which are used in inducing intelligence in the devices. Since location and context are key attributes of pervasive computing applications, the privacy issues need to be handled in a different manner than they are handled in traditional computing. The devices in a pervasive network leave and join in an ad-hoc manner. This device behavior generates a need for new trust models for pervasive computing applications. The trust characteristic is important from the point of view of both the users and the pervasive devices. Thus,

in all aspects, secure authentication and integrity of data communication are an integral part in pervasive computing.

Researchers have been working to eliminate the problems caused by various security, privacy, and trust issues. As part of the Gaia project, Campbell, Al-Muhtadi, and Naldurg (2002), and Campbell and Qian (1998) have identified some pervasive computing and pervasive security characteristics including vanishing boundary, context awareness, scalability, mobility, creation of smart space, unobtrusiveness, and multi-layered authentication. Scalability, creation of smart space, and unobtrusiveness have also been pointed out by Satyanarayan (2001, 2005). Woo (2003) shows how to handle context awareness, customizability, and adaptability of security. Campbell et al. (2002), and Myles, Friday, and Davies (2003) describe context-aware disclosure of information and minimum information disclosure respectively where as Ackerman (2004) deals with feedback and user awareness agendas of pervasive privacy. Wolfe, Ahamed, and Zulkernine (2006) describe the characteristics of pervasive trust as low memory footprint/resource usage, distributed computation, pre-configuration avoidance, capability to customize, circumvention of malicious recommendations, prevention of attacks from variety of attack scenarios, and recalculation of trust.

Thus far, we have discussed the need for security, privacy, and trust of pervasive applications. The rest of the chapter is organized as follows. The next section describes the characteristics of pervasive computing. The three following sections deal with pervasive security, privacy, and trust respectively. Each of these sections includes definitions, an example scenario, characteristics, and recent work in the respective fields. The sixth section sheds some light on several crucial issues interrelating security, privacy, and trust, which is followed by some conclusions.

CHARACTERISTICS OF PERVASIVE COMPUTING

Pervasive computing (Weiser, 1991; Yau, Karim, Wang, Wang, & Gupta, 2002) focuses on combining computing and communications with the surrounding physical environment and making computing and communication transparent to the users in day-to-day activities. Several researchers like Yau et al. (2002), Campbell et al. (2002), Satyanarayan (2001, 2005), and Fan, Akhtar, Chew, Moessner, and Tafazolli (2004) have discussed different attributes of pervasive computing, which have been summarized in the following sub-sections.

Vanishing Boundary

Traditional computing and networking scenarios are comprised of software and hardware to perform computation in the boundaries of buildings that is, strictly within the bricks and walls of a fixed infrastructure. However, pervasive computing distributes the computation beyond any physical boundary. Thus the working environment of pervasive computing includes a number of different devices.

Volatile Environment

Devices can leave and join in the pervasive computing environment arbitrarily and independently producing an extremely volatile network topology. The stability of nodes is completely absent in this scenario. This volatile nature prohibits the usage of more standardized approaches to an array of common problems such as computer security.

Context Awareness

Pervasive computing environments should be able to collect information about the surroundings and use this information in the decision making procedure. As a result, this environment will be able to serve different users, in different situations and times perfectly by adapting itself with the dynamic environment through the use of context knowledge.

Scalability

Hundreds, even thousands, of devices can be present in a pervasive ad-hoc network. These tiny devices can come from several domains, and the interactions of these devices with different users can take place using different protocols. Pervasive computing needs to support this huge heterogeneous network.

Seamless Service

A pervasive computing environment can be used by anybody, in any situation, and at any time. It has to present itself as an "omnipresent" service to the user. The system has to ensure persistent communication between the connected devices. It has to guarantee proper connectivity until any of the users disconnects him/her manually.

Mobility

The users in a pervasive environment are characterized by mobility. They are not static and are moving constantly from one position to another. This characteristic has to be considered in designing pervasive computing applications.

Active Space

In order to embed computational power in the environment and support "context awareness" features, pervasive computing needs to incorporate several devices. These smart devices can collect information like identity, temperature, sound, and so forth, based on the requirement. This feature turns a physical space into an active space that has the monitoring and decision-making capability.

SECURITY FOR PERVASIVE APPLICATIONS

What is Security?

"Computer security is the effort to create a secure computing platform, designed so that agents (users or programs) can only perform actions that have been allowed" (Wikipedia, 2005a). The term "CIA" is commonly used to summarize the purpose of security (ACM Queue, 2005).

- **Confidentiality:** Only authorized users of the information have access to the information, that is, the information is safe from any unauthorized usage.
- **Integrity:** Ensures that any information has not been interpolated by active or passive intruders.
- **Authentication:** Ensures that the correct users of a system are using the system.

In the most primitive sense, computer security allows agents (users or programs) to perform only legitimate actions. Computer security involves numerous facets including network design, encryption, physical security, authorization, authentication, and many more.

Pervasive Security

A paramount security concern of pervasive applications and networks is ensuring network integrity and preventing malicious usage. In the pervasive computing environment, the transparent penetration of everyday life scenarios with computational devices embedded in the environment opens to vulnerabilities and intrusions. As a result, along with the provided services, the whole system has to be secure and impregnable enough to ensure the integrity of the system and information. Essentially, this matter boils down to device authentication. In a traditional wired network, or even in an infrastructure wireless network, a single centralized device or group of devices is responsible for performing this functionality. For instance, public key infrastructure (PKI) provides a method of authentication using identity certificates which are issued by a trusted certificate authority (CA) (Weise, 2001). Due to the nature of pervasive computing environments, authentication schemes involving a centralized approach, like PKI, are not practical. As handheld devices are susceptible to theft, suffer from volatile network connectivity and are limited by low computational power and battery life, some specific well-known security and authentication mechanisms become null and void (Haque & Ahamed, 2006). Again, the pervasive computing devices can come from multiple domains, forming a complex heterogeneous environment and make a secure environment harder to achieve.

An Example Scenario

Dr. Masson, a resident doctor in a hospital, has collected important information about his patient. He has been working for eight hours and is going to sign off. Meantime, Dr. Harrice reaches the hospital. The security system authenticates him and his PDA joins the ad-hoc network of the hospital. The PDA of Dr. Masson recognizes the presence of Dr. Harrice and sends the corresponding information to Dr. Harrice's PDA. As this information is crucial and sophisticated, encrypted data will be sent. The PDA of Dr. Harrice will deal with the decryption phase thus ensuring data security.

Characteristics of Pervasive Security

Several aspects of pervasive security have been addressed by researchers like Campbell et al. (2002), Woo (2003), and Sharmin, Ahmed, Ahamed, Haque, and Khan (2006). The characteristics are summarized in the following:

Unobtrusiveness

The goals of pervasive computing are to reduce the interactions of the users to the minimum level and make the background mechanism transparent to the users. Therefore, the whole security system should be blended with the background in such a way so that the user is minimally disrupted. The enhancement in security will decrease the user interaction. At the same time, security mechanism with enhanced features will consume more bandwidth and increase response delay.

Context Awareness

Context is one of the most important pieces of information that should be incorporated in the decision-making procedure of security mechanism. Context information like location, character, and time play crucial roles in this aspect. Pervasive security systems are comprised of materials like sensors that can collect this contextual information. For example, if the user is in his home or office, then it is less vulnerable than other situations like a public place.

Multi-Layered Authentication

In the pervasive computing environment, authentication varies depending on the context. As a result, one fixed authentication mechanism is not suitable in this environment. The authentication system should include several layers of security architecture. At a specific moment, a specific layer will be chosen based on the role, surrounding environment, and other context-based information.

Customizability

Based on the previous point, it is obvious that we need to shuffle and traverse back and forth among several layers of security architecture. Moreover, we will need time to alter any security

policy. Sometimes, new authentication hardware and software will be needed to blend in with the existing security system. Therefore, the security system should be dynamic and flexible for necessary customization.

Adaptability

The security system has to be adaptable to the environment. It should be able to provide the optimal service within the limitation of available resources. The security system needs to be adaptable in both "pure ad-hoc network" and "managed ad-hoc network".

Interoperability

The pervasive security system is built up of several different types of devices. Again, these smart devices can come from multiple domains forming a complex heterogeneous environment. For example, in a "managed ad-hoc network", a person may intend to communicate with the printer in his office while walking through the lobby. Here, the involvement of the person, authentication mechanism, and the command acceptance mechanism of the lobby and the printer in the office represent three different domains, and the scenario establishes the importance of inter-domain security.

Related Work on Pervasive Security

To materialize the concept of dynamic security, we need to think about three fields (Woo, 2003): (1) A location tracking system for obtaining contextual information which will be used later in run time decision making; (2) An adaptive security framework that can make necessary adjustment in the security policies with the changing dynamic environment; and (3) For implementing decision making algorithm, we need metrics to quantify security level and performance.

In the Gaia project (Campbell & Qian, 1998), an authentication system named "Gaia Authentication" is developed which makes a trade off between the level of security and user interaction. Based on the collected contextual information, this system chooses one of the available authentication processes like wearable devices, smart badges, and so forth. A special value named "net confidence value" ranging from 0 to 1 is assigned to each of the authentication procedures based on the protocol and device used in the authentication. Here, the authentication modules are decoupled into two parts: (1) Authentication mechanism module (AMM) is responsible for implementing the authentication protocol. These modules are independent of authentication devices; (2) Authentication device module (ADM) deals with the authentication device independent of the authentication principle. This system provides the opportunity of adding newly devised authentication devices or protocols in the existing security system like plug ins. CORBA features and universally interoperable core (UIC) are used in the implementation. To obtain an easily customizable security system that can adapt itself with the changing environment, an "access control matrix" is used which is comprised of a set of all subjects, objects, and permissions.

In Cherubim project (Campbell & Qian, 1998), an adaptive agent-based security architecture is developed using the distributed programming environment of Java and the distributed object oriented features of CORBA. To achieve the characteristics of dynamic security, a role-based access control (RBAC) model unifies two frameworks: (1) An authorization framework which is based on "active capability" and deals with several access control mechanisms; (2) A framework that can pick out a run-time security policy based on several pieces of contextual information. This architecture has security agents (SA) running on mobile devices, and it has a central security manager (SM) that deals with all security requests. The system also incorporates active policies (AP)

which indicate the currently available security policies and meta policies which allow necessary customization of policies based on the user and the application.

PRIVACY FOR PERVASIVE APPLICATIONS

What is Privacy?

"Privacy is the ability of an individual or group to stop information about themselves from becoming known to people other than those they choose to give the information to" (Wikipedia, 2005b). Privacy has always been in people's mind, but it is really after the increased use of computers, especially the Internet, that it has become more threatening, and people have started to take it seriously. There are many types of privacy such as territorial privacy, communication privacy, information privacy, and so forth. Information privacy is the one which is getting most attention these days. Transferring data over the Web involves many of these information privacy issues. Privacy has been ignored because many people were unaware of its importance and may be it was not that big of an issue with PCs. However, it is crucial for small handheld devices. These devices are becoming more popular because of their ability to provide unique and helpful services for making our everyday life easier.

Pervasive Privacy

Due to the usage of smart sensors and location based services, privacy has become a critical issue in pervasive computing. In order to use the services, people have to be assured that their privacy is safeguarded. Both the users and the service providers have to work together to achieve it. Privacy, being a very sensitive issue, may place the toughest obstacle in reaching the pinnacle of pervasive computing. The knowledge of privacy is required in sharing information with the right people, at right proportion, and at right time. Pervasive computing applications have to make sure that excessive user information is not disclosed, and it is not being misused or revealed to any unauthorized personnel. In the sectors that deal with sophisticated personal information like healthcare, several useful pervasive computing applications failed to draw the attention of potential users due to users' distrust about the reliability of the application with respect to preserving privacy. Therefore, we need to think about the privacy concerns of the users from the very beginning of the application development phase. Otherwise, a few violations in privacy may lead to permanent distrust of the users towards this promising pervasive technology.

An Example Scenario

Mrs. Shen has been admitted in the gynecology department of "A Medical Centre" for some complex health problems. She was under the medication of Dr. Fin. However, she wants to consult other doctors about her condition. What she would do now? In an usual scenario, Mrs. Shen would call the nurse to obtain a list of other doctors of that field along with their contact numbers. Then she would make phone calls to each of them since it is not possible to visit all of them physically. She would explain her situation to each of the doctors separately. After that, she would choose a doctor and make an appointment with him/her.

As a better alternative approach, she could open her PDA and check the list of doctors that ends with "gyn" (a prefix which indicates that a doctor is a gynecologist). Then she could broadcast a message or chat to one or all of them about her problem and instantly select a doctor and make an appointment with him/her. To maintain her privacy, the application would show just her patient number. When the selected doctor would make a request to the hospital for her file, the application would not pass any personal information

like patient's name, address, and so forth, thus ensuring the privacy of Mrs. Shen.

Characteristics of Pervasive Privacy

A few characteristics of pervasive privacy are mentioned by researchers like Campbell et al. (2002), Ackerman (2004), and Lederer, Hong, Jiang, Dey, Landay, and Mankoff (2003). We have also identified some other attributes. The characteristics with regard to pervasive privacy are listed as follows.

User Awareness

Here, almost invisible and tiny devices are incorporated for collecting user data without the cognition of individuals. Therefore, users need to have prior knowledge on the issues like who will use his/her information and the potential fields where and how the information might be used etc.

Context-Aware Disclosure of Information

Based on contextual variables like identity, location, time, user of the information, and so forth, the level of information shared or disclosed can vary. A user may intend to disclose his/her information to certain organizations only at certain time (Myles et al., 2003). For example, a professor only allows students to know his location during his office hours and does not want them to know when he is in a meeting or not in campus.

Minimum Information Disclosure

To protect privacy, it is necessary to control the amount of outgoing information. Any redundant or excess information should not be revealed because the higher the amount of disclosed information, the higher the probability of privacy-related attacks. For example, if you are using a service that tells you where you are, the service can disclose the information that you are on 20th Street, but it does not need to disclose the information that you are in front of a coffee house.

Feedback

Feedback plays a crucial role in maintaining privacy. Pervasive applications need to inform the users about how personal information is being manipulated and the level of use of this information through feedback. There should be a trade off between the amount of feedback and the obtrusiveness of the application.

Restricted Access

By any means, we need to ensure that private contents are handled through only authorized personnel. At the back end of the system, we have to incorporate rigorous authentication procedure for the users of personal information.

Anonymity

The system should rarely disclose the identity of the person involved. Information of a person or an object might be needed to be shared or manipulated by certain users but no one should know the identity of the person or the object unless it is absolutely necessary.

Related Work on Pervasive Privacy

As part of the Gaia project (Campbell et al., 2002), a privacy preserving system is materialized that can reward the user with location privacy. This privacy aware communication mechanism named Mist (Al-Muhtadi, Campbell, Kapadia, Mickunas, & Yi, 2002; Campbell & Qian, 1998) incorporates hop by hop routing protocol. A unique identifier

known as "handle" is assigned to each of the Mist routers that form the network infrastructure. Based on the handle information of each incoming packet, the Mist routers change the handle of the packet and forward it to the next intermediate router until it reaches the final destination. This protocol keeps the information from the originating and the terminating node transparent to the Mist routers. A specific pervasive device named "Portal" is used to capture contextual information. Portals can recognize the presence of humans or objects by detecting the signals from the badge ID of the people or object through sensors. However, it cannot detect who the person actually is. Each user registers with a special Mist router known as "Lighthouse" which ensures proper forwarding of packets destined and originated from the registered user. Based on the hierarchical level of the Lighthouse, it can recognize the residing area of the user but not the exact location thus preserving the required privacy.

In a workplace study at IMSS General Hospital in Ensenada, Mexico, researchers develop a context-aware mobile system for the hospital environment (Muñoz, Gonzalez, Rodriguez, & Favela, 2003; Muñoz, Rodriguez, Favela, Martinez-Garcia, & Gonzalez, 2003; Rodriguez, Favela, Gonzalez, & Muñoz, 2003) using Jabber open-source IM (instant messaging) server and corresponding XMPP (extensible messaging and presence protocol). They discuss the privacy issues of public displays, and the people involved in the day-to-day hospital scenario like doctors, supervisors, and medical interns (Favela, Rodriguez, Preciado, & Gonzalez, 2004; Tentori, Favela, & Gonzalez, 2005; Tentori, Favela, Gonzalez, & Rodríguez, 2005). The signal strength received from the handheld devices of the users is used in calculating the user location through triangulation. To ensure privacy, they coin a new term called "quality of privacy (QoP)" which is analogous to quality of service (QoS). Here, the application adapts itself through the available contextual information so that it can comply with the user-application agreed QoP value. QoP value is determined based on location, character, type of access (free, restricted, etc.), and the persistency of the user in the environment (hours, days, months, or indefinite). QoP is incorporated in the application through the use of two context-aware filters namely c-filter and s-filter. The c-filter that runs on the client application is employed by the user to define his/her preferred level of QoP. On the other hand, an s-filter running on the server is used to negotiate with the user defined QoP value when the application needs user information.

Several work are proposed for dealing with the issues of privacy in day-to-day life. The principle of minimum asymmetry is proposed by researchers like Lederer et al. (2003) and Lederer, Hong, Dey, and Landay (2004). The underlying theme of minimum asymmetry is to minimize the outgoing user information and at the same time maximize the feedback about the use of disclosed information.

TRUST IN PERVASIVE APPLICATIONS

What is Trust?

Trust is "an aspect of a relationship between two parties, by which a given situation is mutually understood, and commitments are made toward actions in favor of a desired outcome" (Wikipedia, 2005c). According to Gambetta (2000), "trust is the probability that the trusted entity A does something, which is beneficial to the trusting entity B". Trust plays an important role in decision making about whether to reveal any information to the requesting service/device. Trust is determined by the confidence level, and depending on that confidence level, a user decides whether to give out information or not and protects his/her privacy. Trust can be used for every new device

or service which requests your information. For example, you want your staff member to locate you during your office hours so that they can contact you when they want. There is no reason to checking the trust for that service. If you use a service everyday that means that you trust that device, and there is no need to check the trust level. On the other hand, when new services request your information, you need to know whether you can trust that service and can convey the requested information. This can be done when setting up the preferences. First, you check if a preference or permission is already set for that service. If the permission is not set, and if you find that there is no information in the preference, you cannot make a decision and the question of trust comes into the picture. There will be an option added for this type of situation where it will call another trust model to know the trust level. For example, a user can add the following in the preference: From 1 to 10, if the trust level is higher than 7 then trust that device and supply my information, or if a level is between 5-7 then ask the user, and if a level is lower than 5, then do not trust. Once you trust the device and provide information, it will save it so that you do not need to obtain the trust level again. With the growth of handheld devices, a paramount security concern of pervasive applications and networks is ensuring network integrity and preventing malicious use. Essentially, this matter boils down to a flexible, manageable, and configurable trust scheme for the security of pervasive computing applications.

Pervasive Trust

The dependability of one device on another one is very high in an ad-hoc environment. Due to several constraints like storage, battery power, and so forth, almost each task is performed with the help of other devices present in the ad-hoc network. The performance of an ad-hoc network highly relies on the trustworthiness and mutual cooperation of the devices, although the dubiety and incertitude are integral characteristics of such a volatile environment. Therefore, the issue of trust comes up. At present, trust—the fundamental basis of "cooperation"—in pervasive ad-hoc network is under serious threat with the nascent of counterfeiting and malefic force. Several research projects are carried out to build a universally accepted trust model that can ensure reliance of each node on the others in an ad-hoc environment. Recent approaches (Luo, Zerfos, Kong, Lu, & Zhang, 2002; Pirzada & McDonald, 2004; Yi & Kravets, 2002) for pervasive/ad-hoc environments shift the burden of authentication and trust calculation to the community or a subset of the community. In doing so, the security and trust mechanism has fundamentally changed from a pre-emptive, centralized approach to a reactionary, distributed approach.

An Example Scenario

Dr. Ross is presenting an exciting topic in a conference. Dr. Jackson, a conference attendee likes the topic and approach and wants to have the presentation slides. Therefore, Dr. Jackson sends a request to the PDA of Dr. Ross for file transfer. Now the PDA of Dr. Ross calculates the trust value of Dr. Jackson's PDA and sends the file as the requesting PDA possesses the trust value required for that file transfer. Another conference attendee Mr. Pat requests Dr. Ross for some other slides related to healthcare. However, Mr. Pat fails to have enough trust value to get the requested files. As a result, the PDA of Dr. Ross denies sending the files requested by Mr. Pat.

Characteristics of Pervasive Trust

Researchers are trying to build trust models that cope with all the desired characteristics of trust. Pirzada et al. (2004), Wolfe et al. (2006), and Almenarez, Marin, Campo, and Garcia (2004) have introduced some of the following issues with respect to pervasive trust.

Low Memory Footprint/ Resource Usage

Considering the constraints of handheld devices (PDA, smart phone, etc.) like battery life and computational power, any trust service should make efficient use of resources to prevent unnecessary overhead. We have to think of materializing a successful application with all the expected features still capitalizing on a tiny storage.

Distribution of Trust Scheme

Any approach towards a centralized trust scheme is not feasible due to the limitation of the computational capability of the device that will handle the scheme. Therefore, any trust modeling approach must be distributed in some fashion. The highly volatile nature of ad-hoc networks also implies the necessity of a distributed approach.

Avoidance of Pre-Configuration

Some trust modeling approach requires the pre-configuration of all or multiple devices which is actually a time consuming process. This overhead should be avoided as much as possible for a flexible real-world trust model.

Capability to Customize

An administrator can utilize the deployment environment to make the network more secure and functional. The trust scheme should allow for customization. Like security policies, trust schemes should be able to adopt different levels of trust policy based on the context information of the environment.

Circumvention of Malicious Recommendations

"Recommendation" is one of the contexts that plays an important role in the trust value of the node. A recommendation-based trust scheme should contain processes to identify and minimize malicious recommendations.

Prevention of Attacks from a Variety of Attack Scenarios

The trust model should withstand the attack of active and passive intruders, and at the same time it should perform reasonably well against an array of adversary models including a single device, a small number of devices, and a large number of devices.

Recalculation of Trust

The trust value of the nodes in an ad-hoc network is not fixed and the value changes depending on their behavior and other context variables. That is why the trust scheme should involve a periodic update of trust values. Ideally, the time window between updates should be minimal to quickly respond to new threats.

Related Work on Pervasive Trust

Abdul-Rahman and Hailes (1998) introduce the term "distributed trust model", which uses a quantitative scheme for the calculation of trust. Each device holds a trust value ranging from -1 (completely untrustworthy) to 4 (completely trusted) for neighboring nodes. The scheme requires trust to be transitive, and devices make recommendations on another device's behalf. Recommendations are conditional based upon the recommending device's trust level. The trust value assigned by a device is computed based upon the recommended trust values and the trust values of their recommenders.

The resurrecting duckling model forces the network to form a master/slave pairing (Stajano & Anderson, 1999). This model establishes a hierarchical approach which requires pre-configuration. Contrastingly, most distributed approaches like Pirzada et al. (2004) and Luo et al. (2002) involve the device making a judgment based upon its observations and recommendations from neighboring devices. By communicating with another device, the device places a certain level of trust in the other device. Generally, devices allow, disallow, or limit communication based upon the other device's "trust level". Depending upon the approach, trust is some combination of recommendation, the trust in the recommender, direct monitoring, and the importance of traffic, risk or merely the absence of malicious activity.

Pirzada et al. (2004) propose a model based on "effort/return" mechanism for calculating trust. It includes the utility and the importance of the communication in the calculation of trust. This model incorporates trust agent in every device that accumulates necessary data for computing trust independently. This distributed model requires constant monitoring and devaluation of the perceived trust for malicious devices. Additionally, Pirzada et al. (2004) divide the wireless ad-hoc environments into "managed ad-hoc networks" and "pure ad-hoc networks". Managed ad-hoc networks involve situations where assumptions can be made regarding the deployment environment and the presence of other devices. In these situations, an administrator can utilize this information prior to deployment. "Pure ad-hoc" environments consist of situations where no knowledge of the network environment is assumed. These networks require no pre-configuration and make the security and trust mechanism more complex.

A decentralized trust model "PTM" (Almenarez et al., 2004) based on the public key is proposed. It reduces the intervention of users using the autonomy of the devices without incorporating any central fixed infrastructure which was required in public key certification. Here, the initial trust value is formed through prior knowledge which actually builds a "believe space". Later, based on the feedback of the actions, an "evidence space" is formed. The range of trust values has been defined from 0 to 1, where 0, 0.5, and 1 represent "complete distrust", "low trust", and "complete trust", respectively.

GENERAL ISSUES OF SECURITY, PRIVACY, AND TRUST IN PERVASIVE APPLICATIONS

From these discussions, it is obvious that, though security, privacy, and trust issues have quite good solutions in distributed networking or for fixed infrastructure, those solutions are not directly applicable in pervasive computing due to its own inherent set of challenges. In pervasive computing applications, we cannot think of providing every aspect of security, privacy, and trust features due to small memory storage, short wireless range, limited battery power, and so forth. In those applications, the issue of "trade off" becomes crucial. We need to make decisions for the trade off between the pairs of security-usability and privacy-functionality and trust-energy and efficiency. This decision will certainly vary based on applications, user and system requirements, time, locations, and other variables. Here, context awareness presents itself once again with due importance. Context information needs to be used in the decision-making trade off.

If the security and hence authentication mechanism is not rigorous enough to block all hackers, then sophisticated private data will be threatened with unauthorized disclosure and modification. Moreover, we will lose control if the mechanism fails to ensure secure communication. Simultaneously, to ensure authorized use of data security, the mechanism will communicate data to and from only trusted nodes, showing the necessity of a trust model. Thus security, privacy, and trust are

forming a triangle where each node is dependent on the other two.

Wireless sensor network is showing its promise in almost every field including aerospace, civil, and structural engineering, RFID, asset tracking, and so on. One of the hot topics here is pervasive healthcare for the elderly people. Pervasive healthcare has taken steps to design, develop, and evaluate computer technologies for helping citizens to participate closely in taking care of their own health. It is particularly important for seniors as the number of people over age 65 is expected to hit 70 million by 2030 in the United States alone, almost doubling from 35 million in 2000 (Mobiquitous, 2005). This emerging field is moving forward with the goal of vaporizing the materialistic boundaries of the hospitals and bringing hospitals to the home of the elders rather than bringing them to the hospital.

At present, several experiments are being carried out for adopting well-known security, trust, and privacy preserving strategies to fit in the architecture of pervasive computing. For example, the well-known Hopper-Blum authentication protocol has shown to be useful in the authentication of RFID tags (Weis, 2005), where RFID is chosen as a representative of tiny pervasive computing devices. However, a number of issues in pervasive computing are yet to be resolved. For example, a well accepted model that receives consent of a user and provides feedback to the user while keeping the user interaction at the lowest level is yet to be developed. At the same time, many socio-technical issues in case of privacy have to be addressed.

CONCLUSIONS

Pervasive computing has advanced significantly during this decade due to the developments of portable low-cost light-weight devices and emergent short range and low-power wireless communication networks. Although pervasive computing has shown a lot of promise, there are numerous security issues that have not been addressed. These issues include the inherent disadvantages of slow expensive connections, frequent line disconnection, limited host bandwidth, location dependent data, and so forth. In this chapter, we have discussed the need for security, privacy, and trust for pervasive applications. We have also illustrated the characteristics of pervasive computing along with the different features related to pervasive security, privacy, and trust. The chapter also discusses the current state-of-the-art of this field, inter-relationship among security, privacy and trust, and some open issues.

With the emergence of pervasive computing, new terms like m(obile)-business, m-health, pervasive health, QoP (quality of privacy) have been coined. Research work are being continued to avail and adapt every facility and aspect of our day-to-day life in small handheld devices. Along with all the hopes and promises of pervasive computing, we need to address the following issues: (1) What will be the steps if someone loses his/her PDA/smart phone? (2) How to deal with all possible physical and software attacks? (3) Since cryptography-based techniques are not feasible in pervasive environment, should we go for elliptic curve cryptography to overcome low energy constraints? (4) Since in "seamless wireless connectivity", firewall-based security approaches show poor performance, how do we protect ourselves from various attacks such as "denial of service" (CyberTrust, 2005)?

ACKNOWLEDGMENT

The authors appreciate the assistance of Steven Sass for his helpful suggestions to improve this chapter.

REFERENCES

Abdul-Rahman, A., & Hailes, S. (1998). A distributed trust model. *Workshop on New Security Paradigms,* Langdale, Cumbria, UK, September 23-26 (pp. 48-60). New York: ACM Press.

Ackerman, M. S. (2004). Privacy in pervasive environments: Next generation labeling protocols. *Personal and Ubiquitous Computing, 8*(6), 430-439.

ACM Queue. (2005). Purpose of security. *ACM Queue Magazine.* Retrieved December 25, 2005, from http://www.acmqueue.org

Almenarez, F., Marin, A., Campo, C., & Garcia, C. (2004). PTM: A pervasive trust management model for dynamic open environments. *Pervasive Security, Privacy, and Trust.* Retrieved December 2005, from http://jerry.c-lab.de/ubisec/publications/PSPT04_PTM.pdf

Al-Muhtadi, J., Campbell, R., Kapadia, A., Mickunas, D., & Yi, S. (2002). Routing through the mist: Privacy preserving communication in ubiquitous computing environments. *International Conference of Distributed Computing Systems (ICDCS 2002),* Providence, Rhode Island, May 19-22 (pp. 74-83). USA: IEEE CS Press.

Campbell, R. H., Al-Muhtadi, J., Naldurg, P., Sampemane, G., & Mickunas, M. D. (2002). Towards security and privacy for pervasive computing. In *Proceedings of Theories and Systems, Next-NSF-JSPS International Sympsoium,* Tokyo, Japan, November (pp. 1-15).

Campbell, R., & Qian, T. (1998). Dynamic agent-based security architecture for mobile computers. *Second International Conference on Parallel and Distributed Computing and Networks (PDCN'98).* Canada: The International Association of Science and Technology for Development. Retrieved December 2005, from http://srg.cs.uiuc.edu/Security/cherubim/slides/hpcn/index.htm.

Cybertrust. (2005). Pervasive security. *JHU.* Retrieved December 25, 2005, from www.jhuisi. jhu.edu/institute/docs/CyberTrust_81403/pervasive-security-into.pdf

Fan, L., Akhtar, N., Chew, K., Moessner, K., & Tafazolli, R. (2004). Network composition: A step towards pervasive computing. *Fifth IEE International Conference on 3G Mobile Communication Technologies,* London, UK, October 18-20 (pp. 198 -202).

Favela, J., Rodriguez, M., Preciado A., & Gonzalez, V. (2004). Integrating context-aware public displays into a mobile hospital information system. *IEEE Transactions on Information Technology in Biomedicine, 8*(3), 279-286.

Gambetta, D. (2000). Can we trust trust? In D. Gambetta (Ed.), *Trust: Making and breaking cooperative relation* (Chapter 13, pp. 213-237). UK: Department of Sociology, University of Oxford.

Haque, M., & Ahamed, S. (2006). Security in pervasive computing: Current status and open issues. *International Journal of Network Security, 3*(3), 203-214.

Lederer, S., Hong, J. I., Dey, A. K., & Landay, J. A. (2004). Personal privacy through understanding and action: Five pitfalls for designers. *Personal and Ubiquitous Computing, 8*(6), 440-454.

Lederer, S., Hong, J. I., Jiang, X., Dey, A. K., Landay, J. A., & Mankoff, J. (2003). *Towards everyday privacy for ubiquitous computing.* Technical Report UCB-CSD-03-1283, Computer Science Division, University of California, Berkeley. Retrieved December 2005, from http://www.eecs. berkeley.edu/Pubs/TechRpts/2003/5774.html

Luo, H., Zerfos, P., Kong, J., Lu, S., & Zhang, L. (2002). Self-securing ad hoc wireless networks. *Seventh International Symposium on Computers and Communications (ISCC'02),* Italy, July 1-4 (pp. 567-574).

Mobiquitous (2005). Retrieved December 2005, from http://www.mobiquitous.org/2005/challenges.html

Muñoz, M., Gonzalez, V., Rodríguez, M., & Favela, J. (2003). Supporting context-aware collaboration in a hospital: An ethnographic informed design. *Lecture Notes in Computer Science, 2806,* 330-344.

Muñoz, M. A., Rodriguez, M., Favela, J., Martinez-Garcia, A. I., & Gonzalez, V. (2003). Context-aware mobile communication in hospitals. *Computer IEEE, 36*(9), 38-46.

Myles, G., Friday, A., & Davies, N. (2003, January-March). Preserving privacy in environments with location-based applications. *IEEE Pervasive Computing, 2*(1), 56-64.

Pirzada, A., & McDonald, C. (2004). Establishing trust in pure ad-hoc networks. *27th Conference on Australasian Computer Science, 26,* Dunedin, New Zealand (pp. 47-54). Australian Computer Society, Inc.

Rodriguez, M., Favela, J., Gonzalez, V., & Muñoz, M. A. (2003). Agent-based mobile collaboration and information access in a healthcare environment. *eHealth: Application of Computing Science in Medicine and Healthcare,* Cambridge, UK, April 19-21 (pp. 133-148).

Satyanarayanan, M. (2001). Pervasive computing: Vision and challenges. *IEEE Personal Communications, 8*(4), 10-17.

Satyanarayanan, M. (2005). Metrics and benchmarks for pervasive computing. *Pervasive Computing: IEEE, 4*(3), 4-6.

Sharmin, M., Ahmed, S., Ahamed, S. I., Haque, M., & Khan, A. (2006, April). Healthcare aide: Towards a virtual assistant for doctors using pervasive middleware. To appear in *Journal Springer e&i, 123*(4).

Stajano, F., & Anderson, R. (1999). The resurrecting duckling: Security issues for ad-hoc wireless networking. *7th International Workshop on Security Protocols* (pp. 172-194).

Tentori, M., Favela, J., & González, V. (2005). Designing for privacy in pervasive hospital environments. *Ubiquitous Computing & Ambient Intelligence.* Retrieved December 2005, from http://www.isr.uci.edu/~vmgyg/documents/ucami.pdf

Tentori, M., Favela, J., Gonzalez, V., & Rodríguez, M. (2005). Supporting quality of privacy (QoP) in pervasive computing. *Encuentro Internacional de Ciencias de la Computación (ENC),* Pueblo, Mexico, September 26-30 (pp. 58-67). IEEE Computer Society.

Weis, S. A. (2005). Security parallels between people and pervasive devices. *Third International Conference on Pervasive Computing and Communications Workshops (PerCom 2005),* Hawaii, March (pp. 105-109).

Weise, J. (2001). Public key infrastructure overview. *Sun BluePrints Online.* Retrieved December 2005, from http://www.sun.com/blueprints/0801/publickey.pdf

Weiser, M. (1991). The computer for the twenty-first century. *Scientific American, 265,* 66-75.

Wikipedia. (2005a). *Security.* Retrieved December 25, 2005, from http://en.wikipedia.org/wiki/Computer_security

Wikipedia. (2005b). *Privacy.* Retrieved December 25, 2005, from http://en.wikipedia.org/wiki/Privacy

Wikipedia. (2005c). *Trust.* Retrieved December 25, 2005, from http://en.wikipedia.org/wiki/Trust

Wolfe, S. T., Ahamed, S. I., & Zulkernine, M. (2006). A trust framework for pervasive computing environments. *4th ACS/IEEE International Conference on Computer Systems and Applications (AICCSA-06),* Dubai, UAE, March (pp. 312-319). Dubai: IEEE CS Press.

Woo, T. (2003). Dynamic security in pervasive computing. *ECE750 Presentation,* University of Waterloo. Retrieved December 2005, from http://www.vlsi.uwaterloo.ca/~thwoo/ece750/ece750-6report.pdf

Yau, S., Karim, F., Wang, Y., Wang, B., & Gupta, S. K. S. (2002). Reconfigurable context-sensitive middleware for pervasive computing. *IEEE Pervasive Computing, 1*(3), 33-40.

Yi, S., & Kravets, R. (2002). Key management for heterogeneous ad hoc wireless networks. *10th IEEE International Conference on Network Protocols,* Paris, France, November 12-15 (pp. 202-203).

Compilation of References

Abdul-Rahman, A., & Hailes, S. (1998). A distributed trust model. *Workshop on New Security Paradigms,* Langdale, Cumbria, UK, September 23-26 (pp. 48-60). New York: ACM Press.

Abolhasan, M., Wysocki, T., & Dutkiewicz, E. (2004). A review of routing protocols for mobile ad hoc networks. *Ad Hoc Networks, 2*(1), 1-22.

Ackerman, M. S. (2004). Privacy in pervasive environments: Next generation labeling protocols. *Personal and Ubiquitous Computing, 8*(6), 430-439.

ACM Queue. (2005). Purpose of security. *ACM Queue Magazine.* Retrieved December 25, 2005, from http://www.acmqueue.org

Adams, D. A., Nelson, R. R., & Todd, P. A. (1992). Perceived usefulness, ease of use, and usage of information technology: A replication. *MIS Quarterly, 16*(2), 227-247.

Agarwal, R., & Prasad, J. (1998). A conceptual and operational definition of personal innovativeness in the domain of information technology. *Information Systems Research, 9*(2), 204-215.

Aggélou, G. (2004). *Mobile ad hoc networks.* New York: McGraw-Hill Professional.

Aggélou, G. N. (2003). An integrated platform for quality-of-service support in mobile multimedia clustered ad hoc networks. In M. Ilyas (Ed.), *The handbook of ad hoc wireless networks* (pp. 443-465). Boca Raton, FL: CRC Press, Inc.

Aggelou, G., & Tafazolli, R. (1999). RDMAR: A bandwidth-efficient routing protocol for mobile ad-hoc networks. *Proceedings of the 2nd ACM International Workshop on Wireless Mobile Multimedia (WOWMOM*), Seattle, Washington, August 20 (pp. 26-33). ISBN 1-58113-129-1. New York: ACM Press.

Aggélou, G., & Tafazolli, R. (2001). QoS support in 4th generation mobile multimedia ad hoc networks. *Proceedings of the Second International Conference on 3G Mobile Communication Technologies,* London, March 26-28 (pp. 412-416). London: Institute of Electrical Engineers.

Aghito, S. M., & Forchhammer, S. (2006). Context based coding of bi-level images enhanced by digital straight line analysis. *IEEE Transactions on Image Processing, 15*(8), 2120-2130.

Ahmed, T., Mehaoua, A., & Buridant, G. (2001). Implementing MPEG-4 video on demand over IP differentiated services. *Global Telecommunications Conference, GLOBECOM,* San Antonio, TX, November 25-29 (pp. 2489-2493). Piscataway, NJ: IEEE.

Airscanner® Corp. (2005). *Airscanner® Mobile Encrypter V2.2b (BETA) user's manual.* Retrieved July 8, 2005, from http://airscanner.com/downloads/encrypter/EncrypterManual.htm

AirSnort. (2006). Retrieved December 10, 2006, from http://airsnort.shmoo.com

Aizawa, K., Harashima, H., & Saito, T. (1989). Model-based analysis synthesis image coding (MBASIC)

system for a person's face. *Signal Processing: Image Communication, 1*(2), 139-152.

Ajzen, I. (1991). The theory of planned behavior. *Organisational Behavior and Human Decision Process, 52*(2), 179-211.

Akyildiz, I. F., McNair, J., Ho, J., Uzunalioglu, H., & Wang, W. (1999). Mobility management in next-generation wireless systems. *Proceedings of the IEEE, 87*(8), 1349-1351.

Alahuhta, P., Ahola, J., & Hakala, H. (2005). *Mobilising business applications: a survey about the opportunities and challenges of mobile business applications and services in Finland* (Technology Review No. 167/2005). Helsinki: Tekes.

Alba, J. W., & Hutchinson, J. W. (1987). Dimensions of consumer expertise. *Journal of Consumer Research, 13*(3), 411-454.

Aldinger, M. (1994). Multicarrier COFDM scheme in high bitrate radio local area networks. *Proc. of Wireless Computer Networks 94*, Den Haag, Netherlands (pp. 969-973). New York: IEEE.

Al-Gahtani, S. S., & King, M. (1999). Attitudes, satisfaction and usage: Factors contributing to each in the acceptance of information technology. *Behaviour & Information Technology, 18*(4), 277-297.

Ali, M. A., Dooley, L. S., & Karmakar, G. C. (2006). *Object based segmentation using fuzzy clustering.* IEEE International Conference on Acoustics, Speech, and Signal Processing (ICASSP), Toulouse, France, May 15-19.

Almenarez, F., Marin, A., Campo, C., & Garcia, C. (2004). PTM: A pervasive trust management model for dynamic open environments. *Pervasive Security, Privacy, and Trust.* Retrieved December 2005, from http://jerry.c-lab.de/ubisec/publications/PSPT04_PTM.pdf

Al-Mualla, M. E., Canagarajah, C. N., & Bull, D. R. (2001). Simplex minimization for single- and multiple-reference motion estimation. *IEEE Transactions on Circuits and Systems for Video Technology, 11*(12), 1209-1220.

Al-Mualla, M. E., Canagarajah, C. N., & Bull, D. R. (2002). *Video coding for mobile communications: Efficiency, complexity, and resilience.* Amsterdam: Academic Press.

Al-Muhtadi, J., Campbell, R., Kapadia, A., Mickunas, D., & Yi, S. (2002). Routing through the mist: Privacy preserving communication in ubiquitous computing environments. *International Conference of Distributed Computing Systems (ICDCS 2002)*, Providence, Rhode Island, May 19-22 (pp. 74-83). USA: IEEE CS Press.

Amigo Project. (2004, December). *Ambient intelligence for the networked home environment: Summary.* Retrieved May 2, 2006, from http://www.hitech-projects.com/euprojects/amigo/

Anderson, M., Pose, R., & Wallace, C. S. (1986). A password-capability system. *The Computer Journal, 29*(1), 1-8.

Andersson, C. (2001). *GPRS and 3G wireless applications.* NY: John Wiley & Sons, Inc.

Andrews, J. (2005). Using SignLab for formative and summative assessment. *The University of Bristol Learning Technology Support Service Fifth Annual National Conference, Bristol, June 20.* Retrieved April 24, 2006, from http://www.ltss.bris.ac.uk/vleconf05/Speakers/andrews.doc

Andrews, K., & Turner-Stokes, L. (2005). *Rehabilitation in the 21ˢᵗ century: Report of three surveys.* London: Royal Hospital for Neuro-disability.

Appenzeller, G., Baker, M., Lai, K., Maniatis, P., Roussopoulos, M., Swierk, E., & Zhao, X. (1999, July). The mobile people architecture. *Mobile Computing and Communications Review, 3*(3), 36-42.

Ariyoshi, M., Shima, T., Han, J., Karlsson, J., & Urabe, K. (2002). Effect of forward-backward filtering channel estimation in W-CDMA multi-stage parallel interference cancellation receiver. *IEICE Trans. Commun., E85-B*(10), 1898-1905.

Armstrong, L. (1993). Assessing the older communication-impaired person. In J. R. Beech, & L. Harding (Eds.),

Assessment in speech and language therapy (pp. 163-166). London: Routledge.

Ashley, M. (2002). A guide to wireless network security. *Latis Networks*, 1-9.

Aso, T. (2005, May). *Message from the organizers.* Minister of Internal Affairs and Communications, Japan. Tokyo Ubiquitous Network Conference, Tokyo, Japan, May 16-17. Retrieved May 5, 2006, from http://www.wsis-japan.jp/ministeriac_e.html

Atta, R., & Ghanbari, M. (2006). Spatio-temporal scalability-based motion-compensated 3-D subband/DCT video coding. *IEEE Transactions on Circuits and Systems for Video Technology, 16*(1), 43-55.

Auckland City Council. (2005, March). *Transport: Parking–Princess Street pay and display trial.* Retrieved May 4, 2006, from http://www.aucklandcity.govt.nz/auckland/transport/parking/princesst.asp

Ayers, R., & Jansen, W. (2004). *PDA forensic tools: An overview and analysis.* Retrieved July 6, 2005, from http://csrc.nist.gov/publications/nistir/nistir-7100-PDA-Forensics.pdf

Ba, S., & Pavlou, P. A. (2002). Evidence of the effect of trust building technology in electronic markets: price premiums and buyer behavior. *MIS Quarterly, 26*(3), 243-268.

Baker, F., Lindell, B., & Talwar, M. (2000). *RSVP cryptographic authentication.* RFC 2747, IETF. Retrieved May 10, 2006, from http://www.ietf.org/rfc/rfc2747.txt

Balakrishnan, K., Deng, J., & Varhney, P. K. (2005). TWOACK: Preventing selfishness in mobile ad hoc networks. *Proceedings of the IEEE Wireless Communications and Networking Conference (WCNC), 4,* New Orleans, LA, March 13-17 (pp. 2137-2142). IEEE.

Bandyopadhyay, S. K., & Kondi, L. P. (2005). Optimal bit allocation for joint contour-based shape coding and shape adaptive texture coding. *International Conference on Image Processing (ICIP), I,* Genoa, Italy, September 11-14 (pp. 589-592).

Barnsley, M. F. (1988). *Fractals everywhere.* Boston: Academic Press.

Barthel, K. U., Voye, T., & Noll, P. (1993). *Improved fractal image coding.* Picture Coding Symposium, Lausanne, Switzerland, March 17-19.

Barun, T., & Danzeisen, M. (2001). Secure mobile IP communication. In *Proceedings of 26th Annual IEEE Conference on Local Computer Networks (LCN'01),* Tampa, FL, November 14-16 (pp. 586-593). Washington DC, USA: IEEE Computer Society.

Basagni, S., Chlamtac, I., Syrotiuk, V. R., & Woodward, B. A. (1998). A distance routing effect algorithm for mobility (DREAM). *Proceedings of the 4th International Conference on Mobile Computing and Networking (MobiCom),* Dalla, Texas, October 25-30 (pp. 76-84). ISBN 1-58113-035-X. New York: ACM Press.

Bates, R. J. (1995). *Wireless networked communications: Concepts, technology, and implementations.* NY; Singapore: McGraw-Hill, Inc.

Baugher, M., McGrew, S., Naslund, M., Carrara, E., & Norrman, K. (2004). *RTP: The secure real-time transport protocol (SRTP).* RFC 3711, IETF. Retrieved May 9, 2006, from http://www.ietf.org/rfc/rfc3711.txt

Bellenger, D. N., Bernhardt, K. L., & Goldtucker, J. L. (1989). Qualitative research techniques: Focus group interviews. In T. J. Hayes, & C. B. Tathum (Eds.), *Focus group interviews: A reader* (pp. 7-28). Chicago: American Marketing Association.

Bello, P. A. (1963). Characterization of randomly time-variant linear channels. *IEEE Transactions on Communications, 11,* 360-393.

Bellur, B., & Ogier, R. G. (1999). A reliable, efficient topology broadcast protocol for dynamic networks. *Proceedings of the 18th Annual Joint Conference of the IEEE Computer and Communications Societies (InfoCom),* New York, March 21-25 (pp. 178-186). ISBN 0-7803-5417-6, 1. New York: IEEE.

Beltran, F., & Roggendorf, M. (2004, October). *Innovative pricing and charging in next-generation wireless networks* (pp. 1-4). Auckland, New Zealand: PING Research Group.

Beltran, F., & Roggendorf, M. (2004, December). *An incentive-compatible pricing scheme for competitive access to wireless networks* (pp. 1-6). Auckland, New Zealand: PING Research Group.

Beltran, F., & Roggendorf, M. (2005). A simulation model for the dynamic allocation of network resources in a competitive wireless scenario. *Proceedings of the Second International Workshop, MATA 2005*, Montreal, Canada, October 17-19. Lecture Notes in Computer Science 3744. Auckland, New Zealand: PING Research Group.

Bertsekas, D., & Gallager, R. (1987). *Data networks* (pp. 297-333). NJ: Prentice-Hall Inc.

Bhargava S., & Agrawal, D. P. (2001). Security enhancements in AODV protocol for wireless ad-hoc networks. *Proceedings of the Vehicular Technology Conference, 4*, Atlantic City, NJ, October 7-11 (pp. 2143-2147). IEEE.

Bhogal, S. J., Teasell, R. W., & Speechley, M. R. (2003). Intensity of Aphasia therapy, impact on recovery. *Stroke*, (34), 987-993.

Bina, M., & Giaglis, G. M. (2005). *Exploring early usage patterns of mobile data services.* Paper presented at the International Conference on Mobile Business, Sydney, Australia, July 11-13.

Bingham, J. (1990, May). Multicarrier modulation for data transmission: An idea whose time has come. *IEEE Communications Magazine, 28*, 5-14.

Bisdikian, C. (2001). An overview of Bluetooth wireless technology. *IEEE Communications Magazine*, 86-90.

BitDefender. (2007). *BitDefender free edition for Windows CE—AntiVirus Freeware, BitDefender Free Edition for PALM OS—AntiVirus Freeware.* Retrieved September 27, 2007, from http://www.johannrain-softwareentwicklung.de/bitdefender_antivirus_free_edition_pour_windows_ce.htm or http://www.johannrain-softwareentwicklung.de/bitdefender_antivirus_free_edition_pour_palm_os.htm

Black, J. A., & Boal, K. B. (1994). Strategic resources: Traits, configurations, and paths to sustainable competitive advantage. *Strategic Management Journal, 15*, 131-148.

Bluefire Security Technologies. (2005). *Wireless device security products.* Retrieved July 9, 2005, from http://www.bluefiresecurity.com/products.html

Boukerche, A., Hong, S., & Jacob, T., (2003). A two-phase handoff management scheme for synchronizing multimedia units over wireless networks. *Proc. Eighth IEEE International Symposium on Computers and Communication*, Antalya, Turkey, June-July (pp. 1078-1084). Los Alamitos, CA: IEEE Computer Society.

Bozionelos, N. (1996). Psychology of computer use: Prevalence of computer anxiety in British managers and professionals. *Psychological Reports, 78*(3), 995-1002.

Braden, R., Zhang, L., Berson, S., Herzog, S., & Jamin, S. (1997). *Resource ReSerVation Protocol (RSVP).* RFC 2205, IETF. Retrieved May 10, 2006, from http://www.ietf.org/rfc/rfc2205.txt

Brady, N. (1999). MPEG-4 standardized methods for the compression of arbitrarily shaped video objects. *IEEE Transactions on Circuits and Systems for Video Technology, 9*(8), 1170-1189.

Brady, N., Bossen, F., & Murphy, N. (1997). Context-based arithmetic encoding of 2D shape sequences. *International Conference on Image Processing (ICIP), I,* Washington, DC, October 26-29 (pp. 29-32).

Brislen, P. (2005, April). *RoamAD sells metro Wi-Fi to Perth ISP: Kiwi company transitions from carrier to vendor.* Computerworld New Zealand. Retrieved May 3, 2006, from http://computerworld.co.nz/news.nsf/UNID2B8B78E695883CAFCC256FE80026A6D0?OpenDocument&Highlight=2,roamad

Broch, J., Maltz, D., Hu Y.-C., & Jecheva, J. (1998). A performance comparison of multi-hop wireless ad hoc network routing protocols. *Proceedings of the 4th ACM/IEEE International Conference on Mobile Computing and Networking (MobiCom)*, Dallas, Texas, October

25-30 (pp. 85-97). ISBN 1-58113-035-X. New York: ACM Press.

Brüninghaus, K., & Rohling, H. (1997). On the duality of multi-carrier spread spectrum and single-carrier transmission. *Zweites OFDM-Fachgespräch*, Braunschweig, Germany (pp. 210-215). Braunschweig: TU Braunschweig.

Brüninghaus, K., & Rohling, H. (1998). Multi-carrier spread spectrum and its relationship to single carrier transmission. *Proc. of the IEEE VTC'98*, Ottawa, Canada (pp. 2329-2332). New York: IEEE.

Buchegger, S., & Boudec, J.-Y. L. (2002). Nodes bearing grudges: Towards routing security, fairness and robustness in mobile ad hoc networks. *Proceedings of the 10th Euromicro Workshop on Parallel, Distributed and Network-based Processing*, Canary Island, Spain, January 9-11 (pp. 403-410). IEEE Computer Society.

Buchegger, S., & Boudec, J.-Y. L. (2002). Performance analysis of the CONFIDANT protocol. *Proceedings of the 3rd ACM International Symposium on Mobile Ad Hoc Networking and Computing (MobiHoc)*, Lausanne, Switzerland, June 9-11 (pp. 455-465). ISBN 1-58113-501-7. ACM Press.

Budagavi, M., & Gibson, J. D. (2001, February). Multiframe video coding for improved performance over wireless channels. *IEEE Transactions on Image Processing, 10*(2), 252-265.

Bull, D. R., Canagarajah, N. C., & Nix, A. (1999). *Insights into mobile multimedia communications: Signal processing and its applications*. San Diego, CA: Academic Press.

Bulthoff, H. H., Graf, A. B. A., Scholkopt, B., Simoncelli, E. P., & Wichmann, F. A. (2004). Machine learning applied to perception: Decision-images for gender classification. *Advances in Neural Information Processing Systems, 17*. Retrieved April 24, 2006, from http://www.cns.nyu.edu/pub/eero/wichmann04a.pdf

Buracchini, E. (2000). The software radio concept. *IEEE Communication Magazine, 38*, 138-143.

Burton, E., Meeks, N., & Wright, K. (1991). Opportunities for using computers in speech and language therapy: A study of one unit. *British Journal of Disorders in Communication, 26*(2), 207-217.

Calder, B. J., & Staw, B. M. (1975). Self-perception of intrinsic and extrinsic motivation. *Journal of Personality and Social Psychology, 31*(4), 599-605.

Calhoun, P., Johansson, T., Perkins, C., Hiller,T., & McCann, Ed. P. (2005). *Diameter mobile IPv4 application*. IETF, RFC 4004. Retrieved May 8, 2006, from http://www.ietf.org/rfc/rfc4004.txt

Camarillo, G., Handley, M., Johnston, A., Peterson, J., Rosenberg, J., Schooler, E., Schulzrinne, H., & Sparks, R. (2002). RFC 3261 – SIP: Session initiation protocol. *IETF Internet Standard*. Retrieved September 18, 2007, from www.ietf.org/rfc/rfc3261.txt

Campbell, R. H., Al-Muhtadi, J., Naldurg, P., Sampemane, G., & Mickunas, M. D. (2002). Towards security and privacy for pervasive computing. In *Proceedings of Theories and Systems, Next-NSF-JSPS International Sympsoium*, Tokyo, Japan, November (pp. 1-15).

Campbell, R., & Qian, T. (1998). Dynamic agent-based security architecture for mobile computers. *Second International Conference on Parallel and Distributed Computing and Networks (PDCN'98)*. Canada: The International Association of Science and Technology for Development. Retrieved December 2005, from http://srg.cs.uiuc.edu/Security/cherubim/slides/hpcn/index.htm.

Cam-Winget, N., Housley, R., Wagner, D., & Walker, W. (2003). Security flaws in 802.11 data link protocols. *Communication of the ACM, 46*(5).

Capkun, S., Buttyan, L., & Hubaux, J.-P. (2003). Self-organizing public-key management for mobile ad hoc networks. *IEEE Transactions on Mobile Computing, 2*(1), 52-64.

Card, S. K., Moran, T. P., & Newell, A. (1983). *The psychology of human-computer interaction*. Hillsdale, NJ: Lawrence Earlbaum Associates.

Carlsson, C., Hyvonen, K., Repo, P., & Walden, P. (2005). *Adoption of mobile services across different platforms.* Paper presented at the 18th Bled eCommerce Conference, Bled, Slovenia, June 6-8.

Carson, D., Gilmore, A., Gronhaug, K., & Perry, C. (2001). *Qualitative research in marketing.* London: Sage.

Casal, C. R., Burgelman, J. C., & Bohlin, E. (2004). Propects beyond 3G. *Info, 6*(6), 359-362.

Casal, C. R., Schoute, F., & Prasad, R. (1999). A novel concept for fourth generation mobile multimedia communication. *IEEE 50th Vehicular Technology Conference (VTC 1999-Fall), 1,* Amsterdam, The Netherlands, September 19-22 (pp. 381-385). Piscataway, NJ: IEEE Service Center.

Castro, M. D. (1996). *The walnut kernel: A password-capability based operating system.* Ph.D. Thesis, Monash University, Clayton, Australia.

CCITT. (1994). *Facsimile coding schemes and coding functions for group 4 facsimile apparatus.* CCITT Recommendation T.6.

Chae, M., & Kim, J. (2003). What's so different about the mobile Internet? *Communications of the ACM, 46*(12), 240-247.

Chae, M., & Kim, J. (2004). Do size and structure matter to mobile users? An empirical study of the effects of screen size, information structure, and task complexity on user activities with standard web phones. *Behaviour & Information Technology, 23*(3), 165-181.

Chang, R. W. (1966). Synthesis of band-limited orthogonal signals for multichannel data transmission. *Bell Syst. Tech. J., 45,* 1775-1796.

Chapman, N., & Chapman, J. (2000). *Digital multimedia.* London: John Wiley & Sons.

Chau, P. Y. K. (1996). An empirical assessment of a modified technology acceptance model. *Journal of Management Information Systems, 13*(2), 185-204.

Chau, P. Y. K., & Hu, P. J.-H. (2001). Information technology acceptance by individual professionals: a model comparison approach. *Decision Science, 32*(4), 699-719.

Check Point Software Technologies LTD. (2007). *Welcome Pointsec and Reflex Magnetics customers!* Retrieved September 27, 2007, from http://www.check-point.com/pointsec/

Chen, M. J., Chen, L. G., Chiueh, T. D., & Lee, Y. P. (1995). A new block-matching criterion for motion estimation and its implementation. *IEEE Transactions on Circuits and Systems for Video Technology, 5*(3), 231-236.

Chen, T.-W., & Gerla, M. (1998). Global state routing: A new routing scheme for ad-hoc wireless networks. *Proceedings of the IEEE International Conference on Communications (ICC),* Atlanta, Georgia, June 7-11 (pp. 171-175). New York: IEEE.

Cheng, A., & Shang, F. (2005). Priority-driven coding of progressive JPEG images for transmission in real-time applications. *11th IEEE International Conference on Embedded and Real-Time Computing Systems and Applications (RTCSA'05),* Hong Kong, August 17-19 (pp. 129-134). Washington, DC: IEEE Computer Society.

Chiang, C., Wu, H., Liu, W., & Gerla, M. (1997). Routing in clustered multihop, mobile wireless networks. *Proceedings of the IEEE Singapore International Conference on Networks (SICON),* Singapore, April 14-17 (pp. 197-211). New York: IEEE.

Choi, S. J., & Woods, J. W. (1999). Motion-compensated 3-D subband coding of video. *IEEE Transactions on Image Processing, 8*(2), 155-167.

Chou, P. A., Lookabaugh, T., & Gary, R. M. (1989). Entropy constrained vector quantisation. *IEEE Transactions on Acoustics, Speech, and Signal Processing, 37*(1), 31-42.

Clarke, R. J. (1995). *Digital compression of still images and video.* London: Academic Press.

Classen, F., & Meyr, H. (1994). Frequency synchronization algorithms for OFDM systems suitable for communication over frequency selective fading channels.

Proc. IEEE VTC 94, Stockholm, Sweden (pp. 1655-1659). New York: IEEE.

Columbitech AB. (2005). *Columbitech wireless VPN, technical description.* Retrieved September 26, 2007, from http://www.columbitech.com/img/2007/5/7/5300.pdf

Compeau, D. R., & Higgins, C. A. (1995). Computer self-efficacy: Development of a measure and initial test. *MIS Quarterly, 23*(2), 189-211.

Condos, C., James, A., Every, P., & Simpson, T. (2002). Ten usability principles for the development of effective WAP and m-commerce services. *Aslib Proceedings, 54*(6), 345-355.

Coppersmith, D., & Jakobsson, M. (2002). Almost optimal hash sequence traversal. *Conference on Financial Cryptography (FC)*, Southampton, Bermuda, March 11-14, Lecture Notes in Computer Science (pp. 102-119). ISBN: 978-3-540-00646-6. Springer-Verlag.

Corson, M. S., & Ephremides. (1995). A distributed routing algorithm for mobile wireless networks. *ACM/Baltzer Wireless Networks, 1*(1), 61-81.

Cover, T. M., & Thomas, J. A. (1991). *Elements of information theory.* New York: Wiley.

Crawford, A. M. (2002). International media habits on the rise. *Ad Age Global, 2*(11). Retrieved from http://web.ebscohost.com/ehost/detail?vid=3&hid=101&sid=ff86c2ae-e7f7-4388-96b4-7da9c1bc4eb3%40sessionmgr106

Cremonini, M., Damiani, E., De Capitani di Vimercati, S., Samarati, P., Corallo, A., & Elia, G. (2005). Security, privacy, and trust in mobile systems and applications. In M. Pagani (Ed.), *Mobile and wireless systems beyond 3G: Managing new business opportunities* (pp. 312-340). Hershey, PA: IRM Press.

Crochiere, R. E., Webber, S. A., & Flanagan, F. L. (1976). Digital coding of speech in sub-bands. *IEEE International Conference on Acoustics, Speech, and Signal Processing, 1,* April (pp. 233-236).

Csikszentmihalyi, M. (1975). *Beyond boredom and anxiety.* San Francisco: Jossey-Bass.

Cybertrust. (2005). Pervasive security. *JHU.* Retrieved December 25, 2005, from www.jhuisi.jhu.edu/institute/docs/CyberTrust_81403/pervasive-security-into.pdf

Dabholkar, P. A., & Bagozzi, R. P. (2002). An attitudinal model of technology-based self-service: Moderating effects of consumer traits and situational factors. *Journal of Academy of Marketing Science, 30*(3), 184-201.

DARWIN. Retrieved June 23, 2005, from http://developer.apple.com/darwin/projects/streaming/

Das, S. R., Castaneda, R., & Yan, J. (2000). Simulation-based performance evaluation of routing protocols for mobile ad hoc networks. *ACM/Baltzer Mobile Networks and Applications (MONET), 5*(3), 179-189.

Daubechies, I. (1990). The wavelet transform, time-frequency localization and signal analysis. *IEEE Transactions on Information Theory, 36*(5), 961-1005.

Davis, F. D. (1989). Perceived usefulness, perceived ease of use, and user acceptance in information technology. *MIS Quarterly, 13*(3), 319-340.

Davis, F. D., Bagozzi, R. P., & Warshaw, P. R. (1989). User acceptance of computer technology: A comparison of two theoretical models. *Management Science, 35*(8), 982-1002.

Davis, F. D., Bagozzi, R. P., & Warshaw, P. R. (1992). Extrinsic and intrinsic motivation to use computers in the workplace. *Journal of Applied Social Psychology, 22,* 1111-1132.

Deci, E. L., & Ryan, R. M. (1985). *Intrinsic motivation and self-determination in human behavior.* New York: Plenum Press.

del Galdo, E. M., & Nielsen, J. (Eds.) (1996). *International user interfaces.* London: John Wiley & Sons.

Denzin, N. K. (1989). *The research act: A theoretical introduction to sociological methods (3rd ed.).* Englewood Cliffs, N. J.: Prentice Hall.

Deo, S., Nichols, D. M., Cunningham, S. J., Witten, I. H., & Trujillo, M. F. (2004). Digital library access for illiterate users. *Proceedings of the 2004 International*

Research Conference on Innovations in Information Technology, Dubai (UAE), October (pp. 506-516). United Arab Emirates: UAE University.

Dholakia, R. R., & Dholakia, N. (2004). Mobility and markets: Emerging outlines for m-commerce. *Journal of Business Research, 57*(12), 1391-1396.

Diener, E., & Suh, E. (1997). Measuring quality of life: Economic, social and subjective indicators. *Social Indicators Research, 40*(1-2), 189-216.

Dietz, M., & Meltzer, S. (2002, July). CT-aacPlus: A state-of-the-art audio coding scheme. *EBU Technical Review*, (291), 1-7. Retrieved from http://www.ebu.ch/en/technical/trev/trev_291-dietz.pdf and http://www.ebu.ch/en/technical/trev/trev_index-digital.html

Dietze, C. (2005). The smart card in mobile communication: Enabler of next-generation (NG) services. In M. Pagani (Ed.), *Mobile and wireless systems beyond 3G: Managing new business opportunities* (pp. 221-253). Hershey, PA: IRM Press.

Dillon, A., & Morris, M. (1996). User acceptance of information technology: theories and models. *Journal of American Society for Information Science, 31*, 3-32.

Distasi, R., Nappi, M., & Riccio, D. (2006). A range/domain approximation error-based approach for fractal image compression. *IEEE Transactions on Image Processing, 15*(1), 89-97.

Dittmann, J., Wohlmacher, P., & Ackermann, R. (2001). Conditional and user specific access to services and resources using annotation watermarks. In R. Steinmetz, J. Dittman, & M. Steinebach (Eds.), *Communications and multimedia security issues of the new century* (pp. 137-142). Norwell, MA: Kluwer Academic Publishers.

Domenech, A. L. (2003). *Port-based authentication for wireless LAN access control.* Report of Graduation Degree. Department of Electrical Engineering Eindhoven University of Technology (TU/e), Eindhoven, Netherlands. Retrieved September 24, 2007, from http://people.spacelabs.nl/~alex/Port_Based_Authentication_for_Wireless_LAN_Access_Control.pdf

Douceur, J. (2002). The Sybil attack. *1st International Workshop Peer-to-Peer Systems (IPTPS)*, Cambridge, MA, March 7-8, Lecture Notes in Computer Science (pp. 251-260). ISBN 3-540-44179-4. Springer-Verlag.

Douglas, D. (2004). *Windows mobile-based devices and security: Protecting sensitive business information.* Retrieved July 1, 2005, from http://download.microsoft.com/download/4/7/c/47c9d8ec-94d4-472b-887d-4a9ccf194160/6.%20WM_Security_Final_print.pdf

Drew, Jr., W. (2003, February). Wireless networks: New meaning to ubiquitous computing. *The Journal of Academic Librarianship, 29*(2), 1-11.

Dube, R., Rais, C. D., Kuang-Yeh, W., & Tripathi, S. K. (1997). Signal stability-based adaptive routing (SSA) for ad-hoc mobile networks. *IEEE Personal Communications, 4*(1), 36-45.

Dufaux, F., & Moscheni, F. (1995). Motion estimation techniques for digital TV: A review and a new contribution. *Proceedings of the IEEE, 83*(6), 858-876.

Dufaux, F., & Nicholson, D. (2004). JPWL: JPEG 2000 for wireless applications. Photonic devices and algorithms for Computing VI. In K. M. Iftekharuddin, & A. A. S. Awwal (Eds.), *Proceedings of the SPIE, 5558*, 309-318.

Duncan, M. V., Akhtari, M. S., & Bradford, P. G. (2004). Visual security for wireless handheld devices. In *JOSHUA Journal of Science & Health at The University of Alabama, 2*/2004. The University of Alabama.

Dunlop, H., Cunningham, S. J., & Jones, M. (2002). A digital library of conversational expressions: Helping profoundly disabled users communicate. *Proceedings of the 2nd ACM/IEEE-CS Joint Conference on Digital Libraries* (JCDL), Portland (Oregon, USA), July 14-18 (pp. 273-274). New York: ACM Press.

Dutta, A., Chen, J., Madhani, S., McAuley, A., NakaJima, N., & Schulzrinne, H. (2001). Implementing a testbed for mobile multimedia. In *Proceedings of the IEEE Conference on Global Communications (GLOBECOM)*, San Antonio, Texas, November 25-29 (pp. 1944-1949). IEEE.

Dutta, A., Das, S., Li, P., McAuley, A., Ohba, Y., Baba, S., & Schulzrinne, H. (2004). *Secured mobile multimedia communication for wireless Internet.* ICNSC 2004, Taipei.

Dutta, A., Zhang, T., Madhani, S., Taniuchi, K., Fujimoto, K., Katsube, Y., Ohba, Y., & Schulzrinne, H. (2004). Secure universal mobility for wireless Internet. In *Proceedings of the 2nd ACM International Workshop on Wireless Mobile Applications and Services on WLAN Hotspots (WMASH)*, Philadelphia, PA, October 1 (pp. 71-80). New York: ACM Press.

Ebrahimi, T. (1997). *MPEG-4 video verification model version 8.0.* International Standards Organization, ISO/IEC JTC1/SC29/WG11 MPEG97/N1796.

Eden, M., & Kocher, M. (1985). On the performance of a contour coding algorithm in the context of image coding. Part i: Contour segment coding. *Signal Processing, 8,* 381-386.

Electronic Communication Privacy Act. (2006). Retrieved December 10, 2006, from http://www.usiia.org/legis/ecpa.html

Elliot, G., & Phillips, N. (2004). *Mobile commerce and wireless computing systems.* Harlow: Pearson Education Limited.

Elliot, S., & Loebbecke, C. (2000). Interactive, interorganizational innovations in electronic commerce. *Information Technology & People, 13*(1), 46-66.

ePanorama. (2006). Retrieved February 2007, from http://www.epanorma.net/links/tele_mobile.html

Ethereal. (2006). Retrieved December 10, 2006, from http://www.ethereal.com

ETHEREAL. Retrieved June 30, 2005, from www.ethereal.com

ETSI EN 302 304. (2004). *Digital video broadcasting (DVB): Transmission system for handheld terminals (DVB-H).* Retrieved September 26, 2007, from http://pda.etsi.org/pda/queryform.asp

Fan, L., Akhtar, N., Chew, K., Moessner, K., & Tafazolli, R. (2004). Network composition: A step towards pervasive computing. *Fifth IEE International Conference on 3G Mobile Communication Technologies,* London, UK, October 18-20 (pp. 198 -202).

Favela, J., Rodriguez, M., Preciado A., & Gonzalez, V. (2004). Integrating context-aware public displays into a mobile hospital information system. *IEEE Transactions on Information Technology in Biomedicine, 8*(3), 279-286.

Featherstone, I., & Zhang, N. (2003). Towards a secure videoconferencing system for mobile users. In the *5th European Personal Mobile Communications Conference,* Glasgow, Scotland, April 22-25 (pp. 477-481). London, UK: IEEE.

Ferraro, C. I. (2003). *Choosing between IPsec and SSL VPNs.* Retrieved July 10, 2005, from http://searchsecurity.techtarget.com/qna/0,289202,sid14_gci940324,00.html

Figge, S. (2004). Situation-dependent services: A challenge for mobile operators. *Journal of Business Research, 57*(12), 1416-1422.

Fishbein, M., & Ajzen, I. (1975). *Belief, attitude, intention and behaviour: An introduction to theory and research.* Reading, MA: Addison-Wesley.

Foundation for Assistive Technology (FAST). (2004, April). *Reporting on assistive technology in a rapidly changing world* (pp. 11-14). Retrieved April 24, 2006, from http://www.fastuk.org/RAPID.pdf

Frankfort-Nachmias, C., & Nachmias, D. (1996). *Research methods in the social sciences (5th ed.).* New York: St. Martin's Press.

Freeman, H. (1961). On the encoding of arbitrary geometric configurations. *IRE Trans. Electronic Computers, EC-10,* 260-268.

Fu, S., Atiquzzaman, M., Ma, L., & Lee, Y. (2005, November). Signaling cost and performance of SIGMA: A seamless handover scheme for data networks. *Journal* of *Wireless Communications and Mobile Computing,* *5*(7), 825-845.

Fu, S., Ma, L., Atiquzzaman, M., & Lee, Y. (2005). Architecture and performance of SIGMA: A seamless mobility architecture for data networks. *40th IEEE International Conference on Communications (ICC),* Seoul, Korea, May 16-20 (pp. 3249-3253). Institute of Electrical and Electronics Engineers Inc.

Funk, J. L. (2005). The future of the mobile phone Internet: An analysis of technological trajectories and lead users in the Japanese market. *Technology in Society,* *27*(1), 69-83.

Gafni, E., & Bertsekas, D. (1981). Distributed algorithms for generating loop-free routes in networks with frequently changing topology. *IEEE Transactions on Communications,* *29*(1), 11-18.

Galanxhi-Janaqi, H., & Nah, F. F.-H. (2004). U-commerce: Emerging trends and research issues. *Industrial Management & Data Systems,* *104*(9), 744-755.

Galda, D., Rohling, H., Costa, E., Haas, H., & Schulz, E. (2002). A low complexity transmitter structure for the OFDM-FDMA uplink. *Proc. IEEE VTC'02 Spring,* Birmingham, Alabama, May (pp. 1024-1028). New York: IEEE.

Gambetta, D. (2000). Can we trust trust? In D. Gambetta (Ed.), *Trust: Making and breaking cooperative relation* (Chapter 13, pp. 213-237). UK: Department of Sociology, University of Oxford.

Garcia-Luna-Aceves, J. J., & Spohn, M. (1999). Source-tree routing in wireless networks. *Proceedings of the 7th Annual International Conference on Network Protocols (ICNP),* Toronto, Canada, October 31-November 3 (pp. 273-282). ISBN 0-7695-0412-4. Los Alamitos, CA: IEEE Computer Society Press.

Garfield, L. (2004). *Infosync: Reporting from the digital.* Retrieved May 2007, from http://www.infosyncworld.com/news/n/5048.html

Gartner. (2005). *Smart phones are favoured as thin clients for mobile workers.* Retrieved May 2007, from http://www.nokia.com/NOKIA_COM_1/About_Nokia/Press/White_Papers/pdf_files/Whitepaper_TheMyth-sofMobility.pdf

Gefen, D., Karahanna, E., & Straub, D. W. (2003). Trust and TAM in online shopping: an integrated model. *MIS Quarterly, 27*(1), 51-90.

Gehrmann, C., Persson, J., & Smeet, B. (2004). *Bluetooth security.* Norwood, MA: Artech House, Inc.

Georgievski, M., & Sharda, N. (2005). Enhancing user experience for networked multimedia systems. *Proceedings of the 4th International Conference on Information Systems Technology and its Applications (ISTA2005),* Massey University, Palmerston North, New Zealand, May 23-25 (pp. 73-84). Bonn: Lecture Notes in Informatics (LNI), Gesellschaft für Informatik (GI).

Georgievski, M., & Sharda, N. (2005). Implementation and usability of user interfaces for quality of service management. *Tencon'05: Proceedings of the Annual technical Conference of IEEE Region 10,* Australia, November 21-24. New Jersey: IEEE.

Gerla, M., Hong, X., & Guangyu, P. (2000). Landmark routing for large ad hoc wireless networks. *Proceedings of the IEEE Global Telecommunications Conference (GLOBECOM), 3,* San Francisco, November 27-December 1 (pp. 1702-1706). New York: IEEE.

Gerla, M., Lee Y.-Z., Park, J.-S., & Yi, Y. (2005). On demand multicast routing with unidirectional links. *IEEE Wireless Communications and Networking Conference, 4,* 2162-2167.

Gerstheimer, O., & Lupp, C. (2004). Needs versus technology: The challenge to design third-generation mobile applications. *Journal of Business Research, 57*(12), 1409-1415.

Ghanbari, M. (1989). Two-layer coding of video signals for VBR networks. *IEEE Journal on Selected Areas in Communications, 7*(5), 771-781.

Ghanbari, M. (1991). Subband coding algorithms for video applications: Videophone to HDTV-conferencing.

IEEE Transactions on Circuits and Systems for Video Technology, 1(2), 174-183.

Ghanbari, M. (1999). Video coding: An introduction to standard codecs. *IEE Telecommunications Series, 42.*

Ghanbari, M., & Seferides, V. (1993). Cell-loss concealment in ATM video codecs. *IEEE Transactions on Circuits and Systems for Video Technology, 3*(3), 238-247.

Girod, B., & Farber, N. (1999). Feedback-based error control for mobile video transmission. *Proceedings of the IEEE: Special Issue on Video for Mobile Multi-media, 97*(10), 1707-1723.

Girod, B., Aaron, A., Rane, S., & Rebollo-Monendero, D. (2005). Distributed video coding. *Proceedings of the IEEE, 93*(1), 71-83.

Glass, S., Hiller, T., Jacobs, S., & Perkins, C. (2000). *Mobile IP authentication, authorization, and accounting requirements.* IETF, RFC 2977. Retrieved May 8, 2006, from http://www.ietf.org/rfc/rfc2977.txt

Glykas, M., & Chytas, P. (2004). Technology assisted speech and language therapy. *International Journal of Medical Informatics, 73,* 529-541.

Goff, T., Moronski, J., Phatak, D. S., & Gupta, V. (2000). Freeze-TCP: A true end-to-end TCP enhancement mechanism for mobile environments. *IEEE INFOCOM,* Tel Aviv, Israel, March 26-30 (pp. 1537-1545). NY: IEEE.

Goldberger, J., & Greenspan, H. (2006). Context-based segmentation of image sequences. *IEEE Transactions on Pattern Analysis and Machine Intelligence, 28*(3), 463-468.

Gonzalez, L. V., Ge, Y., & Lamont, L. (2005). HOLSR: A hierarchical proactive routing mechanism for mobile ad hoc networks. *IEEE Communication Magazine, 43*(7), 118-125.

Gonzalez-Castano, F. J., & Garcia-Reinoso, J. (2002). Bluetooth location networks. *IEEE Global Telecommunications Conference (GlobeComm) 2002,* Taipei, Taiwan, November 17-21 (pp. 233-237). New York: IEEE.

Goodman, J., Dickinson, A., & Syme, A. (2004). Gathering requirements for mobile devices using focus groups with older people. *Designing a More Inclusive World, Proceedings of the 2nd Cambridge Workshop on Universal Access and Assistive Technology (CWUAAT),* Cambridge, UK, March. Retrieved April 24, 2006, from http://www.computing.dundee.ac.uk/projects/UTOPIA/publications/navigation_workshop.pdf

Greenspan, H., Goldberger, J., & Mayer, A. (2004). Probabilistic space-time video modeling via piecewise GMM. *IEEE Transactions on Pattern Analysis and Machine Intelligence, 26*(3), 384-396.

Grifin, P. (2002). Security flaws shut down Telecom's mobile email. *New Zealand Herald,* April 28.

Gross, J., Karl, H., Fitzek, F., & Wolisz, A. (2003). Comparison of heuristic and optimal subcarrier assignment algorithms. *Proc. of Intl. Conf. on Wireless Networks (ICWN),* Las Vegas, Nevada (pp. 249-255). Las Vegas: CSREA Press.

Grundström, C., & Wilkinson, I. F. (2004). The role of personal networks in the development of industry standards: A case study of 3G mobile telephony. *Journal of Business and Industrial Marketing, 19*(4), 283-293.

GSM Security Portal. (2005). Retrieved July 11, 2005, from http://www.gsm-security.net/

Guangyu, P., Geria, M., & Hong, X. (2000). LANMAR: Landmark routing for large scale wireless ad hoc networks with group mobility. *Proceedings of the 1st Annual Workshop on Mobile and Ad Hoc Networking and Computing (MobiHoc),* Boston, MA, August 11 (pp. 11-18). Piscataway, NJ: IEEE Press.

Guidance Software, Inc. (2005). *EnCase Forensic Version 5.* Retrieved July 8, 2005, from http://www.guidancesoftware.com/products/ef_index.asp

Gunes, M., Sorges, U., & Bouazizi, I. (2002). ARA: The ant-colony based routing algorithm for MANETs. *Proceedings of the International Conference on Parallel Processing (ICPP) Workshops,* Vancouver, Canada, August 18-21 (pp. 79-85). Washington, DC: IEEE Computer Society.

Gupte, S., & Singhal, M. (2003). Secure routing in mobile wireless ad hoc networks. *Ad Hoc Networks, 1*(1), 151-174.

H'otter, M. (1990). Object-oriented analysis-synthesis coding based on moving two-dimensional objects. *Signal Processing, 2,* 409-428.

Haas, Z. J. (1997). A new routing protocol for the reconfigurable wireless networks. *Proceedings of the 6th IEEE International Conference on Universal Personal Communications (ICUPC),* 2, San Diego, October 12-16 (pp. 562-566). IEEE Computer Society.

Haas, Z. J., & Pearlman, M. R. (1998). The performance of a new routing protocol for the reconfigurable wireless networks. *Proceedings of the IEEE International Conference on Communications (ICC),* 1, Atlanta, Georgia, June 7-11 (pp. 156-160). New York: IEEE.

Hackett, S. (2001). *Aglie communication: m.Net Australia consortium wins federal funding for 3G/WLL test bed and applications development environment.* Retrieved May 2007, from http://www.agile.com.au/press/press-29-05-2001.htm

Halsall, F. (2001). *Multimedia communications: Applications, networks, protocols and standards.* Harlow, England; NY: Addision-Wesley.

Hammond, K. (2001). B2C e-commerce 2000-2010: What experts predict. *Business Strategy Review, 12*(1), 43-50.

Hanzo, L. et al. (2003). *OFDM and MC-CDMA for broadband multi-user communications, WLANs and broadcasting.* New York: Wiley.

Hanzo, L., & Streit, J. (1995, August). Adaptive low-rate wireless videophone schemes. *IEEE Trans. Circuits Syst. Video Technol., 5*(4), 305-318.

Haque, M., & Ahamed, S. (2006). Security in pervasive computing: Current status and open issues. *International Journal of Network Security, 3*(3), 203-214.

Harris, C. (2004). Progressing from paper towards technology. *Communication Matters, 18*(2), 33-37.

Hart, J., & Hannan, M. (2004). The future of mobile technology and mobile wireless computing. *Campus-Wide Information Systems, 21*(5), 201-204.

Haskell, B. G., Mounts, F. W., & Candy, J. C. (1972). Interframe coding of videotelephone pictures. *Proceedings of the IEEE, 60*(7), 792-800.

Havinga, P. J. M. (2000). *Mobile multimedia systems. Ph.D. thesis, University of Twente, Netherlands, ISBN 90-365-1406-1.*

Helsinki University of Technology. (2006). *InfraHIP project portal.* Retrieved May 10, 2006, from http://infrahip.hiit.fi

Higaki, H., & Umeshima, S. (2004). Multiple-route ad-hoc on-demand distance vector (MRAODV) routing protocol. *Proceedings of the 18th International Parallel and Distributed Processing Symposium (IPDPS),* Santa Fe, New Mexico, April 26-30. New York: IEEE.

Hong, X., Xu, K., & Gerla, M. (2002). Scalable routing protocols for mobile ad hoc networks. *Kluwer Wireless Networks, 16*(4), 11-21.

Horton, R. P., Buck, T., Waterson, P. E., & Clegg, C. W. (2001). Explaining intranet use with the technology acceptance model. *Journal of Information Technology, 16,* 237-249.

Hou, T.-C., & Li, V. (1986). Transmission range control in multihop packet radio networks. *IEEE Transactions on Communications, 34*(1), 38-44.

Housley, R., & Arbaugh, W. (2003). Security problems in 802.11-based networks. *Communications of the ACM, 46*(5).

Hu, Y.-C., & Johnson, D. (2000). Caching strategies in on-demand routing protocols for wireless ad hoc networks. *Proceedings of the 6th International Conference on Mobile Computing and Networking (MobiCom),* Boston, MA, August 6-11 (pp. 231-242). ISBN 1-58113-197-6. New York: ACM Press.

Hu, Y.-C., & Johnson, D. (2001). Implicit source routes for on-demand ad hoc network routing. *Proceedings of the 2nd ACM International Symposium on Mobile Ad Hoc*

Networking and Computing (MobiHoc), Long Beach, CA, October 4-5 (pp. 1-10). ISBN 1-58113-428-2. New York: ACM Press.

Hu, Y.-C., Johnson, D. B., & Perrig, A. (2002). SEAD: Secure efficient distance vector routing for mobile wireless ad hoc networks. *Proceedings of the 4th IEEE Workshop on Mobile Computing Systems and Applications (WMCSA)*, Callicoon, NY, June 20-21 (pp. 3-13). ISBN 0-7695-1647-5. IEEE.

Hu, Y.-C., Perrig, A., & Johnson, D. B. (2002). Ariadne: A secure on-demand routing protocol for ad-hoc networks. *Proceedings of the 8th Annual International Conference on Mobile Computing and Networking (MobiCom)*, Atlanta, Georgia, September 23-28 (pp. 12-23). ISBN 1-58113-486-X. ACM Press.

Hu, Y.-C., Perrig, A., & Johnson, D. B. (2003). Packet leashes: A defense against wormhole attacks in wireless ad hoc networks. *Proceedings of the 22nd IEEE Computer and Communications Societies (InfoCom)*, San Francisco, CA, April 1-3 (pp. 1976-86). IEEE.

Hui, S. Y., & Yeung, K. H. (2003). Challenges in the migration to 4G mobile systems. *IEEE Communication Magazine, 41*(12), 54-59.

Hung, S.-Y., Ku, C.-Y., & Chang, C.-M. (2003). Critical factors of WAP services adoption: An empirical study. *Electronic Commerce Research and Applications, 2*(1), 42-60.

HUT. Retrieved June 1, 2005, from http://www.cs.hut.fi/research/dynamics/

Hwu, J.-S., Chen, R.-J., & Lin, Y.-B. (2006). An efficient identity-based cryptosystem for end-to-end mobile security. *IEEE Transaction on Wireless Communications, 5*(9), 2586-2593.

IFC. (2005). *IFC: Universal mobile telecommunications system (UMTS) proto.* Retrieved May 2007, from http://www.iec.org/online/tutorials/umts/topic01.html

Igbaria, M., Parasuraman, S., & Baroudi, J. J. (1996). A motivational model of microcomputer usage. *Journal of Management Information Systems, 13*(1), 127-143.

Ilgun, K., Kemmerer, R. A., & Porras, P. A. (1995). State transition analysis: A rule-based intrusion detection approach. *IEEE Transactions on Software Engineering, 21*(3), 181-199.

Illgner, R., & Lappe, D. (1995). Mobile multimedia communications in a universal telecommunications network. *Proc. SPIE Conf. Visual Communication Image Processing*, Taipei, Taiwan, May 23-26 (pp. 1034-1043). USA: SPIE.

Intel. (2004). *Intel wireless trusted platform: Security for mobile devices.* White Paper. Retrieved September 26, 2007, from http://whitepapers.zdnet.co.uk/0,1000000651,260091578p,00.htm

Intoto Inc. (2005). *iGateway SSL-VPN.* Retrieved July 9, 2005, from http://www.intoto.com/product_briefs/iGateway%20SSL%20VPN.pdf

Ishii, K. (2004). Internet use via mobile phone in Japan. *Telecommunications Policy, 28*(1), 43-58.

Islam, M. M., Pose, R., & Kopp, C. (2004). A link layer security protocol for suburban ad-hoc networks. *Australian Telecommunication Networks and Applications Conference (ATNAC)*, Sydney, Australia, December 8-10 (pp. 174-177). Australia.

Islam, M. M., Pose, R., & Kopp, C. (2005). Challenges and a solution to support QoS for real-time traffic in multi-hop ad-hoc networks. *Proceedings of the 2nd IEEE and IFIP International Conference on Wireless and Optical Communications Networks (WOCN)*, Dubai, UAE, March 6-8. ISBN 0-7803-9019-9. New York: IEEE.

Islam, M. M., Pose, R., & Kopp, C. (2005). Link layer security for SAHN protocols. *3rd IEEE PerCom Workshops on PWN*, Kauai Island, Hawaii, March 8-12 (pp. 279-283). IEEE Computer Society.

Islam, M. M., Pose, R., & Kopp, C. (2005). An intrusion detection system for suburban ad-hoc networks. *IEEE Tencon*, Melbourne, Australia, November 12-14. ISBN: 0-7803-9312-0. IEEE Press.

Islam, M. M., Pose, R., & Kopp, C. (2005). MAC layer support for real-time traffic in a SAHN. *International*

Conference on Information Technology: Coding and Computing (ITCC), Las Vegas, NV, April 4-6 (pp. 639-645). Washington, DC: IEEE Computer Society.

Islam, M. M., Pose, R., & Kopp, C. (2005). Making SAHN-MAC independent of single frequency channel and omnidirectional antennas. *IASTED International Conference on Networks and Communication Systems (NCS 2005)*, Krabi, Thailand, April 18-20, 6pp. ISBN: 0-88986-490-X. ACTA Press.

Islam, M. M., Pose, R., & Kopp, C. (2005). Suburban ad-hoc networks in information warfare. *Proceedings 6th Australian InfoWar Conference*, Geelong, Australia, November 25-26 (pp. 71-80). ISBN: 1-74156-028-4. Australia: Deakin University.

Islam, R. (2005). *Enhanced security in mobile IP communication*. MSc Thesis, Department of Computer and Systems Sciences, Royal Institute of Technology, Stockholm, Sweden.

ISO. (1992). *Coded representation of picture and audio information—Progressive bi-level image compression*. ISO Draft International Standard 11544.

ISO/IEC 14496-10 & ITU-T Rec. (2003). *H.264, Advanced video coding*.

ISO/IEC 14496-2. (2001). *Coding of audio-visual objects – Part 2: Visual*. Annex F.

ITU [International Telecommunication Union]. (2005). *New initiatives programmes, Feb 2005*. Retrieved May 3, 2006, from http://www.itu.int/osg/spu/ni/

ITU [International Telecommunication Union]. (2005, April). *ITU workshop on ubiquitous network societies: The case of the Republic of Korea* (pp. 10-45). Retrieved May 10, 2006, from http://www.itu.int/osg/spu/ni

ITU [International Telecommunication Union]. (2005, April). *ITU workshop on ubiquitous network societies: The case of the Japan* (pp. 13-45). Retrieved May 10, 2006, from http://www.itu.int/osg/spu/ni

ITU [International Telecommunication Union]. (2005, April). *ITU workshop on ubiquitous network societies:*

The case of the Italian Republic (pp. 25-50). Retrieved May 10, 2006, from http://www.itu.int/osg/spu/ni

ITU [International Telecommunication Union]. (2005, April). *ITU workshop on ubiquitous network societies: The case of the Republic of Singapore* (pp. 21-39). Retrieved May 10, 2006, from http://www.itu.int/osg/spu/ni

ITU-T Recommendation H.261. (1993). *Video CODEC for audiovisual services at px64 kbit/s.*

ITU-T Recommendation H.263. (1998). *Video coding for low bit rate communication, Version 2.*

Jacobs, E. W., Fisher, Y., & Boss, R. D. (1992). Image compression: A study of the iterated transform method. *Signal Processing, 29*(3), 251-263.

Jacquet, P., Muhlethaler, P., Clausen, T., Laouiti, A., Qayyum, A., & Viennot, L. (2001). Optimized link state routing protocol for ad hoc networks. *Proceedings of the IEEE National Multi-Topic Conference (INMIC)*, Lahore, Pakistan, December 28-30 (pp. 62-68). New York: IEEE.

Jacquin, A. E. (1992). Image coding based on a fractal theory of iterated contractive image transformations. *IEEE Transactions on Image Processing, 1*(1), 18-30.

Jain, A. K. (1989). *Fundamentals of digital image processing*. Englewood Cliffs, NJ: Prentice-Hall.

Jain, J., & Jain, A. (1981). Displacement measurement and its application in interframe image coding. *IEEE Transactions on Communication, COMM-29*(12), 1799-1808.

Jansen, W. A. (2003). Authenticating users on handheld devices. In the *15th Annual Canadian Information Technology Security Symposium (CITSS)*, Ottawa, Canada, May 12-15. Ottawa, Canada: Communications Security Establishment. Retrieved September 24, 2007, from http://csrc.nist.gov/mobilesecurity/publications.html#MD

Jarvenpaa, S. L., Lang, K. R., Takeda, Y., & Tuunainen, V. K. (2003). Mobile commerce at crossroads. *Communications of the ACM, 46*(12), 41-44.

Jiang, J. J., Hsu, M. K., Klein, G., & Lin, B. (2000). E-commerce user behaviour model: An empirical study. *Human Systems Management, 19*(4), 265-276.

Jiang, M., Li, J., & Tay, Y. C. (1999). *Cluster based routing protocol*. Draft-ietf-manet-cbrp-spec-01.txt, Internet Draft, August.

Joa-Ng, M., & Lu, I.-T. (1999). A peer-to-peer zone-based two-level link state routing for mobile ad-hoc networks. *IEEE Journal on Selected Areas in Communications, 17*(8), 1415-1425.

Johnson, D., & Maltz, D. (1996). Dynamic source routing in ad hoc wireless networks. In T. Imielinski, & H. Korth (Eds.), *Mobile computing* (vol. 353, pp. 151-181). USA: Kluwer Academic Publishers.

Johnson, D., & Maltz, D. (1996). Dynamic source routing in ad hoc wireless networks. In T. Imielinski, & H. Korth (Eds.), *Mobile computing* (vol. 353, pp. 151-181). USA: Kluwer Academic Publishers.

Johnston, J. D. (1980). A filter family designed for use in quadratic mirror filter banks. *IEEE International Conference on Acoustics, Speech, and Signal Processing (ICASSP)* (pp. 291-294).

Jung, K. U. (2004, December). *Community, S&A research group, research group notice board: Ubiquitous campus in Korea*. Ubiquitous IT Korea Forum. Retrieved May 12, 2006, from http://www.ukoreaforum.or.kr/bbs/view.php?code=c_rg01&page=1&number=2&keyfield=&key=

Kaiser, S. (1998). *Multi-carrier CDMA mobile radio systems: Analysis and optimization of detection, decoding and channel estimation*. Fortschritt-Berichte VDI, Reihe 10, Nr. 531, VDI-Verlag, Düsseldorf, Germany.

Kaiser, S. (2002). OFDM code-division multiplexing in fading channels. *IEEE Trans. on Communications, 50*, 1266-1273.

Kamaci, N., Altunbasak, Y., & Mersereau, R. M. (2005). Frame bit allocation for the H.264/AVC video coder via Cauchy-density-based rate and distortion models. *IEEE Transactions on Circuits and Systems for Video Technology, 15*(8), 994-1006.

Kampmann, M. (2002). Automatic 3-D face model adaptation for model-based coding of videophone sequences. *IEEE Transactions on Circuits and Systems for Video Technology, 12*(3), 172-182.

Karlsson, G., & Vetterli, M. (1988). Three-dimensional subband coding of video. *IEEE International Conference on Acoustics, Speech, and Signal Processing (ICASSP)*, New York, April (pp. 1100-1103).

Karmakar, G. C. (2002). *An integrated fuzzy rule-based image segmentation framework*. PhD Thesis. Gippsland School of Computing and Information Technology. Monash University: Australia.

Karygiannis, T., & Owens, L. (2002). *Wireless Network Security*. National Institute of Standard and technology, US Department of Commerce, Special Publication 800-48.

Kasera, K. K., & Ramanathan, R. (1997). A location management protocol for hierarchically organized multihop mobile wireless networks. *Proceedings of the IEEE 6th International Conference on Universal Personal Communications (ICUPC)*, 1, San Diego, CA, October 12-16 (pp. 158-162). IEEE Communications Society.

Kaspersky Lab Products & Services. (2007). *Kaspersky anti-virus mobile*. Retrieved September 24, 2007, from http://www.kaspersky.com/antivirus_mobile

Kaspersky Lab. (2005). *Worm.SymbOS.Cabir.a*. Retrieved July 12, 2005, from http://www.viruslist.com/en/viruslist.html?id=1689517

Katsaggelos, A. K., Kondi, L. P., Meier, F. W., Ostermann, J., & Schuster, G. M. (1998). MPEG-4 and rate-distortion-based shape-coding techniques. *Proceedings of the IEEE, 86*(6), 1126-1154.

Katto, J., Ohki, J., Nogaki, S., & Ohta, M. (1994). A wavelet codec with overlapped motion compensation for very low bit-rate environment. *IEEE Transactions on Circuits and Systems for Video Technology, 4*(**3**), 328-338.

Katz, M. L., & Shapiro, C. (1986). Technology adoption in the presence of network externalities. *Journal of Political Economy, 94*(4), 822-841.

Kaufaris, M. (2002). Applying the technology acceptance model and flow theory to online consumer behaviour. *Information Systems Research, 13*(2), 205-223.

Kaup, A. (1998). Object-based texture coding of moving video in MPEG-4. *IEEE Transactions on Circuits and Systems for Video Technology, 9*(1), 5-15.

Keil, M., Beranek, P. M., & Konsynski, B. R. (1995). Usefulness and ease of use: Field study evidence regarding task considerations. *Decision Support Systems, 13*(1), 75-91.

Khalifa, M., & Cheng, S. K. N. (2002). *Adoption of mobile commerce: Role of exposure.* Paper presented at the 35th Hawaii International Conference on System Sciences, Hilton Waikoloa Village, Hawaii, January 7-10 (pp. 46-52). IEEE Computer Society.

Khansari, M., Jalai, A., Dubois, E., & Mermelstein, P. (1996, February). Low bit-rate video transmission over fading channels for wireless microcellular system. *IEEE Trans. Circuits Syst. Video Technol., 6*(1), 1-11.

Kharif, O. (2005). May I see your voice, please? In *BusinessWeek online*, April 20. Retrieved July 11, from http://www.businessweek.com/technology/content/apr2005/tc20050420_1036_tc024.htm?campaign_id=rss_techn

Kim, C., & Hwang, J.-N. (2002). Fast and automatic video object segmentation and tracking for content-based applications. *IEEE Transactions on Circuits and Systems for Video Technology, 12*(2), 122-129.

Kindberg, T., & Fox, A. (2002). System software for ubiquitous computing. *IEEE Pervasive Computing*, January, 70-81.

King, J. L., Gurbaxani, V., Kraemer, K. L., McFarlan, F. W., Raman, K. S., & Yap, C. S. (1994). Institutional factors in information technology innovation. *Information Systems Research, 5*(2), 139-169.

Kirton, M. (1976). Adopters and innovators: a description and measure. *Journal of Applied Psychology, 61*(5), 622-629.

Kismet. (2006). Retrieved December 10, 2006, from http://kismetwireless.net

Kitamura, M. (2002, September 1). *NRI papers: Using ubiquitous networks to create new services based on the commercial and public infrastructure* (pp. 1-14). NRI Papers No. 54. Japan: Nomura Research Institute.

Kivinen, T., & Tschofenig, H. (2006). *Design of the IKEv2 Mobility and Multihoming (MOBIKE) Protocol, draft-ietf-mobike-design-08.txt.* RFC 4621, IETF. Retrieved March 3, 2006, from http://www.ietf.org/internet-drafts/draft-ietf-mobike-design-08.txt

Klasen, L. (2002). Migrating an online service to WAP: Case study. *The Electronic Library, 20*(3), 195-201.

Kleijen, M., Wetzels, M., & de Ruyter, K. (2004). Consumer acceptance of wireless finance. *Journal of Financial Services Marketing, 8*(3), 206-217.

Kleinrock, L., & Stevens, K. (1971). *Fisheye: A lens like computer display transformation.* Technical Report, UCLA Computer Science Department.

Knutsen, L., Constantiou, I. D., & Damsgaard, J. (2005). *Acceptance and perceptions of advanced mobile services: Alterations during a field study.* Paper presented at the International Conference on Mobile Business, Sydney, Australia, July 11-13.

Ko, Y. B., & Vaidya, N. H. (1998). Location-aided routing (LAR) in mobile ad hoc networks. *Proceedings of the 4th Annual ACM/IEEE International Conference on Mobile Computing and Networking (MobiCom)*, Dallas, Texas, October 25-30 (pp. 66-75). ISBN 1-58113-035-X. New York: ACM Press.

Kondi, L. P. (2005). Transactions letters. A rate-distortion optimal hybrid scalable/multiple description video codec. *IEEE Transactions on Circuits and Systems for Video Technology, 15*(7), 921-927.

Kondi, L. P., Melnikov, G., & Katsaggelos, A. K. (2001). Jointly optimal coding of texture and shape. *International Conference on Image Processing (ICIP), 3,* Thessaloniki, Greece, October 7-10 (pp. 94-97).

Kondi, L. P., Melnikov, G., & Katsaggelos, A. K. (2004). Joint optimal object shape estimation and encoding. *IEEE Transactions on Circuits and Systems for Video Technology, 14*(4), 528-533.

Kondi, L.P., Meier, F. W., Schuster, G. M., & Katsaggelos, A. K. (1998) Joint optimal object shape estimation and encoding. *SPIE Visual Communication and Image Processing*, San Jose, California, USA, January (pp. 14-25).

Kopp, C. (1997). *An I/O and stream inter-process communications library for a password capability system.* Masters Thesis, Monash University, Clayton, Australia.

Kopp, C., & Pose, R. (1998). Bypassing the home computing bottleneck: The suburban area network. *3rd Australasian Computer Architecture Conference (ACAC), 20*(4), Perth, Australia (pp. 87-100). ISBN: 981-3083-93-X. Springer-Verlag.

Kopp, C., & Pose, R. (1998). Bypassing the home computing bottleneck: The suburban area network. *3rd Australasian Computer Architecture Conference (ACAC), 20*(4), 87-100, Perth, Australia, February 2-3. ISBN: 981-3083-93-X. Springer-Verlag.

Kunt, M., Ikonomopoloulos, A., & Kocher, R. (1985). Second generation image coding techniques. *Proceedings of the IEEE, 73*(4), 549-574.

Kwon, T. T., Gerla, M., Das, S., & Das, S. (2002). Mobility management for VoIP service: Mobile IP vs. SIP. *IEEE Wireless Communications, 9,* 66-75.

Langa, F. (2002, August). Langa letter: A real-life GPS road test. *InformationWeek.* Retrieved May 11, 2006, from http://www.informationweek.com/story/showArticle.jhtml?articleID=6502601&pgno=1

Le Maistre, R. (2002). *US to top 3G chart in 2010 European editor, Unstrung.* Retrieved May 2007, from http://www.unstrung.com/document.asp?doc_id=24887

Lederer, S., Hong, J. I., Dey, A. K., & Landay, J. A. (2004). Personal privacy through understanding and action: Five pitfalls for designers. *Personal and Ubiquitous Computing, 8*(6), 440-454.

Lederer, S., Hong, J. I., Jiang, X., Dey, A. K., Landay, J. A., & Mankoff, J. (2003). *Towards everyday privacy for ubiquitous computing.* Technical Report UCB-CSD-03-1283, Computer Science Division, University of California, Berkeley. Retrieved December 2005, from http://www.eecs.berkeley.edu/Pubs/TechRpts/2003/5774.html

Lee, B.-G., Choi, D.-H., Kim, H.-G., Sohn, S.-W., & Park, K.-H. (2003). Mobile IP and WLAN with AAA authentication protocol using identity-based cryptography. *10th International Conference on Telecommunications ICT 2003, 1,* Colmar, France, February 23-March 1 (pp. 597-603). USA: IEEE.

Lee, C. H., Lee, D., & Kim, J. W. (2004). Seamless MPEG-4 video streaming over Mobile-IP enabled wireless LAN. *Proceedings of SPIE, Multimedia Systems and Applications*, Philadelphia, Pennsylvania, October (pp. 111-119). USA: SPIE.

Lee, M. S. Y., McGoldrick, P. J., Keeling, K. A., & Doherty, J. (2003). Using ZMET to explore barriers to the adoption of 3G mobile banking services. *International Journal of Retail & Distribution Management, 31*(6), 340-348.

Lee, S., Cho, D., Cho, Y., Son, S., Jang, E., Shin, J., & Seo, Y. (1999). Binary shape coding using baseline-based method. *IEEE Transactions on Circuits and Systems for Video Technology, 9*(1), 44-58.

Lemay, R. (2006). *Perth, Adelaide get optus, Vodafone 3g.* Retrieved May 2007, from http://www.zdnet.com.au/news/communications/soa/perth-adelaide-get-optus-vodafone-3g/0,130061791,139261668,00.htm

Leonard-Barton, D., & Deschamps, I. (1988). Managerial influence in the implementation of new technology. *Management Science, 34*(10), 1252-1265.

Lewis, J., Tucker, W., & Blake, E. (2003). SoftBridge: A multimodal instant messaging bridging system. *Proceedings of the Southern African Telecommunication Networks and Applications Conference (SATNAC) 2003*, Southern Cape, South Africa, September 7-10 (pp. 255-256). South Africa: SATNAC.

Li, H., & Forchheimer, R. (1994). Two-view facial movement estimation. *IEEE Transactions on Circuits and Systems for Video Technology, 4*(3), 276-287.

Li, J., & Lei, S. (1997). Rate-distortion optimized embedding. In *Proc. Picture Coding Symposium*, Berlin, Germany, September (pp. 201-206).

Li, S., & Li, W. (1995). Shape-adaptive discrete wavelet transforms for arbitrarily shaped visual object coding. *IEEE Transactions on Circuits and Systems for Video Technology, 10*(5), 725-743.

Li, W., & Salari, W. (1995). Successive elimination algorithm for motion estimation. *IEEE Transactions on Image Processing, 4*(1), 105-107.

Lin, H., & Wang, Y. (2005). *Predicting consumer intention to use mobile commerce in Taiwan*. Paper presented at the International Conference on Mobile Business, Sydney, Australia, July 11-13.

Linde, Y., Buzo, A., & Gary, R. M. (1980). An algorithm for vector quantization. *IEEE Transactions on Communication, 28*(1), 84-95.

Linnartz, J. P. (2000). Synchronous MC-CDMA in dispersive, mobile rayleigh channels. *Proc. of 2nd IEEE Benelux Signal Processing Symposium (SPS-2000)*, Hilvarenbeek, The Netherlands (pp. 1-4). New York: IEEE.

Liu, Q., Reihaneh, S.-N., & Sheppard, N. P. (2003). Digital rights management for content distribution. In *Proceedings of the Australasian Information Security Workshop Conference on ACSW Frontiers 2003, 21*, Adelaide, Australia, February 1 (pp. 49-58). Darlinhurst, Australia: Australian Computer Society, Inc.

Liu, T., & Choudary, C. (2004). Content-aware streaming of lecture videos over wireless networks. *IEEE Sixth International Symposium on Multimedia Software Engineering (ISMSE'04)*, Miami, FL, December 13-15 (pp. 458-465). Washington, DC: IEEE Computer Society.

LKSCTP. Retrieved June 1, 2005, from http://lksctp.sourceforge.net

Lu, J., Yu, C., Liu, C., & Yao, J. E. (2003). Technology acceptance model for wireless Internet. *Internet Research: Electronic Networking Applications and Policy, 13*(3), 206-222.

Lucent Technologies. (2000). *ORiNOCO Manager Suite Users Guide*.

Lundberg, J., & Candolin, C. (2003). Mobility in the host identity protocol (HIP). In *Proceedings of the International Symposium on Telecommunications (IST2003)*, Isfahan, Iran, August (pp. 754-757). Iran: Iran Telecom Research Center.

Luo, H., & Lu, S. (2000). *Ubiquitous and robust authentication services for ad hoc wireless networks*. Technical Report, TR-200030, Department of Computer Science, UCLA.

Luo, H., Zerfos, P., Kong, J., Lu, S., & Zhang, L. (2002). Self-securing ad hoc wireless networks. *Seventh International Symposium on Computers and Communications (ISCC'02)*, Italy, July 1-4 (pp. 567-574).

Lynn, L. H., Aram, J. D., Reddy, N. M., & Ostermann, J. (1997). Methodologies used for evaluation of video tools and algorithms in MPEG-4. *Signal Processing: Image Communication, 9*(4), 343-365.

Maclagan, M. (2000). Where are we going in our language? New Zealand English today. *New Zealand Journal of Speech-Language Therapy, 53-54*, 14-20.

Man, H., de Queiroz, R., & Smith, M. (2002). Three-dimensional subband coding techniques for wireless video communications. *IEEE Transactions on Circuits and Systems for Video Technology, 12*(3), 386-397.

Marina, M. K., & Das, S. R. (2001). Performance of route caching strategies in dynamic source routing. *Proceed-*

ings of the International Workshop on Wireless Networks and Mobile Computing (WNMC) in Conjunction with International Conference on Distributed Computing Systems (ICDCS), Phoenix, Arizona, April 16-19 (pp. 425-432). Los Alamitos, CA: IEEE Computer Society.

Marina, M. K., & Das, S. R. (2001). On-demand multipath distance vector routing in ad hoc networks. *Proceedings of the International Conference for Network Protocols (ICNP)*, Riverside, CA, November 11-14 (pp. 14-23). Washington, DC: IEEE Computer Society.

Marina, M. K., & Das, S. R. (2003). *Ad-hoc on-demand multipath distance vector routing.* Technical Report, Computer Science Department, Stony Brook University.

Marti, S., Giuli, T. J., Lai, K., & Baker, M. (2000). Mitigating routing misbehavior in mobile ad-hoc networks. *Proceedings of the 6th International Conference on Mobile Computing and Networking (MobiCom)*, Boston, MA, August 6-11 (pp. 255-265). ISBN 1-58113-197-6. ACM Press.

Massey, A. P., Khatri, V., & Ramesh, V. (2005). *From the Web to the wireless Web: Technology readiness and usability.* Paper presented at the 38th Hawaii International Conference on System Sciences, Hilton Waikoloa Village, Hawaii, January 3-6 (p. 32b). IEEE Computer Society.

Mathwick, C., Malhotra, N., & Rigdon, E. (2001). Experiental value: Conceptualization, measurement and application in the catalog and Internet shopping environment. *Journal of Retailing, 77*(1), 39-56.

McCullough, J. (2004). *185 wireless secrets: Unleash the power of PDAs cell phones, and wireless networks.* Indianapolis, IN: Wiley Publishing, Inc.

Meier, F. W., Schuster, G. M., & Katsaggelos, A. K. (2000). A mathematical model for shape coding with B-splines. *Signal Processing: Image Communications, 15*(7-8), 685-701.

Meyer, U., & Wetzel, S. (2004). On the impact of GSM encryption and man-in-the-middle attacks on the security of interoperating GSM/UMTS networks. In the *15th IEEE International Symposium on Personal, Indoor and Mobile*

Radio Communications (PIMRC 2004), 4, Barcelona, Spain, September 5-8 (pp. 2876-2883). USA: IEEE. Retrieved July 12, 2005, from http://www.cdc.informatik.tu-darmstadt.de/~umeyer/UliPIMRC04.pdf

Michiardi, P., & Molva, R. (2002). CORE: A collaborative reputation mechanism to enforce node cooperation in mobile ad hoc networks. *Proceedings of the IFIP TC6/TC11 Sixth Joint Working Conference on Communications and Multimedia Security*, Portoroz, Slovenia, September 26-27 (pp. 107-121). ISBN 1-4020-7206-6. Berlin: Springer.

Michiardi, P., & Molva, R. (2002). Simulation-based analysis of security exposures in mobile ad-hoc networks. *Proceedings of the European Wireless Conference.*

Ministry of Internal Affairs and Communications, Japan. (2005, May).*Towards realisation of ubiquitous networking society.* Tokyo Ubiquitous Networking Conference, Tokyo, Japan, May 16-17. Retrieved May 12, 2006, from http://www.wsis-japan.jp/about_e.html

Ministry of Internal Affairs and Communications, Japan. (2005, May). *Tokyo Ubiquitous Networking Conference: Program (Session 5) Ubiquitous Network Society*, Tokyo, Japan, May 16-17. Retrieved May 12, 2006, from: http://www.wsis-japan.jp/session5_e.html

Minton, G. C., & Scheneider, F. W. (1980). *Differential psychology.* Prospect Heights, IL: Waveland Press.

MIP. Retrieved June 1, 2005, from opensource.nus.edu.sg/projects/mobileip/mip.html

MITA. (2002). *Mobile Internet technical architecture: The complete package.* Nokia, Finland, ISBN 951-826-669-7.

Mizoguchi, M. et al. (1998). A fast burst synchronization scheme for OFDM. *Proc ICUPC 98*, Florence, Italy (pp. 125-129). New York: IEEE.

MNET. Retrieved June 1, 2005, from http://mosquitonet.stanford.edu/

Mobile Broadcast/Multicast Service (MBMS). (2004, August). White Paper, TeliaSonera. Retrieved July 11,

2005, from http://www.medialab.sonera.fi/workspace/MBMSWhitePaper.pdf

Mobile. (2007). *Mobile computing*. Retrieved May 2007, from http://searchmobilecomputing.techtarget.com/sDefinition/0,,sid40_gci506042,00.html

MobileCloak™ Web Portal. (2005). Retrieved July 11, 2005, from http://www.mobilecloak.com

Mobiquitous (2005). Retrieved December 2005, from http://www.mobiquitous.org/2005/challenges.html

Moon, J.-W., & Kim, Y.-G. (2001). Extending the TAM for a World-Wide-Web context. *Information & Management, 38*(4), 217-230.

Moreau, C. P., Lehmann, D. R., & Markman, A. B. (2001). Entrenched knowledge structures and consumer response to new products. *Journal of Marketing Research, 38*(1), 14-29.

Morgan, D. L., & Krueger, R. A. (1993). When to use focus groups and why. In D. L. Morgan (Ed.), *Successful focus groups* (pp. 1-19). London: Sage Publications.

Moskowitz, R., & Nikander, P. (2006). *Host Identity Protocol (HIP) Architecture*. RFC 4423, IETF.

MPEG. Retrieved June 1, 2005, from http://mpeg4ip.sourceforge.net/faq/index.php

Muller, T. (1999). *Bluetooth security architecture: Version 1.0*. Bluetooth White Paper, Document # 1.C.116/1.0. Retrieved September 26, 2007, from http://www.bluetooth.com/NR/rdonlyres/C222A81E-D9F9-48CA-91DE-9C81F5C8B94F/0/Security_Architecture.pdf

Muñoz, M. A., Rodriguez, M., Favela, J., Martinez-Garcia, A. I., & Gonzalez, V. (2003). Context-aware mobile communication in hospitals. *Computer IEEE, 36*(9), 38-46.

Muñoz, M., & Rubio, C. G. (2004). A new model for service and application convergence in B3G/4G networks. *IEEE Wireless Communications, 35*(5), 539-549.

Muñoz, M., Gonzalez, V., Rodríguez, M., & Favela, J. (2003). Supporting context-aware collaboration in a hospital: An ethnographic informed design. *Lecture Notes in Computer Science, 2806,* 330-344.

Murakami, T. (2004, August 1). *NRI papers: Ubiquitous networking: Business opportunities and strategic issues* (pp. 1-24). NRI Papers No. 79. Japan: Nomura Research Institute.

Murphy, A. L., Picco, G. P., & Roman, G. C. (2001, May). Time: A middleware for physical and logical mobility. *Proceedings of the 21st International Conference on Distributed Computing Systems.*, April (pp. 524-533).

Murthy, M., & Garcia-Luna-Aceves, J. J. (1995). A routing protocol for packet radio networks. *Proceedings of the 1st Annual International Conference on Mobile Computing and Networking (MobiCom)*, Berkeley, CA, November 13-15 (pp. 86-95). ISBN 0-89791-814-2. New York: ACM Press.

Myers, D. (2004). *Mobile video telephony*. New York: McGraw-Hill Professional.

Myles, G., Friday, A., & Davies, N. (2003, January-March). Preserving privacy in environments with location-based applications. *IEEE Pervasive Computing, 2*(1), 56-64.

Nagumo, T. (2002, June 1). *NRI papers: Innovative business models in the era of ubiquitous networks* (pp. 1-13). NRI Papers No. 49. Japan: Nomura Research Institute.

Nanda, S., & Pearlman, W. S. (1992). Tree coding of image subbands. *IEEE Transactions on Signal Processing, 1*(2), 133-147.

Narula, A., & Lim, J. S. (1993). Error concealment techniques for an all-digital high-definition television system. In *Proc. SPIE Conf. Visual Commun. and Image Proc.*, Chicago, IL, May (pp. 304-315).

Nasir, A., & Mah-Rukh, M.-R. (2006). Internet mobility using SIP and MIP. *Proc. Third International Conference on Information Technology: New Generations ITNG'06*, Las Vegas, NV, April 10-12 (pp. 334-339). USA: IEEE Computer Society.

Navakitkanok, P., & Aramvith, S. (2004). Improved rate control for advanced video coding (AVC) standard under low delay constraint. *International Conference on Information Technology: Coding and Computing (ITCC'04)*, *2*, Las Vegas, NV, April 5-7 (p. 664). Washington, DC: IEEE Computer Society.

NetStumbler. (2006). Retrieved December 10, 2006, from http://www.netstumbler.com

Ngan, K. N., & Chooi, W. L. (1994). Very low bit rate video coding using 3D subband approach. *IEEE Transactions on Circuits and Systems for Video Technology, 4*(3), 309-316.

NHS. (2004). *Allied health professionals*. Retrieved April 24, 2006, from http://www.nhscareers.nhs.uk

Nikaein, N., Bonnet, C., & Nikaein, N. (2001). HARP: Hybrid ad hoc routing protocol. *Proceedings of the International Symposium on Telecommunications (IST)*, Tehran, Iran, September 1-3.

Nikaein, N., Labiod, H., & Bonnet, C. (2000). DDR: Distributed dynamic routing algorithm for mobile ad hoc networks. *Proceedings of the 1st Annual Workshop on Mobile and Ad Hoc (MobiHoc)*, Boston, MA, August 11 (pp. 19-27). Piscataway, NJ: IEEE Press.

Nikander, P. (2001). *Denial-of-service, address ownership, and early authentication in the IPv6 world*. Retrieved May 10, 2006, from http://www.tml.tkk.fi/~pnr/publications/cam2001.pdf

Nikander, P., Ylitalo, J., & Wall, J. (2003). Integrating security, mobility, and multi-homing in a HIP way. *Proceedings of Network and Distributed Systems Security Symposium (NDSS'03)*, San Diego, USA, February (pp. 87-99). Reston, USA: Internet Society.

Nokia. (2006). *State of workforce mobility*. Retrieved May 2007, from http://www.nokia.com/NOKIA_COM_1/About_Nokia/Press/White_Papers/pdf_files/Whitepaper_TheMythsofMobility.pdf

Novak, T. P., Hoffman, D. L., & Yung, Y. (2000). Measuring the customer experience in online environments: A structural modeling approach. *Marketing Science, 19*(1), 22-42.

O'Connell, K. J. (1997). Object-adaptive vertex-based shape coding method. *IEEE Transactions on Circuits and Systems for Video Technology, 7*(1), 251-255.

Ogier, R., Templin, F., & Lewis, M. (2004, February). *RFC 3684 on topology dissemination based on reverse-path forwarding (TBRPF)*. Draft-ietf-manet-tbrpf-11.txt, Internet Draft, Network Working Group.

Ohya, T., & Miki, T. (2005). Enhanced-reality multimedia communications for 4G mobile networks. *1st International Conference on Multimedia Services Access Networks (MSAN '05)*, Orlando, FL, June 13-15 (pp. 69-72). Piscataway, NJ: IEEE Service Center.

Øien, G. E. (1993). *L_2-optimal attractor image coding with fast decoder convergence*. PhD thesis. Trondheim, Norway.

Olzak, T. (2005). *Wireless handheld device security*. Retrieved July 11, 2003, from http://www.securitydocs.com/pdf/3188.PDF

Ong, C. S., Nahrstedt, K., & Yuan, W. (2003). Quality of protection for mobile multimedia applications. In *Proceedings of IEEE International Conference on Multimedia and Expo (ICME2003)*, Baltimore, MD, July 7-9 (pp. 137-140). IEEE. Retrieved July 11, 2005, from http://cairo.cs.uiuc.edu/publications/papers/ICME03-chui.pdf

Onoe, Y., Atsumi, Y., Sato, F., & Mizuno, T. (2001). A dynamic delayed ack control scheme on Mobile IP networks. *International Conference on Computer Networks and Mobile Computing*, Los Alamitos, CA, October 16-19 (pp. 35-40). Los Alamitos, CA: IEEE Computer Society.

Ordentlich, E., Weinberger, M., & Seroussi, G. (1998). A low-complexity modeling approach for embeddded coding of wavelet coefficients. *IEEE Data Compression Conference (DCC)*, Snowbird, Utah, March 30-April 1 (pp. 408-417).

Orr, E. (2004). *3G-324M helps 3G live up to its potential.* Retrieved May 2007, from http://www.wsdmag.com/Articles/Print.cfm?ArticleID=7742

Pagani, M. (2004). Determinants of adoption of third generation mobile multimedia services. *Journal of Interactive Marketing, 18*(3), 46-59.

Pan, Y., Lee, M., Kim, J. B., & Suda, T. (2004, May). An end-to-end multipath smooth handoff scheme for streaming media. *IEEE Journal on Selected Areas in Communications, 22*(4), 653-663.

Papadimitratos, P., & Haas, Z. J. (2002). Secure routing for mobile ad hoc networks. *Proceedings of the SCS Communication Networks and Distributed Systems Modelling and Simulation Conference (CNDS)*, San Antonio, Texas, January 27-31. USA: Society for Computer Simulation.

Papadimitratos, P., & Haas, Z. J. (2003). Secure link state routing for mobile ad hoc networks. *Proceedings of the Symposium on Applications and the Internet Workshops (SAINT)*, Orlando, FL, January 27-31 (pp. 27-31). USA: IEEE Computer Society.

Papadimitratos, P., & Haas, Z. J. (2003). Secure message transmission in mobile ad hoc networks. *Ad Hoc Networks, 1*(1), 193-209.

Paraben Corporation. (2007). *Device seizure v1.2: Cell phone & PDA forensic software.* Retrieved September 27, 2007, from http://www.paraben-forensics.com/catalog/product_info.php?cPath=25&products_id=405

Parasuraman, A. (2000). Technology readiness index: A multiple item scale to measure readiness to embrace new technologies. *Journal of Service Research, 2*(4), 307-320.

Park, V. D., & Corson, M. S. (1997). A highly adaptive distributed routing algorithm for mobile wireless networks. *Proceedings of the 16th Annual Joint Conference of the IEEE Computer and Communications Societies (InfoCom)*, 3, Kobe, Japan, April 7-11 (pp. 1405-1413). ISBN 0-8186-7780-5. New York: IEEE.

Patanapongpibul, L., & Mapp, G. (2003). A client-based handoff mechanism for Mobile IPv6 wireless networks. *Proc. Eighth IEEE International Symposium on Computers and Communications*, Antalya, Turkey, June-July (pp. 563-568). Los Alamitos, CA: IEEE Computer Society.

Patton, M. Q. (1990). *Qualitative evaluation and research methods (2nd ed.).* London: Sage Publications.

Pätzold, M. (2002). *Mobile fading channels.* New York: Wiley.

PC Guardian Technologies. (2005). *PDASecure—Powerful security. Simple to use. Superior service & support.* Retrieved July 9, 2005, from http://www.pcguardiantechnologies.com/PDASecure/PDASecure_Brochure.pdf

Pearson, D. E. (1995). Developments in model-based video coding. *Proceedings of the IEEE, 83*(6), 892-906.

Pei, G., Gerla, M., & Chen, T. (2000). Fisheye state routing in mobile ad hoc networks. *Proceedings of the ICDCS Workshop on Wireless Networks and Mobile Computing*, Taipei, Taiwan, April 10 (pp. 71-78). IEEE Computer Society.

Pei, G., Gerla, M., Hong, X., & Chiang, C. C. (1999). A wireless hierarchical routing protocol with group mobility. *Proceedings of the IEEE Wireless Communications and Networking Conference (WCNC)*, 3, New Orleans, September 21-24 (pp. 1538-1542). Piscataway, NJ: IEEE Press.

Peikari, C., & Fogie, S. (2003). *Maximum wireless security.* SAMS Publishing.

Peled, A., & Ruiz, A. (1980). Frequency domain data transmission using reduced computational complexity algorithms. *Proc. IEEE ICASSP,* Denver, Colorado (pp. 964-967). New York: IEEE.

Perelson, S., & Botha, R. (2004). An investigation into access control for mobile devices. In H. S. Venter, J. H. P. Eloff, L. Labuschagne, & M.M. Eloff (Eds.), *ISSA 2004 Enabling Tomorrow Conference, Peer-Reviewed Proceedings of the ISSA 2004 Enabling Tomorrow*

Conference, Information Security South Africa (ISSA), Gallagher Estate, Johannesburg, South Africa, June 30-July 2. CDROM, ISBN 1-86854-522-9. ISSA. Retrieved July 11, 2005, from http://www.nmmu.ac.za/rbotha/Pubs/docs/LC_017.pdf

Perkins, C. (1996). IP mobility support. *IETF RFC 2002*, October.

Perkins, C. (2002). *IP mobility support for IPv4, IETF*. RFC 3344. Retrieved May 8, 2006, from http://www.ietf.org/rfc/rfc3344.txt

Perkins, C. E., & Royer, E. M. (1999). Ad-hoc on-demand distance vector routing. *Proceedings of the 2nd IEEE Workshop on Mobile Computing Systems and Applications (WMCSA)*, New Orleans, LA, February 25-26 (pp. 90-100). IEEE Computer Society.

Perkins, C., & Bhagwat, P. (1994). Highly dynamic destination-sequenced distance-vector routing (DSDV) for mobile computers. *Proceedings of the ACM Conference on Communications Architectures, Protocols and Applications (SIGCOMM)*, London, UK, August 31-September 2 (pp. 234-244). New York: ACM Press.

Perkins, E. C. (2002). RFC 3220 – IP mobility support for IPv4. *IETF Internet Standard*. Retrieved September 18, 2007, from www.ietf.org/rfc/rfc3220.txt

Perrig, A., Canetti, R., Tygar, J. D., & Song, D. (2000). Efficient authentication and signing of multicast streams over lossy channels. *Proceedings of the IEEE Symposium on Security and Privacy*, Berkeley, CA, May 14-17 (pp. 56-73). IEEE Computer Society.

Phillips, A., Nelson, B., Enfinger, F., & Steuart, C. (2005). *Guide to computer forensics and investigations* (2nd ed.). USA: Course Technology Press.

Phobe.com. (2003). *Yer latin lesson for today: Today's latin lesson*. Retrieved May 14, 2006, from http://www.phobe.com/octopusmotor/latin.html

Pirzada, A., & McDonald, C. (2004). Establishing trust in pure ad-hoc networks. *27th Conference on Australasian Computer Science, 26*, Dunedin, New Zealand (pp. 47-54). Australian Computer Society, Inc.

Podilchuk, C., Jayant, N., & Farvardin, N. (1995). Three dimensional subband coding of video. *IEEE Transactions on Image Processing, 4*(2), 125-39.

Pose, R. (2001). Password-capabilities: Their evolution from the password-capability system into walnut and beyond. In G. Heiser (Ed.), *6th Australasian Computer Systems Architecture Conference (ACSAC)*, 23, Gold Coast, Australia, January 29-30 (pp. 105-113). IEEE Computer Society.

Privat, G. (2005, April). *Ubiquitous network & smart devices: New telecom services & evolution of human interfaces*. ITU Ubiquitous Network Societies Workshop, April 6. Retrieved May 20, 2006, from http://www.itu.int/osg/spu/ni/ubiquitous/Presentations/8_privat_applications.pdf

Pulkkis, G., Grahn, K., Karlsson, J., Martikainen, M., & Daniel, D. E. (2005). Recent developments in WLAN security. In M. Pagani (Ed.), *Mobile and wireless systems beyond 3G: Managing new business opportunities* (pp. 254-311). Hershey, PA: IRM Press

Puri, R., & Ramchandran, K. (2002). PRISM: A new robust video coding architecture based on distributed compression principles. *Allerton Conference on Communication, Control, and Computing*, Allerton, IL, October.

Queensland Aphasia Groups. (2001). *Web developer's guidelines*. Retrieved April 24, 2006, from http://www.shrs.uq.edu.au/cdaru/aphasiagroups/Web_Development_Guidelines.html

Queensland Aphasia Groups. (2001). *What is aphasiafriendly?* Retrieved April 24, 2006, from http://www.shrs.uq.edu.au/cdaru/aphasiagroups/Aphasia_Friendly.html

Rabbani, M., & Jones, P. W. (1991). *Digital image compression techniques*. Bellingham, Washington: SPIE Optical Engineering Press.

Rabin, M. O. (1989). Efficient dispersal of information for security, load balancing, and fault tolerance. *Journal of ACM, 36*(2), 335-348.

Radhakrishnan, S., Racherla, G., Sekharan, C. N., Rao, N. S. V., & Batsell, S. G. (1999). DST- A routing protocol for ad hoc networks using distributed spanning trees. *Proceedings of the IEEE Wireless Communications and Networking Conference, (WCNC), 3*, New Orleans, September 21-24 (pp. 1543-1547). Piscataway, NJ: IEEE Operations Center.

Radhakrishnan, S., Racherla, G., Sekharan, C. N., Rao, N. S. V., & Batsell, S. G. (2003). Protocol for dynamic ad-hoc networks using distributed spanning trees. *Kluwer Wireless Networks, 9*(6), 673-686.

Raghunathan, A., Ravi, S., Hattangady, S., & Quisquater, J. (2003). Securing mobile appliances: New challenges for the system designer. *Proceedings of the Design, Automation and Test in Europe Conference and Exhibition, 1* (pp. 10176).

Ramabadran, T., & Gaitonde, S. (1988). A tutorial on CRC application. *IEEE Micro, 8*(4), 62-75.

Ramanujan, R., Ahamad, A., Bonney, J., Hagelstrom, R., & Thurber, K. (2000). Techniques for intrusion-resistant ad hoc routing algorithms (TIARA). *Proceedings of the 21st Century Military Communications Conference (MILCOM)*, Los Angeles, CA, October 22-25 (pp. 660-664). ISBN 0-7803-6521-6. Piscataway, NJ: IEEE.

Ramasubramanian, V., Haas, Z. J., & Sirer, E. G. (2003). SHARP: A hybrid adaptive routing protocol for mobile ad hoc networks. *Proceedings of the 4th ACM International Symposium on Mobile Ad Hoc Networking and Computing (MobiHoc)*, Annapolis, MD, June 1-3 (pp. 303-314). ISBN 1-58113-684-6. New York: ACM Press.

Rao, M., & Mendoza, L. (Eds.) (2005). *Asia unplugged: The wireless and mobile media boom in the Asia-Pacific*. New Delhi: Response Books (A Division of Sage Publications).

Ratliff, J. M. (2002). NTT DoCoMo and its i-mode success: Origins and implications. *California Management Review, 44*(3), 55-71.

Ray, S. K. (2006). Fourth generation (4G) networks: Roadmap-migration to the future. *IETE Technical Review, 23*, 253-265.

Reaz, A. S., Atiquzzaman, M., & Fu, S. (2005). Performance of DNS as location manager. *IEEE Globecom*, St. Louis, MO, November 28-December 2 (pp. 359-363). USA: IEEE Computer Society.

Redmill, D. W. (1994). *Image and video coding for noisy channels*. PhD thesis. University of Cambridge. Signal Processing and Communications Laboratory.

Redmill, D. W., & Kingsbury, N. G. (1996). The EREC: An error resilient technique for coding variable-length blocks of data. *IEEE Transactions on Image Processing, 5*(4), 565-574.

Repo, P., Hyvonen, K., Pantzar, M., & Timonen, P. (2004). *Users intenting ways to enjoy new mobile services: The case of watching mobile videos*. Paper presented at the 37th Hawaii International Conference on System Sciences, Hawaii, January 5-8 (p. 40096.3). IEEE Computer Society.

Rhodes, S. R. (1983). Age-related differences in work attitudes and behavior: A review of conceptual analysis. *Psychological Bulletin, 93*(2), 328-367.

Richardson, I. E. (2003). *H.264 and MPEG-4 video compression*. Chichester: John Wiley & Sons.

Rivest, R. L., Shamir, A., & Adleman, L. (1978). A method for obtaining digital signatures and public-key cryptosystems. *Communication ACM, 21*(2), 120-126.

RoamAD. (2005, January). *Products & services: Introduction*. Retrieved May 16, 2006, from http://www.roamad.com/prodservintro.htm

Robbins, J. (2005). *An explanation of computer forensics*. Retrieved July 11, 2005, from http://www.computerforensics.net/forensics.htm

Robins, F. (2003). The marketing of 3G. *Marketing Intelligence & Planning, 21*(6), 370-378.

Rodriguez, M., Favela, J., Gonzalez, V., & Muñoz, M. A. (2003). Agent-based mobile collaboration and information access in a healthcare environment. *eHealth: Application of Computing Science in Medicine and Healthcare*, Cambridge, UK, April 19-21 (pp. 133-148).

Roehm, M. L., & Sternthal, B. (2001). The moderating effect of knowledge and resources on the persuasive impact of analogies. *Journal of Consumer Research, 28*(2), 257-272.

Rogers, E. M. (1995). *Diffusion of innovations.* New York: Free Press.

Rohling, H., & Grünheid, R. (1997). Performance comparison of different multiple access schemes for the downlink of an OFDM communication system. *Proc. IEEE VTC'97,* Phoenix, Arizona (pp. 1365-1369). New York: IEEE.

Rohling, H., Galda, D., & Schulz, E. (2004). An OFDM based cellular single frequency communication network. *Proc. of the Wireless World Research Forum '04,* Beijing, China (pp. 254-258). Zurich: WWRF.

Rokou, F. P., & Rokos, Y. (2004). Integral laboratory for creating and delivery lessons on the Web based on a pedagogical content repurposing approach. *Fourth IEEE International Conference on Advanced Learning Technologies (ICALT'04),* Joensuu, Finland, August 30-September 1 (pp. 732-734). Washington, DC: IEEE Computer Society.

Romsey Associates Ltd. (2005). *Details—PDALok V1.0...,PDALok—The technology behind the security software.* Retrieved July 3, 2005, from http://www.pdalok.com/pda_security_products/PDALok_details.htm

Rosenberg, J., Schulzrinne, H., Camarillo, G, Johnston, A., Peterson, J., Sparks, R., Handley, M., & Schooler, E. (2002). *SIP: Session initation protocol.* RFC 3261, IETF. Retrieved May 9, 2006 from http://www.ietf.org/rfc/rfc3261.txt

Rosenberg, R. S. (2004). *The social impact of computers (3rd ed.).* USA: Elsevier Academic Press.

ROSISTEM. (2005). *Biometric education » Fingerprint.* Retrieved July 9, 2005, from http://www.barcode.ro/tutorials/biometrics/fingerprint.html

Rossotto, C. M., Kerf, M., & Rohlfs, J. (2000). Competition in mobile telecommunications: Sector growth, benefits for the incumbent and policy trends. *Info, 2*(1), 67-73.

Royal College of Speech and Language Therapists (RCSLT). (2004). *What do speech and language therapists do?* Retrieved January 15, 2006, from http://www.rcslt.org/whatdo.shtml

Royer, E. M., & Toh, C.-K. (1999). A review of current routing protocols for ad hoc mobile wireless networks. *IEEE Personal Communications, 6*(2), 46-55.

Rueppel, T. (1992). *Stream ciphers.* SIMM.

Ruiz, P. M. (2002). *Beyond 3G: Fourth generation wireless networks.* Retrieved May 2007, from http://internetng.dit.upm.es/ponencias-jing/2002/ruiz/ruiz.PDF

Saaksjarvi, M. (2003). Consumer adoption of technological innovations. *European Journal of Innovation Management, 6*(2), 90-100.

Said, A., & Pearlman, W. (1996). A new, fast, and efficient image codec based on set partitioning in hierarchical trees. *IEEE Transactions on Circuits and Systems for Video Technology, 6*(3), 243-250.

Salama, P., Shroff, N. B., Coyle, E. J., & Delp, E. J. (1995). Error concealment techniques for encoded video streams. *IEEE International Conference on Image Processing (ICIP),* Washington, DC, October 23-26 (pp. 9-12).

Salsano, S., Veltri, L., & Papalilo, D. (2002). SIP security issues: The SIP authentication procedure and its processing load. *IEEE Network, 16*(6), 38-44.

Saltzberg, B. R. (1967). Performance of an efficient parallel data transmission system. *IEEE Trans. on Communications, 15,* 805-811.

Salzman, M., Palen, L., & Harper, R. (2001). *Mobile communications: Understanding users, adoption and design.* CHI 2001, Seattle, WA, March 31-April 5.

Santa-Cruz, D., Grosbois, R., & Ebrahimi, T. (2002). JPEG 2000 performance evaluation and assessment. *Signal Processing: Image Communication, 17*(1), 113-130.

Santivez, C., Ramanathan, R., & Stavrakakis, I. (2001). Making link-state routing scale for ad hoc networks. *Proceedings of the 2nd ACM International Symposium on Mobile Ad Hoc Networking Computing (MobiHoc)*, Long Beach, CA, October 4-5 (pp. 22-32). ISBN 1-58113-428-2. New York: ACM Press.

Sanzgiri, K., Dahill, B., Levine, B. N., Shields, C. & Belding-Royer, E. M. (2002). A secure routing protocol for ad hoc networks. *Proceedings of the 10th IEEE International Conference on Network Protocols (ICNP)*, Paris, France, November 12-15 (pp. 78-89). IEEE Computer Society.

Sanzgiri, K., Dahill, B., Levine, B. N., Shields, C., & Belding-Royer, E. M. (2005). Authenticated routing for ad hoc networks. *IEEE Journals on Selected Areas in Communications, 23*(3), 598-610.

Satyanarayanan, M. (1996). Fundamental challenges in mobile computing. *Fifteenth ACM Symposium on Principles of Distributed Computing, 1*(1), Philadelphia, PA, May, 7pp.

Satyanarayanan, M. (2001). Pervasive computing: Vision and challenges. *IEEE Personal Communications, 8*(4), 10-17.

Satyanarayanan, M. (2005). Metrics and benchmarks for pervasive computing. *Pervasive Computing: IEEE, 4*(3), 4-6.

Sawada, M., Tani, N., Miki, M., & Maruyama, Y. (1998). Advanced mobile multimedia services and applied network techniques. *IEEE International Conference on Universal Personal Communications, 1*, 79-85.

Schnorr, C. P. (1991). Efficient signature generation by smart cards. *Journal of Cryptology, 4*(3), 161-174.

Schulzrinne, H., Casner, S., Frederick, R., & Jacobson, V. (2003). *RTP: A transport protocol for real-time applications*. RFC 3550, IETF. Retrieved May 8, 2006, from http://www.ietf.org/rfc/rfc3550.txt

Schuster, G. M., & Katsaggelos, A. K. (1997). *Rate-distortion based video compression: Optimal video frame compression and object boundary encoding*. Boston: Kluwer Academic Publishers.

Schuster, G. M., & Katsaggelos, A. K. (1998). An optimal boundary encoding scheme in the rate distortion sense. *IEEE Transactions on Image Processing, 7*(1), 13-26.

Schuster, G. M., & Katsaggelos, A. K. (2006). Motion compensated shape error concealment. *IEEE Transactions on Image Processing, 15*(2), 501-510.

Schuster, G. M., Katsaggelos, A. K., & Xiaohuan, L. (2004). Shape error concealment using Hermite splines. *IEEE Transactions on Image Processing, 13*(6), 808-820.

Secker, A., & Taubman, D. S. (2004). Highly scalable video compression with scalable motion coding. *IEEE Transactions on Image Processing, 13*(8), 1029-1041.

Sekaran, U. (2000). *Research methods for business: A skill building approach*. New York: John Wiley and Sons.

Seol, S., Kim, M., Yu, C., & Lee., J. H. (2002). Experiments and analysis of voice over MobileIP. *13th IEEE International Symposium on Personal, Indoor and Mobile Radio Communications (PIMRC)*, Lisboa, Portugal, September 15-18 (pp. 977-981). Piscataway, NJ: IEEE.

Setec Portal. (2005). Retrieved July 11, 2005, from http://www.setec.fi

sfr Gesellschaft für Datenverarbeitung mbH. (2005). *The patented technology in all of our visual key products*. Retrieved July 3, 2005, from http://www.viskey.com/viskeypalm/index.html and http://www.viskey.com/tech.html

Shamir, A. (1979). How to share a secret. *Communications of ACM, 22*(11), 612-613.

Shapiro, J. M. (1993). Embedded image coding using zerotrees of wavelet coefficients. *IEEE Transactions on Signal Processing, 41*(12), 3445-3462.

Sharda, N. (1999). *Multimedia information networking*. New Jersey: Prentice Hall.

Sharda, N., & Georgievski, M. (2002). A holistic quality of service model for multimedia communications. *International Conference on Internet and Multimedia Systems and Applications (IMSA2002)*, Kaua'i, Hawaii, August 12-14 (pp. 282-287). Calgary, Alberta, Canada: ACTA Press.

Sharmin, M., Ahmed, S., & Ahamed, S. I. (2006). MARKS (middleware adaptability for resource discovery, knowledge usability and self-healing) for mobile devices of computing environments. *Proceedings of Third International Conference on Information Technology: New Generation (ITNG 2006)*, Las Vegas, NE, April (pp. 306-313).

Sharmin, M., Ahmed, S., Ahamed, S. I., Haque, M., & Khan, A. (2006, April). Healthcare aide: Towards a virtual assistant for doctors using pervasive middleware. To appear in *Journal Springer e&i, 123*(4).

Shen, M., Li, G., & Liu, H. (2004). Design tradeoffs in OFDMA traffic channels. *Proc. of IEEE ICASSP '04*, Montreal, Canada (pp. 757-760). New York: IEEE.

Sheppard, B. H., Hartwick, J., & Warshaw, P. R. (1988). The theory of reasoned action: A meta-analysis of past research with recommendations for modifications and future research. *Journal of Consumer Research, 15*(3), 325-343.

Shin, C. C., & Johnson, D. M. (1978). Avowed happiness as an overall assessment of quality of life. *Social Indicators Research, 5*, 475-492.

Shirani, S., Kossentini, F., & Ward, R. (2000). A concealment method for video communications in an error-prone environment. *IEEE Journal on Selected Areas in Communications, 18*(6), 1122-1128.

Siau, K., Lim, E. P., & Shen, Z. (2001). Mobile commerce: Promises, challenges, and research agenda. *Journal of Databases Management, 12*(2), 4-13.

Sikora, T., & Makai, B. (1995). Shape-adaptive DCT for generic coding of video. *IEEE Transactions on Circuits and Systems for Video Technology, 5*(3), 59-62.

Sikora, T., Bauer, S., & Makai, B. (1995). Efficiency of shape adaptive transforms for coding of arbitrarily shaped image segments. *IEEE Transactions on Circuits and Systems for Video Technology, 5*(3), 254-258.

Singh, S., Woo, M., & Vaidya, N. H. (1998). Power-aware routing in mobile ad hoc networks. *Proceedings of the 4th ACM/IEEE International Conference on Mobile Computing and Networking (MobiCom)*, Dallas, Texas, October 25-30 (pp. 181-190). ISBN 1-58113-035-X. New York: ACM Press.

Slepian, J. D., & Wolf, J. K. (1973). Noiseless coding of correlated information sources. *IEEE Transactions on Information Theory, IT-19*, 471-480.

Smith, J. R., & Jabri, M. A. (2004). The 3G-324M protocol for conversational video telephony. *IEEE MultiMedia, 11*(3), 102-105.

Soares, L. D., & Pereira, F. (2004). Spatial shape error concealment for object-based image and video coding. *IEEE Transactions on Image Processing, 13*(4), 586-599.

Soares, L. D., & Pereira, F. (2006). Temporal shape error concealment by global motion compensation with local refinement. *IEEE Transactions on Image Processing, 15*(6), 1331-1348.

Sohel, F. A., Dooley, L. S., & Karmakar, G. C. (2005). A dynamic Bezier curve model. *International Conference on Image Processing (ICIP), II*, Genoa, Italy, September (pp. 474-477).

Sohel, F. A., Dooley, L. S., & Karmakar, G. C. (2005). *A novel half-way shifting Bezier curve model*. IEEE Region 10 Conference (Tencon), Melbourne, Australia, November.

Sohel, F. A., Dooley, L. S., & Karmakar, G. C. (2006). Accurate distortion measurement for generic shape coding. *Pattern Recognition Letters, 27*(2), 133-142.

Sohel, F. A., Dooley, L. S., & Karmakar, G. C. (2006). *Variable width admissible control point band for vertex based operational-rate-distortion optimal shape coding algorithms.* International Conference on Image Processing (ICIP), Atlanta, GA, October.

Sohel, F. A., Dooley, L. S., & Karmakar, G. C. (2007). New dynamic enhancements to the vertex-based rate-distortion optimal shape coding framework. *IEEE Transactions on Circuits and Systems for Video Technology, 7*(10).

Sohel, F. A., Karmakar, G. C., & Dooley, L. S. (2005). An improved shape descriptor using Bezier curves. *First International Conference on Pattern Recognition and Machine Intelligence (PReMI). Lecture Notes in Computer Science, 3776,* Kolkata, India, December (pp. 401-406).

Sohel, F. A., Karmakar, G. C., & Dooley, L. S. (2006). *Dynamic sliding window width selection strategies for rate-distortion optimal vertex-based shape coding algorithms.* International Conference on Signal Processing (ICSP), Guilin, China, November 16-20.

Sohel, F. A., Karmakar, G. C., & Dooley, L. S. (2007). Fast distortion measurement using chord-length parameterisation within the vertex-based rate-distortion optimal shape coding framework. *IEEE Signal Processing Letters, 14*(2), 121-124.

Sohel, F. A., Karmakar, G. C., & Dooley, L. S. (2007). Spatial shape error concealment utilising image-texture. *IEEE Transactions on Image Processing* (revision submitted).

Sohel, F. A., Karmakar, G. C., & Dooley, L. S. (2007). *Bezier curve-based character descriptor considering shape information.* IEEE/ACIS International Conference on Computer and Information Science (ICIS), Melbourne, Australia, July.

Sohel, F. A., Karmakar, G. C., Dooley, L. S. & Arkinstall, J. (2007). Quasi Bezier curves integrating localised information. *Pattern Recognition* (in press).

Sohel, F. A., Karmakar, G. C., Dooley, L. S., & Arkinstall, J. (2005). Enhanced Bezier curve models incorporating local information. *IEEE International Conference on Acoustics, Speech, and Signal Processing (ICASSP), IV,* Philadelphia, PA, March 18-23 (pp. 253-256).

Stajano, F., & Anderson, R. (1999). The resurrecting duckling: Security issues for ad-hoc wireless networking. *7th International Workshop on Security Protocols* (pp. 172-194).

Stamp, M. (2003). Digital rights management: The technology behind the hype. Journal of Electronic Commerce Research, 4(3), 202-212. Long Beach, CA: California State University Long Beach. Retrieved May 12, 2006, from http://home.earth-link.net/~mstamp1/DRMpaper.pdf

Stedman, R., Gharavi, H., Hanzo, L., & Steele, R. (1993, February). Transmission of subband-coded images via mobile channels. *IEEE Trans. Circuit Syst. Video Technol., 3,* 15-27.

Stewart, R. (2005, June). *Stream control transmission protocol (SCTP) dynamic address configuration.* IETF DRAFT, draft-ietf-tsvwgaddip-sctp-12.txt.

Stubblefield, A., Ioannidis, J., & Rubin, A. (2002). Using the Fluhrer, Mantin and Shamir attack to break WEP. *Proceedings of the Network and Distributed Systems Security Symposium* (pp. 17-22).

Sun, J., Howie, D., Koivisto, A., & Sauvola, J. (2001). A hierarchical framework model of mobile security. *Proc. 12th IEEE International Symposium on Personal, Indoor and Mobile Radio Communication, 1,* San Diego, CA, September 30-October 3 (pp. 56-60). USA: IEEE.

Sun, J., Howie, D., Koivisto, A., & Sauvola, J. (2001). Design, implementation, and evaluation of Bluetooth security. *Proc. IEEE International Conference on Wireless LANs and Home Networks,* Singapore, December 5-7 (pp. 121-130). USA: IEEE. Retrieved July 11, 2005, from http://www.mediateam.oulu.fi/publications/pdf/87.pdf

Sun, S., Haynor, D., & Kim, Y. (2003). Semiautomatic video object segmentation using v-snakes. *IEEE Transactions on Circuits and Systems for Video Technology, 13*(1), 75-82.

Sun, Y., Ahmad, I., Li, D., & Zhang, Y.-Q. (2006). Region-based rate control and bit allocation for wireless video transmission. *IEEE Transactions on Multimedia, 8*(1), 1-10.

Sundaresan, H. (2003). *OMAP^{TM} platform security features*. White Paper. Retrieved July 1, 2005, from http://focus.ti.com/pdfs/wtbu/omapplatformsecuritywp.pdf

Symantec Corporation. (2005). *Symantec AntiVirus™ for handhelds annual service edition, Symantec anti-Virus for handhelds safeguards Palm and Pocket PC mobile users*. Retrieved July 9, 2005, from http://www.symantec.com/sav/handhelds/ and http://www.symantec.com/press/2003/n030825.html

Symantec. (2003). *Wireless handheld and smartphone security*. Retrieved September 26, 2007, from http://whitepapers.zdnet.co.uk/0,1000000651,260085794p,00.htm?r=1

Tabesh, A., Bilgin, A., Krishnan, K., & Marcellin, M. W. (2005). JPEG2000 and motion JPEG2000 content analysis using codestream length information. *Proceedings of The Data Compression Conference (DCC'05)*, Snowbird, UT, March 29-31 (pp. 329-337). Washington, DC: IEEE Computer Society.

Talwar, V., & Nahrstedt, K. (2000). Securing RSVP for multimedia applications. *Multimedia Security Workshop ACM Multimedia*, Los Angeles, CA, October 30-November 3 (pp. 153-156). USA: ACM Press.

Tang, C.-W., Chen, C.-H., Yu, Y.-H., & Tsai, C.-J. (2006). Visual sensitivity guided bit allocation for video coding. *IEEE Transactions on Multimedia, 8*(1), 11-18.

Tao, H., Sawhney, H. S., & Kumar, R. (2002). Object tracking with Bayesian estimation of dynamic layer representations. *IEEE Transactions on Pattern Analysis and Machine Intelligence, 24*(1), 75-89.

Taubman, D. (2000). High performance scalable image compression with EBCOT. *IEEE Transactions on Image Processing, 9*(7), 1158-1170.

Taubman, D. S., & Marcellin, M. W. (2002). *JPEG2000: Image compression fundamentals, standards and practice*. Boston: Kluwer Academic Publishers.

Taubman, D. S., & Zakhor, A. (1994). Multirate 3-D subband coding of video. *IEEE Transactions on Image Processing, 3*(4), 572-88.

Taubman, D., & Marcellin, M. (2002). *JPEG2000: Image compression fundamentals, standards and practice*. Netherlands: Kluwer Academic Publishers.

Taylor, L. (2004). *Handheld security, part III: Evaluating security products*. Retrieved July 1, 2005, from http://www.firewallguide.com/pda.htm

Taylor, L. (2004-2005). *Handheld security, part I-V*. Retrieved September 25, 2007, from http://www.pdastreet.com/articles/2004/12/2004-12-6-Handheld-Security-Part.html

Taylor, S., & Todd, P. A. (1995). Understanding information technology usage: A test of competing models. *Information Systems Research, 6*(2), 144-176.

Taylor, S., & Todd, P. A. (1995). Assessing IT usage: The role of prior experience. *MIS Quarterly, 19*(4), 561-570.

Tekalp, A. M. (1995). *Digital video processing*. Prentice Hall Signal Processing Series. Prentice Hall, Englewood Cliffs: NJ.

Telecom New Zealand. (2004, June). *Product finder: Internet and data, Telecom's wireless hotspot service*. Retrieved May 18, 2006, from http://www.telecom.co.nz/content/0,3900,204163-1487,00.html

Tentori, M., Favela, J., & González, V. (2005). Designing for privacy in pervasive hospital environments. *Ubiquitous Computing & Ambient Intelligence*. Retrieved December 2005, from http://www.isr.uci.edu/~vmgyg/documents/ucami.pdf

Tentori, M., Favela, J., Gonzalez, V., & Rodríguez, M. (2005). Supporting quality of privacy (QoP) in pervasive computing. *Encuentro Internacional de Ciencias de la Computación (ENC)*, Pueblo, Mexico, September 26-30 (pp. 58-67). IEEE Computer Society.

Teo, T. S. H., & Pok, S. H. (2003). Adoption of WAP-enabled mobile phones among Internet users. *Omega: The International Journal of Management Science, 31*(6), 483-498.

Terry, D. J. (1993). Self-efficacy expectancies and the theory of reasoned action. In D. C. Terry, C. Gallois, &

M. McCamish (Eds.), *The theory of reasoned action: Its application to AIDS-preventive behaviour* (pp. 135-152). Oxford: Pergamon.

Thai, B., Wan, R., Seneviratne, A., & Rakotoarivelo, T. (2003, February). Integrated personal mobility architecture: A complete personal mobility solution. *ACM Mobile Networks and Applications (MONET) Special Issue: Personal Environment Mobility in Multi-Provider and Multi-Segment Network, 8,* 27-36.

Tham, J. Y., Ranganath, S., Ranganath, M., & Kassim, A. A. (1998). A novel unrestricted center-biased diamond search algorithm for block motion estimation. *IEEE Transactions on Circuits and Systems for Video Technology, 8*(4), 369-377.

The Familiar Project. (2005). *The familiar project.* Retrieved July 10, 2005, from http://familiar.handhelds. org/

Thompson, R. L., Higgins, C. A., & Howell, J. M. (1991). Personal computing: Toward a conceptual model of utilization. *MIS Quarterly, 15*(1), 125-143.

Thompson, R., Higgins, C., & Howell, J. (1994). Influence of experience on personal computer utilization: Testing a conceptual model. *Journal of Management Information Systems, 11*(1), 167-187.

Thomson, S., & Narten, T. (1998, December). *IPv6 stateless address autoconfiguration.* IETF RFC 2462.

Toh, C. K. (1996). A novel distributed routing protocol to support ad-hoc mobile computing. *Proceedings of the 15th IEEE International Performance, Computing, and Communications Conference (IPCCC),* Phoenix, Arizona, March 27-29 (pp. 480-486). New York: IEEE.

Toh, C. K. (1997). Associativity-based routing for ad hoc mobile networks. *Journal of Wireless Personal Communications, 4*(2), 103-139.

Toivonen, T., & Heikkilä, J. (2004). Fast full search block motion estimation for H.264/avc with multilevel successive elimination algorithm. In *Proc. International Conference on Image Processing (ICIP), 3,* Singapore, October (pp. 1485-1488).

Topaz Systems Inc. (2005). *Electronic signature pad with interactive LCD display.* Retrieved July 7, 2005, from http://www.topazsystems.com/products/specs/TL462.pdf

Topiwala, P. N. (1998). *Wavelet image and video compression.* Boston: Kluwer Academic Publishers.

Toufik, I., & Knopp, R. (2004). Channel allocation algorithms for multi-carrier systems. *Proc. of the IEEE VTC '04,* Los Angeles, CA, September (pp. 1129-1133). New York: IEEE.

Trust Digital. (2005). *Trust digital 2005™.* Retrieved July 5, 2005, from http://www.trustdigital.com/downloads/productsheet_trust2005.pdf

Trusted Computing Group. (2004). *Security in mobile phones.* Retrieved July 1, 2005, from http://www.trustedcomputinggroup.org/downloads/whitepapers/TCG-SP-mobile-sec_final_10-14-03_V2.pdf

Tsuchiya, P. F. (1988). The landmark hierarchy: A new hierarchy for routing in very large networks. *Computer Communication Review, 18*(4), 35-42.

UMTS World. (2003). *3G applications.* Retrieved May 2007, from http://www.umtsworld.com/applications/applications.htm

UMTS. (2003). *Mobile evolution: Shaping the future.* Retrieved August 28, 2005, from http://www.umts-forum.org/servlet/dycon/ztumts/umts/Live/en/umts/Resources_Papers_index

Vaishampayan, V. A. (1993). Design of multiple description scalar quantizers. *IEEE Transaction on Information Theory, 39*(3), 821-834.

van de Sandt-Koenderman, M., Wiegers, J., & Hardy, P. (2005, May). A computerised communication aid for people with aphasia. *Disability Rehabilitation, 27*(9), 529-533.

Van Dyke, C., & Koc, C. K. (2003, February). On ubiquitous network security and anomaly detection. *Proceedings of the Applications and the Internet Workshops,* January 27-31 (pp. 374-378). Electrical & Computer Engineering, Oregon State University.

van Steenderen, M. (2002). Business applications of WAP. *The Electronic Library, 20*(3), 215-223.

Venkatesh, V., & Davis, F. D. (2000). A theoretical extension of the technology acceptance model: four longitudinal field studies. *Management Science, 46*(2), 186-204.

Venkatesh, V., & Morris, M. G. (2000). Why don't men ever stop to ask for directions? Gender, social influence, and their role in technology acceptance and usage behavior. *MIS Quarterly, 24*(1), 115-139.

Venkatesh, V., Morris, M. G., Davis, G. B., & Davis, F. D. (2003). User acceptance of information technology: Toward a unified view. *MIS Quarterly, 27*(3), 425-478.

Veridt, LLC. (2005). *Veridt verification and identification, product information.* Retrieved September 27, 2007, from http://www.veridt.com/Products.html

Vigna, G., Gwalani, S., Srinivasan, K., Belding-Royer, E. M., & Kemmerer, R. A. (2004). An intrusion detection tool for AODV-based ad hoc wireless networks. *Proceedings of the Annual Computer Security Applications Conference (ACSAC)*, Miami Beach, FL, December 10-14 (pp. 16-27). IEEE Computer Society.

Villamil, F. (2005). *Firewall wizards: Looking for PDA firewall.* Retrieved July 10, 2005, from http://seclists.org/lists/firewall-wizards/2004/Jan/0031.html

Vodafone New Zealand. (2005, March). *Business services: Paying with your mobile – TXT-a-Park.* Retrieved May 20, 2006, from http://www.vodafone.co.nz/promos/txt-a-park/txt_a_park.jsp?item=txt_a_park

VQEG. (1998). Final report from the Video Quality Expert Group on the validation of objective models of video quality assessment.

Wade, N., & Swanston, M. (2001). *Visual perception: An introduction (2nd ed.).* London: Psychology Press.

Wahlqvist, M. et al. (1997). Capacity comparison of an OFDM based multiple access system using different dynamic resource allocation. *Proc. IEEE VTC'97*, Phoenix, Arizona (pp. 1664-1668). New York: IEEE.

Walker, J. (2001, May). *Unsafe at any key size: An analysis of the WEP encapsulation.* IEEE 802.11 doc 00-362, grouper.ieee.org/groups/802/11.

Wallace, G. K. (1991). The JPEG still picture compression standard. *Communications of the ACM, 34*(4), 30-44.

Wan, P.-J., Calinescu, G., & Yi, C.-W. (2004). Minimum-power multicast routing in static ad hoc wireless networks. *IEEE/ACM Transactions on Networking, 12*(3), 507-514.

Wang, H., Raman, B., Biswas, R., Chuah, C., Gummadi, R., Hohlt, B., Hong, X., Kiciman, E., Mao, Z., Shih, J., Subramanian, L., Zhao, B., Joseph, A., & Katz, R. (2000, August). Iceberg: An Internet-core network architecture for integrated communications. *IEEE Personal Communications (2000): Special Issue on IP-based Mobile Telecommunication Networks, 7*(4), 10-19.

Wang, H., Schuster, G. M., & Katsaggelos, A. K. (2005). Rate-distortion optimal bit allocation for object-based video coding. *IEEE Transactions on Circuits and Systems for Video Technology, 15*(9), 113-1123.

Wang, L., Zhang, L. F., Shu, Y. T., Dong, M., & Yang, O. W. W. (2001). Adaptive multi-path source routing in wireless ad-hoc networks. *Proceedings of the IEEE International Conference on Communications (ICC)*, St. Petersburg, Russia, June 11-15 (pp. 867-871). New York: IEEE.

Wang, Q., & Abu-Rgheff, M. A. (2002). Integrated mobile IP and SIP approach for advanced location management. *4th International Conference on 3G Mobile Communication Technologies, 2003*, London, UK, June 25-27 (pp. 205-209). New York: IEEE.

Wang, W. et al. (2003). Impact of multiuser diversity and channel variability on adaptive OFDM. *Proc. IEEE VTC 2003 Fall*, Orlando, Florida, October (pp. 547-551). New York: IEEE.

Wang, Y., & Zhu, Q. (1998). Error control and concealment for video communication: A review. *Proceedings of the IEEE, 86*(5), 974-997.

Wang, Y.-S., Wang, Y.-M., Lin, H.-H., & Tang, T.-I. (2003). Determinants of user acceptance of Internet banking: An empirical study. *International Journal of Service Industry Management, 14*(5), 501-519.

Watson, D. (2002, September). *Auckland wireless net lures Asians: A firm planning a large wireless network around downtown Auckland is already claiming interest from Asia and the Pacific islands in its distance-boosting technology.* Computerworld New Zealand. Retrieved May 20, 2006, from http://computerworld.co.nz/news. nsf/UNID/CC256CED0016AD1ECC256C3C007A629 9?OpenDocument&Highlight=2,roamad

Weatherall, J., & Jones, A. (2002, February). Ubiquitous networks and its applications. *IEEE Wireless Communications, 9*(1), 18-29.

Weinstein, S. B., & Ebert, P. M. (1971). Data transmission by frequency-division multiplexing using the discrete fourier transform. *IEEE Transactions on Communication Technology, 19*, 628-634.

Weis, S. A. (2005). Security parallels between people and pervasive devices. *Third International Conference on Pervasive Computing and Communications Workshops (PerCom 2005),* Hawaii, March (pp. 105-109).

Weise, J. (2001). Public key infrastructure overview. *Sun BluePrints Online.* Retrieved December 2005, from http://www.sun.com/ blueprints/0801/publickey.pdf

Weiser, M. (1991). The computer for the twenty-first century. *Scientific American, 265*, 66-75.

Weiser, M. (1993, July). Some computer science problems in ubiquitous computing. *Communications of the ACM, 36*(7), 75-84.

Welch, T. A. (1984). A technique for high performance data compression. *IEEE Computer, 17*(6), 8-19.

Wellington City Council. (2005, March). *Technology & software, TXT-a-Park.* Retrieved May 21, 2006, from http://www.wellington.govt.nz/innovation/details/tx-tapark.html

Wen, J., & Villasenor, J. D. (1997). A class of reversible variable length codes for robust image and video coding. In *Proc. IEEE International Conference on Image Processing (ICIP), 2,* Washington, DC, October (pp. 65-68).

WEPcrack. (2006). Retrieved December 10, 2006, from http://www.wirelessdefence.org/contents/WEPcrack-main.htm

WEPCrack. (2007). Retrieved September 26, 2007, from http://wepcrack.sourceforge.net

Wesel, E. K. (1998). *Wireless multimedia communications: Networking video, voice, and data.* Addison Wesley.

Wi-Fi Alliance Portal. (2005). Retrieved July 11, 2005, from http://www.wi-fi.org

Wikipedia, the free encyclopedia. Retrieved July 11, 2005, from http://www.wikipedia.org

Wikipedia. (2005). *Secure sockets layer (SSL) and transport layer security (TLS).* Retrieved July 10, 2005, from http://en.wikipedia.org/wiki/Secure_Sockets_Layer

Wikipedia. (2005, May). *Article: Ubiquitous computing.* Wikipedia: The Free Encyclopaedia. Retrieved May 5, 2006, from http://mobileman.projects.supsi.ch/glossary. html

Wikipedia. (2005). *Security.* Retrieved December 25, 2005, from http://en.wikipedia.org/wiki/Computer_security

Wikipedia. (2005). *Privacy.* Retrieved December 25, 2005, from http://en.wikipedia.org/wiki/Privacy

Wikipedia. (2005). *Trust.* Retrieved December 25, 2005, from http://en.wikipedia.org/wiki/Trust

Wikipedia. (2006). *Wikipedia, the free encyclopedia.* Retrieved October 2006, from http://en.wikipedia. org/wiki/

Wikipedia. *Computer forensics.* Retrieved July 11, 2005, from http://en.wikipedia.org/wiki/Computer_forensics

Wireless Developer Network. (2002, August). *Wireless developer network – Daily news: RoamAD announces 802.11b breakthrough with metropolitan-wide Wi-Fi network.* Wireless Developer Network. Retrieved

May 9, 2006, from http://www.wirelessdevnet.com/news/2002/238/news1.html

Witten, I., Neal, R., & Cleary, J. (1987). Arithmetic coding for data compression. *Communication of the ACM, 30*(6), 520-540.

Wolfe, S. T., Ahamed, S. I., & Zulkernine, M. (2006). A trust framework for pervasive computing environments. *4th ACS/IEEE International Conference on Computer Systems and Applications (AICCSA-06)*, Dubai, UAE, March (pp. 312-319). Dubai: IEEE CS Press.

Woo, S.-C. M., & Singh, S. (2001). Scalable routing protocol for ad hoc networks. *Journal of Wireless Networks, 7*(5), 513-529.

Woo, T. (2003). Dynamic security in pervasive computing. *ECE750 Presentation*, University of Waterloo. Retrieved December 2005, from http://www.vlsi.uwaterloo.ca/~thwoo/ece750/ece750-6report.pdf

Woods, J., & O'Neil, S. (1986). Subband coding of images. *IEEE Trans. Acoustics, Speech, and Signal Processing, 34*(5), 1278-1288.

Wu, H. R., & Rao, K. R. (2006). *Digital video image quality and perceptual coding*. Boca Raton, FL: CRC Press: Taylor and Francis.

Wu, T., Wu, S., Huang, H., & Gong, F. (1999). Securing QoS: Threats to RSVP messages and their countermeasures. *Proceedings of IWQoS'99 – Seventh International Workshop on Quality of Service*, UCL, London, May 31-June 4 (pp. 62-64). USA: IEEE.

Wu, W., Banerjee, N., Basu, K., & Das, S. K. (2003). Network assisted IP mobility support in wireless LANs. *Second IEEE International Symposium on Network Computing and Applications, NCA'03*, Cambridge, MA, April 16-18 (pp. 257-264). Los Alamitos, CA: IEEE Computer Society.

Wyner, A. D., & Ziv, J. (1976). The rate-distortion function for source coding with side information at the decoder. *IEEE Transactions on Information Theory, IT-22*(1), 1-10.

Xydis, T. G., & Blake-Wilson, S. (2002). *Security comparision: Bluetooth™ Communications vs. 802.11*. Retrieved July 11, 2005, from http://ccss.isi.edu/papers/xydis_bluetooth.pdf

Xylomenos, G., & Polyzos, G. C. (2001). Quality and service support over multi-service wireless Internet links. *Computer Networks, 37*(5), 601-615.

Yamaguchi, N., Ida, T., & Watanabe, T. (1997). A binary shape coding method using modified MMR. In *Proc. Special Session on Shape Coding (ICIP97), I*, Washington, DC, October (pp. 504-508).

Yamauchi, T., & Markman, A. B. (2000). Inference using categories. *Journal of Experimental Psychology: Learning, Memory, and Cognition, 26*(3), 776-795.

Yang, H., Luo, H., Ye, F., Lu, S., & Zhang, L. (2004). Security in mobile ad hoc networks: Challenges and solutions. *IEEE Wireless Communications, 11*(1), 38-47.

Yang, H., Luo, H., Ye, F., Lu, S., & Zhang, L. (2004). Security in mobile ad hoc networks: Challenges and solutions. *IEEE Wireless Communications, 11*(1), 38-47.

Yau, S., Karim, F., Wang, Y., Wang, B., & Gupta, S. K. S. (2002). Reconfigurable context-sensitive middleware for pervasive computing. *IEEE Pervasive Computing, 1*(3), 33-40.

Yau, S., Wang, Y., & Karim, F. (2002). Development of situation aware application software for ubiquitous computing environments. *26th Annual International Computer Software and Application Conference* (pp. 233-238).

Yegin, A., Ohba, Y., Penno, R., Tsirtsis, G., & Wang, C. (2005). *Protocol for carrying authentication for network access (PANA) requirements*. RFC 4058, IETF.

Yen, D. C., & Chou, D. C. (2000). Wireless communications: Applications and managerial issues. *Industrial Management & Data Systems, 100*(9), 436-443.

Yi, S., & Kravets, R. (2002). *Key management for heterogeneous ad hoc wireless networks*. Technical Report, UIUCDCS-R-2002-2290, Department of Computer Science, University of Illinois at Urbana-Champaign.

Yi, S., & Kravets, R. (2002). Key management for heterogeneous ad hoc wireless networks. *10ᵗʰ IEEE International Conference on Network Protocols*, Paris, France, November 12-15 (pp. 202-203).

Yi, S., Naldurg, P., & Kravets, R. (2001). Security-aware ad hoc routing for wireless networks. *Proceedings of the 2ⁿᵈ ACM International Symposium on Mobile Ad Hoc Networking and Computing (MobiHoc)*, Long Beach, CA, October 4-5 (pp. 299-302). ISBN 1-58113-428-2. ACM Press.

Yin, R. K. (1994). *Case study research: Design and methods.* Beverley Hills: Sage.

Youssef, M., Agrawala, A., & Shankar, U. (2003). WLAN location determination via clustering and probability distributions. *IEEE International Conference on Pervasive Computing and Communications (PerCom) 2003*, Dallas, Texas, March 23-26 (pp. 143-150). New York: IEEE.

Yuhan, A. H. (2003). *Ubiquitous computing and its network requirements.* Samsung Advanced Institute of Technology. Retrieved May 13, 2006, from http://www.t-engine.org/aucnc2003/yuhan.pdf

Zander, J., & Kim, S. L. (2001). *Radio resource management for wireless networks.* London: Artech House Publishers, Mobile Communications Series.

Zapata, M. G., & Asokan, N. (2002). Securing ad hoc routing protocols. *Proceedings of the 3ʳᵈ ACM Workshop on Wireless Security (WiSe)*, Atlanta, Georgia, September 28 (pp. 1-10). ISBN 1-58113-585-8. ACM Press.

Zeithaml, V. A., & Gilly, M. C. (1987). Characteristics affecting the acceptance of retailing technologies: A comparison of elderly and nonelderly consumers. *Journal of Retailing, 63*(1), 49-68.

Zeng, W., Zhuang, X., & Lan, J. (2004). Network friendly media security: Rationales, solutions, and open issues. *2004 International Conference on Image Processing ICIP '04, 1*, Singapore, October 24-27 (pp. 565-568). USA: IEEE.

Zhang, L., Lambertsen, G., & Yamada, T. (2003). Security scenarios within IP-based mobile multimedia metropolitan area network. In *Proceedings of Global Telecommunications Conference, GLOBECOM '03, 3*, San Francisco, CA, December 1-5 (pp. 1522-1526). Piscataway, NJ: IEEE Operations Center.

Zhang, Y., & Lee, W. (2000). Intrusion detection in wireless ad-hoc networks. *Proceedings of the 6ᵗʰ International Conference on Mobile Computing and Networking (MobiCom)*, Boston, MA, August 6-11 (pp. 275-283). ISBN 1-58113-197-6. ACM Press.

Zhang, Y., Lee, W., & Huang, Y. (2003). Intrusion detection techniques for mobile wireless networks. *Kluwer Wireless Networks, 9*(5), 545-556.

Zhou, L., & Haas, Z. J. (1999). Securing ad hoc networks. *IEEE Networks, 13*(6), 24-30.

Zhu, Q.-F., Wang, Y., & Shaw, L. (1993). Coding and cell-loss recovery in DCT-based packet video. *IEEE Transactions on Circuits and Systems for Video Technology, 3***(3)**, 238-247.

Zimmermann, R. (2001). Lokalisierung Mobiler Geraete (german). *Seminar Mobile Computing ETH Zuerich 2001.* Technical report. Retrieved September 18, 2007, from www.vs.inf.ethz.ch/edu/SS2001/MC/beitraege/07-location-rep.pdf

Ziv, J., & Lempel, A. (1977). A universal algorithm for sequential data compression. *IEEE Transactions on Information Theory, IT-23*(3), 337-343.

About the Contributors

Gour C. Karmakar received the BSc Engg degree in computer science and engineering from Bangladesh University of Engineering and Technology in 1993 and Masters and PhD degrees in information technology from the Faculty of Information Technology, Monash University in 1999 and 2003 respectively. He was with the Computer Division, Bangladesh Open University from 1994 to 1998. He is currently a lecturer at the Gippsland School of Information Technology, Monash University (since December 2002). He has published over 50 peer-reviewed research publications and was jointly awarded the IEEE Communications Society sponsored Outstanding Paper Prize at the *1st International Conference on Next Generation Wireless Systems (ICNEWS'06)*. His research interests include image and video processing, mobile and wireless sensor networks.

Laurence S. Dooley received his BSc (Hons), MSc, and PhD degrees in electrical engineering from the University of Wales, Swansea in 1981, 1983, and 1987, respectively. From 1999 to 2006, he was a professor of multimedia technology in the Faculty of Information Technology, Monash University, Australia. His major research interests are in multimedia signal processing, mobile communications, image/video object segmentation, bioinformatics, wireless sensor networks, and R&D commercialisation strategies for regional small business. He has edited one book and published well over 170 international scientific peer-reviewed journals, book chapters, and conference papers, and was jointly awarded the IEEE Communications Society sponsored Outstanding Paper Prize at the *1st International Conference on Next Generation Wireless Systems (ICNEWS'06)* this year. He has successfully supervised 13 PhD/Masters research students as well as being in receipt of many external grants from both government and industry to support his multi-faceted research projects. He is a senior member of the IEEE, a Chartered Engineer (C.Eng), and a corporate member of the BCS.

* * *

Sheikh I. Ahamed is an assistant professor in the Department of Math., Stat., and Computer Science at Marquette University, USA. He is a member of the IEEE, ACM, and the IEEE Computer Society. Dr. Ahamed received the BSc in computer science and engineering from the Bangladesh University of Engineering and Technology, Bangladesh in 1995. He completed his PhD in computer science from Arizona State University, U.S. in 2003. His research interests are security in ad-hoc networks, middleware for ubiquitous/pervasive computing, sensor networks, and component-based software development. He serves regularly on international conference program committees in software engineering and pervasive computing such as COMPSAC 04, COMPSAC 05, COMPSAC 06, and ITCC 05. He is the program co-chair of the *International Workshop on Security, Privacy, and Trust for Pervasive Computing Applications (SPTPA 2006)*. Dr. Ahamed can be contacted at iq@mscs.mu.edu; http://www.mscs.mu.edu/~iq.

Mohammed Atiquzzaman received the MSc and PhD degrees in electrical engineering from the University of Manchester, England. Currently he is a faculty member in the School of Computer Science at University of Oklahoma. He is the co-editor-in-chief of *Computer Communications Journal*, and serves on the editorial boards of *IEEE Communications Magazine, Telecommunications Systems Journal, Wireless and Optical Networks Journal*, and *Real Time Imaging Journal*. He is the co-author of the book, *TCP/IP over ATM Networks*. Dr. Atiquzzaman was the technical co-chair of 2003 Workshop on High Performance Switching and Routing, and SPIE Quality of Service over Next-Generation Data Networks Conference (2001, 2002, 2003). He will serve as the co-chairs of symposiums in IEEE Globecom 2006 and ICC 2007, the two flagship conferences of the IEEE Communications Society. Dr. Atiquzzaman's current research interests are in wireless, satellite and mobile networks, quality of service for next generation Internet, broadband networks, multimedia over high speed networks, and image processing. His research has been supported by state and federal agencies like NSF (USA), NASA (USA), U.S. Air Force, Ohio Board of Regents (USA), and DITARD (Australia). He has over 130 refereed publications in these interest areas, most of which can be accessed at http://www.cs.ou.edu/~atiq.

Sally-Jo Cunningham received degrees in computer science and Asian studies from the University of Tennessee, and a PhD in computer science from Louisiana State University. She has been a lecturer with the University of Waikato, New Zealand since 1990. Her research interests include machine learning, digital libraries, computer education, and computer applications in textiles.

Kaj J. Grahn, Dr. Tech., is presently senior lecturer in telecommunications at the Department of Business Administration, Media, and Technology at Arcada Polytechnic, Helsinki, Finland. His current research interests include mobile and wireless networking and network security.

Jairo Gutiérrez is a senior lecturer in the Information Systems and Operations Management Department at the University of Auckland. He teaches data communications and computer networking, and has supervised the research projects of more than 35 postgraduate students over the past nine years. He is the editor-in-chief of the *International Journal of Business Data Communications and Networking* and has served as a reviewer for several leading academic publications. His current research topics are in network management systems, viable business models for mobile commerce, programmable networks, and quality of service issues associated with Internet protocols. He is also the co-ordinator for the University of the Cisco Networking Academy Program. Dr. Gutiérrez received a Systems and Computer Engineering degree from The University of The Andes (Colombia, 1983), a Masters degree in computer science from Texas A&M University (1985), and a PhD in information systems from The University of Auckland in 1997.

Munirul M. Haque is a graduate student and member of Ubicomp Research Lab in the Department of Math. Stat., and Computer Science at Marquette University, USA. He has received his BSc degree in computer science and engineering from Bangladesh University of Engineering and Technology (BUET), Bangladesh in 2003. His field of interest encompasses pervasive security, trust model in pervasive computing, and wireless sensor network. His contact address is md.haque@mu.edu; http://www.mscs. mu.edu/~mhaque.

Catherine Harris qualified in speech therapy at Leicester and is a member of the Royal College of Speech and Language Therapists. She has worked as a specialist SLT in augmented communication since 1995 at Portsmouth and now in Gloucester and served as a non-executive director of Southampton Primary Healthcare Trust from 2001-2004. In conjunction with industry and charities such as the Motor Neurone Disease Association, she has developed therapy programmes for dysarthria and dyspraxia using augmented and alternative communication devices.

Sally Rao Hill, B.Bus. (Hons) PhD CPM, is a lecturer of marketing at the School of Commerce at the University of Adelaide. Her research interests include relationship marketing, Internet marketing, and services marketing. She is an active researcher and has published in the *Journal of Business and Industrial Marketing, European Journal of Marketing, Qualitative Market Research: An International Journal, Australasian Marketing Journal, International Journal of Internet Marketing and Advertising* and the *Journal of Internet Business.* She has won the best paper award in an international conference.

Muhammad Mahmudul Islam, B.Sc. Engg.(Hons), has received the BSc Engg in computer science and engineering with first class honours in 2002 from Bangladesh University of Engineering and Technology (BUET). Currently, he is working towards a PhD degree on designing and analysis of lower level protocols for suburban ad-hoc networks (SAHN) from the Faculty of Information Technology, Monash University, Australia. His research interests include wireless ad-hoc and sensor networks, computer system security, and parallel and distributed computer systems architecture. He has published 15 research articles.

Laura Jefferies has a First Class Honours degree in multimedia and is studying for a PhD in the use of accessible 2D and 3D multimedia in speech and language therapy.

Joarder Kamruzzaman received his BSc and MSc in electrical and electronic engineering from Bangladesh University of Engineering & Technology, Dhaka, Bangladesh (1986 & 1989, respectively), and PhD in information systems engineering from Muroran Institute of Technology, Japan (1993). Currently he is a faculty member in the Faculty of Information Technology, Monash University, Australia. His research interest includes computational intelligence, computer networks and bioinformatics. He has published more than 90 refereed papers in international journals and conference proceedings. He is currently serving as a program committee member of a number of international conferences.

Odej Kao is an associated professor in the Department of Computer Science, University of Paderborn, Germany. He received a Masters degree in computer science in 1995 and a PhD degree in 1997, both from the TU Clausthal, Germany. From 1998 to 2002, he worked on his advanced PhD (Habilitation) and joined thereafter the University of Paderborn as a professor for distributed and operating systems. Since 2003 he has also been the managing director of the Paderborn Centre for Parallel Computing. His research area includes parallel and distributed computing, distributed multimedia systems, multimedia retrieval, and resource management in grid environments.

Jonny Karlsson has a BSc in information technology from Arcada Polytechnic, Helsinki, Finland. Since May 2002, he has been working in Arcada Polytechnic as a course assistant and course teacher in programming and network security-related courses, and as a research assistant. His current research interests include wireless and mobile network security.

Carlo Kopp, BE(Hons), MSc, PhD, graduated in electrical engineering from the University of Western Australia with First Class Honours, in 1984. In 1996 he completed a research Masters degree in computer science at Monash University in Melbourne, and in September, 2000, he completed his PhD thesis. The thesis dealt with the properties of airborne high capacity long range microwave ad-hoc networks, and the adaptation of active electronically steered arrays for these applications. His career interests have been broad, involving a professional engineering and computer science track, and also concurrently a track as a defence analyst. He has held a wide range of positions in the computer and communications industries, primarily as a design engineer, computer programmer, systems integrator, and consultant. His best known project in the computer industry was the design and development of the first Australian manufactured Unix SPARC workstation computer, in 1993. His other professional and academic interests include air warfare strategy, doctrine, and information warfare. His work in this area has been published by the Royal Australian Air Force and the United States Air Force, and he was one of the authors of the information theoretical basis of information warfare. His research interests encompass inter-process communications, OS and machine performance, real-time/embedded systems, file systems and I/O subsystems/performance, ad-hoc networking protocols, survivable network topologies, high speed/optical communications, information warfare, and air power theory. At this time he lectures on computing topics at Monash SCSSE, and is a research fellow in Regional Military Strategy at the Monash Asia Institute.

Stefan Lietsch is a PhD student at the Paderborn Center for Parallel Computing, University of Paderborn, Germany. He received a Masters degree in computer science in 2005 from the University of Paderborn, Germany. He has worked in the areas of user supporting communication systems and communication in wireless ad-hoc networks. His current research interests include distributed computing, high performance visualization, and parallel medical image reconstruction.

Michael Mathew has completed his Bachelors at Monash University from the business faculty in 1999. Then he went on to do his Masters in information management and systems. While doing this, he worked as a lecturer at Monash and Holmesglen TAFE. His industry experience comes from working in communications with PABX systems and mobile systems. Currently, he is completing a Masters in IT. The research looks at the analysis and the development of new mobile systems integrating 4G networks. The research is industry specific and fully funded by the Australian communication giant Telstra. The research is focused on giving Telstra's clients the edge with their communication systems.

Kevin Park is a technical consultant with Hewlett Packard New Zealand Ltd. (Auckland). He has previous experience in software development and network administration. He has a Bachelors (Honours) degree in commerce (Information Systems Major) and Bachelors degree in science (Computer Science Major). His current research interests are in network management, quality of service, and network security. He can be reached at kevin.park@infosys.geek.nz.

Ronald Pose's (BSc[Hons], PhD [Monash University]), PhD thesis, *A Capability-based Tightly-Coupled Multiprocessor,* was supervised by Prof. C. S. Wallace. Formerly a research scientist at Telecom Australia Research Laboratories, where he worked on public key cryptography and authentication and certification techniques for secure communications, in 1988, Dr. Pose was invited to join the Computer Science Department at Monash University. Dr. Pose's current research interests include virtual reality and telerobotics technology, computer architecture, parallel and distributed computer systems architecture, secure operating systems, reconfigurable computer systems, multiprocessor interconnection networks, wireless ad-hoc networks, spread-spectrum microwave communication, and computer system security. He currently has PhD students working on computer security analysis, multi-user virtual reality, and wireless ad-hoc networking. He has published over 70 research papers in refereed journals and conferences, supervised a number of research students and served on international advisory committees for several international conferences.

Göran Pulkkis, Dr. Tech., is presently a senior lecturer in computer science and engineering at the Department of Business Administration, Media, and Technology at Arcada Polytechnic, Helsinki, Finland. His current research interests include network security, applied cryptographic, and quantum informatics.

Nina Reeves has a First degree in mathematics and computing, an MSc in information science, and a PhD in human computer interaction. Her research interests include on-line help systems, digital libraries, and ethnographic methods of enquiry. She is a member of the British HCI Group and the Higher Education Academy and has been a senior lecturer and course leader in multimedia computing at the University of Gloucestershire since 1995.

Hermann Rohling is with the Hamburg University of Technology, Germany, where he has developed an international reputation in the field of OFDM transmission techniques. Previously, Prof. Rohling was with the AEG Research Institute, Ulm as a researcher working in the area of digital signal processing for radar and communications applications. His research interests have included wideband mobile communications especially based on multicarrier transmission techniques (OFDM) for future broadband systems (4G), signal theory, digital radar signal processing, detection, estimation, and differential GPS for high precision navigation. Prof. Rohling is a member of Informationstechnische Gesellschaft (ITG), German Institute of Navigation (DGON), and a fellow of IEEE.

Nalin Sharda gained BTech and PhD degrees from IIT Delhi. Presently, he teaches and leads research in multimedia and Internet communications at the School of Computer Science and Mathematics, Victoria University, Australia. He has presented seminars, lectures, and tutorials in Austria, Australia, Germany, Hong Kong, India, Japan, Pakistan, Sweden, Switzerland, UAE, and USA. His publications include the textbook titled, *Multimedia Information Networking,* and around 100 papers and handbook chapters. He has published on networked multimedia systems in the *Handbook of Multimedia Computing*, the *Handbook of Internet Computing*, the *Communications of the AIS*, *IEEE & ACM Journals*, and numerous international conferences.

Surendra Sivagurunathan holds a bachelor's degree on computer science from the Univerity of madras, India. He graduated from the University of Oklahoma in 2005 with a Masters degree in computer science. Since 2003, he has worked for the Telecom and Network Research Laboratory at the University of Oklahoma on variety wireless networking projects, including SIGMA (Seamless IP-diversity based Generalized Mobility Architecture) sponsored by NASA, experimental evaluation of SIGMA over Mobile IP, HTTP over SCTP, multimedia transmission over SCTP, and various wireless network-related project. He is currently employed at YouSendIt Inc., California.

Ferdous A. Sohel received his BSc in computer science and engineering (CSE) from Bangladesh University of Engineering and Technology in 2002. From 2002 until 2004, he was a lecturer in the Dept. of CSE, International Islamic University, Chittagong (Dhaka Campus), Bangladesh, and since 2004, he has been studying for a PhD degree in the Faculty of Information Technology, Monash University, Australia. His current research interests include the family of parametric curves, object and content analysis, contour-based shape coding, and error concealment for interactive multimedia. He has published 20 peer-reviewed papers on his research to date.

Nhat Dai Tran is a BSc student in information technology at Arcada Polytechnic, Helsinki, Finland. He is currently working at Arcada Polytechnic as a research assistant. His research interests include wireless and mobile network security.

Indrit Troshani, PhD, MSc (CBIS), GradCertEd (TertTeach), BBA (Hons), and MACS, teaches Internet commerce, electronic commerce, and information systems at both undergraduate and postgraduate levels at the School of Commerce, University of Adelaide. His research interests include adoption and diffusion of network innovations (e.g., XBRL) and mobile services (e.g., 3G), as well as software development process improvement. He has contributed to the body of knowledge in electronic commerce by co-authoring refereed journal and international conference publications. He is also a member of the Australian Computer Society.

Mohammad Zulkernine is a faculty member of the School of Computing of Queen's University, Canada, where he is leading the Queen's Reliable Software Technology (QRST) research group. He received his BSc degree in computer science and engineering from Bangladesh University of Engineering and Technology in 1993. Dr. Zulkernine received an MEng degree in computer science and systems engineering from Muroran Institute of Technology, Japan in 1998. He received his PhD from the Department of Electrical and Computer Engineering of the University of Waterloo, Canada in 2003, where he belonged to the university's Bell Canada Software Reliability Laboratory. Dr. Zulkernine's research focuses on software engineering (software reliability and security), specification-based automatic intrusion detection, and software behavior monitoring. His research works are funded by a number of provincial and federal research organizations of Canada, while he is having an industry research partnership with Bell Canada. He is a member of the IEEE, ACM, and the IEEE Computer Society. Dr. Zulkernine is also cross-appointed in the Department of Electrical and Computer Engineering of Queen's University, and a licensed professional engineer of the province of Ontario, Canada. He can be reached at mzulker@cs.queensu.ca; http://www.cs.queensu.ca/~mzulker/.

Index